Scientific practice and ordinary action

Scientific practice and ordinary action

Ethnomethodology and social studies of science

MICHAEL LYNCH
Brunel University

CAMBRIDGE UNIVERSITY PRESS
Cambridge, New York, Melbourne, Madrid, Cape Town, Singapore, São Paulo

Cambridge University Press
The Edinburgh Building, Cambridge CB2 2RU, UK

Published in the United States of America by Cambridge University Press, New York

www.cambridge.org
Information on this title: www.cambridge.org/9780521431521

First published 1993
First paperback edition 1997

A catalogue record for this publication is available from the British Library

Library of Congress Cataloguing in Publication data

Lynch, Michael, 1948–

Scientific practice and ordinary action : ethnomethodology and
social studies of science / Michael Lynch.

p. cm

Includes bibliographical references and index.

ISBN 0–521–43152–2

1. Science – Social aspects. 2. Science – Methodology.
3. Sociology – Methodology. 4. Ethnomethodology. I. Title.
Q175.5.L9 1993
306.4′5–dc20 92–34526
CIP

ISBN-13 978-0-521-43152-1 hardback
ISBN-10 0-521-43152-2 hardback

ISBN-13 978-0-521-59742-5 paperback
ISBN-10 0-521-59742-0 paperback

Transferred to digital printing 2005

To Nancy

Contents

Acknowledgments

Although I wrote this book, I cannot take full credit (or, for that matter, blame) for its contents. The book explores the possibility of developing what I am calling a *postanalytic* approach to the study of scientific practices. As should be obvious throughout, this orientation is strongly influenced (perhaps infected) by Harold Garfinkel's ethnomethodological approach to situated practical action and practical reasoning. For the past twenty years, I have had the benefit of reading numerous unpublished drafts of Garfinkel's writings and attending many lectures and seminars in which he and his students discussed and demonstrated novel ways to investigate the production of social order. The specific references I have made in this book to published and unpublished writings can cover only a small part of what I learned from Garfinkel, his colleagues, and his students, including Eric Livingston, Albert (Britt) Robillard, George Girton, Ken Morrison, Ken Liberman, Richard Fauman, Doug Macbeth, Melinda Baccus, and Stacy Burns. My initial efforts to understand ethnomethodology were aided immeasurably by close friends and colleagues, including David Weinstein, Alene Terasaki, Bill Bryant, and Nancy Fuller, with whom I shared a preoccupation with the question "*What in the world* was Harold talking about?" Garfinkel also read an earlier draft of this book and gave me specific and helpful comments on it.

My understanding of different approaches to ethnomethodology and conversation analysis also relied on what I learned from seminars, informal data sessions, and discussions with Melvin Pollner, Gail Jefferson, Emanuel Schegloff, Anita Pomerantz, and Harvey Sacks. Although I am critical of some of their work in this volume, I hope this will not obscure my appreciation of their achievements. More recently, my understanding of ethnomethodology and related matters benefited from discussions and collaborative projects with Jeff Coulter, Wes Sharrock, Bob Anderson, George Psathas, David Bogen, Dusan Bjelic, Graham Button, Lucy Suchman, John O'Neill, Eileen Crist, Kathleen Jordan, Jeff Stetson, Ed Parsons, and Edouard Berryman. I am especially indebted to Jeff Coulter for his strong encouragement and support, for reading and commenting on an earlier draft of this manuscript, and for teaching me most of what I know about Wittgenstein's later writings. My debt to David Bogen is both pervasive and detailed, especially in Chapter 6, which includes arguments, examples, and some revised passages from coauthored publications and conference presentations. Although the relevant passages are written "in my hand," there is no separating what those passages say from what I have gained from our many conversations and collaborative studies.

My access to the issues and critical debates in the social studies of science has been helped by collaborations, editorial advice, critical exchanges, and many enjoyable conversations with John Law, Steve Woolgar, Sam Edgerton, Gus Brannigan, Andy Pickering, Trevor Pinch, Steven Shapin, Joan Fujimura, Bruno Latour, David Edge, Susan Leigh Star, Harry Collins, and others whom I have neglected to mention. I am also grateful to David Bloor for his role in a critical exchange, parts of which I have incorporated into Chapter 5. Although it came late in the course of my preparation of this manuscript, I also benefited from a visiting appointment in the science studies program at the University of California at San Diego in 1991–92. I was especially informed by the debates and discussions among students and faculty in the core seminar in history, philosophy, and sociology of science that I cotaught with Robert Marc Friedman, Jerry Doppelt, and Chandra Mukerji.

Much of what I wrote in this book was dredged up from a computer hard disk on which I had deposited files, drafts, and notes for various projects and papers. As a result, I intermingled the contents of this book with parts of several papers that were published separately. While doing so, I selected passages, examples, and arguments that were relevant to the overall aims of this book, and I reshaped them accordingly.

I would also like to thank the editors and staff at the New York office of Cambridge University Press for expediting the publication of this book, and I am especially grateful to the three anonymous reviewers contacted by Cambridge University Press who gave me helpful advice and constructive criticism. Finally, I am indebted to Nancy Richards for her loving support, patience, and tolerance while living for countless hours with an asocial writer of social texts.

Introduction

Nobody doubts the significance of science in modern society. Science is often held responsible for spurring the technological transformations, the rises in population, and the shifts in economic production and sources of inequality that characterize the modern landscape. At the same time, nobody seems to have figured out just what science is, and how it is distinguished from other modes of knowledge. Debates persist in the philosophy, history, and sociology of science about how science differs from more commonplace modes of reasoning and practical action. Many participants in these debates have grown doubtful about whether it even makes sense to speak of science as a coherent method, separate from the economic interests, material culture, and specialized skills that distinguish the different subfields of biology, chemistry, astronomy, physics, and the like. The once unquestionable conviction that science must be different from "mere" political opinion, untested speculation, and commonsense belief has recently taken a beating, and the defenders of science are nowadays asked to account for how science is not patriarchal or to explain how it is not an extension of Western colonialism.

In this volume I do not intend to add fuel to such debates so much as to suggest how we might develop more differentiated conceptions of the sciences, scientific methods, and the relationship between scientific and commonsense knowledge. I do not try to solve the problem of defining "science" or the problem of demarcating science from other modes of reasoning and practical action. Instead, I suggest a way to investigate the sciences and to respecify[1] the topics that so often come up in discussions of science, topics like "observation," "representation," "measurement, "proof,"

[1] This term is taken from Harold Garfinkel, "Respecification: evidence for locally produced, naturally accountable phenomena of order, logic, reason, meaning, method, etc., in and as of the essential haecceity of immortal ordinary society (I) – an announcement of studies," pp. 10–19, in G. Button, ed., *Ethnomethodology and the Human Sciences* (Cambridge University Press, 1991). Briefly, I understand a "respecification" of these topics to mean not a redefinition of the meaning of terms but a way of investigating the different activities in which "order," "logic," "meaning," and so forth are locally and practically relevant.

and "discovery." This agenda derives from my interest in two specialized modes of investigation – ethnomethodology and the sociology of scientific knowledge – that are usually considered to be subfields of sociology.

Considered as "parts" of sociology, ethnomethodology and the sociology of science are relatively minor fields. Ethnomethodology is commonly said to be the study of "micro" social phenomena – the range of "small" face-to-face interactions taking place on street corners and in families, shops, and offices – and the sociology of science is said to investigate one of the several modern social institutions. Neither is given much space in conventional sociology textbooks. In the heartland of sociology, far more attention is given to the "larger" social and historical forces that give rise to and maintain systems of economic production, labor markets, bureaucratic organizations, religious and political ideologies, and social classes. Ethnomethodology and sociology of science also are marginal to the cutting edge of social science methodology. Neither area is noted for using the most recently developed quantitative methods of data analysis.[2] More often, they use "soft" modes of research, such as historical case study, ethnography, interviewing, and textual criticism.

Ethnomethodology and the sociology of science also happen to be the two fields in which I work, so naturally I am inclined to argue for their importance, and I do so in this book. Although I believe that professional sociologists should pay more attention to the two areas, my primary objective is not to persuade sociologists to allot more space on the program to them. Rather, I am more interested in arguing for the transdisciplinary relevance of ethnomethodology and the sociology of science. I propose that they are of interest not because of the "parts" of society they investigate, but because of their overridingly *epistemic* focus. They offer distinctive empirical approaches to investigating the production of knowledge, and they enable a refinement of contemporary discussions on the nature and consequences of scientific and technological rationality.

Sociology and transdisciplinary critical discourse

Sociology currently faces an interesting set of circumstances. With the emergence of a transdisciplinary critical discourse in numerous historical, philosophical, and literary fields, many academic scholars and researchers

[2] There was a time when sociologists of science helped develop applications for sociometric methods of network analysis. Sociologists such as Nicholas Mullins, Diana Crane, Derek De Solla Price, and many others developed bibliometric maps of "invisible colleges" in various scientific fields, by systematically representing the patterns of citations between research reports. For example, see Y. Elkana, J. Lederberg, R. K. Merton, A. Thackray, and H. Zuckerman, eds., *Toward a Metric of Science: The Advent of Science Indicators* (New York: Wiley, 1978).

have begun to appreciate the thematic importance of social practices. For lack of a better term, I use the phrase *transdisciplinary critical discourse* to speak of the various antifoundationalist and "post-ist" movements – poststructuralist, postmodernist, postconventionalist – in philosophy, law, literary studies, and social science. These are associated with various appropriations and criticisms of the writings of Foucault, Habermas, Derrida, Gadamer, Rorty, Barthes, Deleuze, Lyotard, and, from an earlier generation, Wittgenstein, Heidegger, Merleau-Ponty, Benjamin, and Dewey.

An interest in "epistemology" is often said to unite the various lines of antifoundationalist research and debate, although it can fairly be said that the legacy of Wittgenstein and Heidegger might best be characterized as antiepistemological. In any event, with the eruptions of feminist and other politicized modes of textual criticism in every humanities and social science discipline (and, to an extent, in biology, archaeology, and some of the other natural sciences as well),[3] textual criticism has merged with social criticism, and (anti-)epistemology has become deeply textual and sociological.[4] Sociology's traditional topical concerns – race, class, gender, power, ideology, technology, symbolic communication, and the social conditioning of language – have been taken up in countless discussions and debates throughout the humanities and human sciences.

At the same time, participants in these debates rarely seem to think that it would be worthwhile to consult the pages of the *American Sociological Review* and related professional journals. This is understandable, since the latest sociological models of status attainment and the advances in rational-choice theory are worse than irrelevant; they are *symptoms* of the very mode of discourse criticized by antifoundationalist philosophers and literary theorists. Moreover, vernacular concepts like *race, class,* and *gender* are featured in highly contentious public discourses, so that a strategy of de-politicizing

[3] See Sandra Harding, "Is there a feminist method?" *Hypatia* 2 (1987): 17–32; Donna Haraway, "Situated knowledges: the science question in feminism and the privilege of partial perspective," *Feminist Studies* 14 (1988): 575–99; Evelyn Fox Keller, *Reflections on Gender and Science* (New Haven, CT: Yale University Press, 1984); Alison Wylie, "The constitution of archaeological evidence: gender politics and science," in P. Galison and D. Stump, eds., *Disunity and Contextualism: New Directions in the Philosophy of Science Studies* (Stanford, CA: Stanford University Press 1996), pp. 311–43 and Athena Beldecos, Sarah Bailey, Scott Gilbert, Karen Hicks, Lori Kenschaft, Nancy Niemczyk, Rebecca Rosenberg, Stephanie Schaertel, and Andrew Wedel (The Biology and Gender Study Group), "The importance of feminist critique for contemporary cell biology," *Hypatia* 37 (1988): 172–87.

[4] What I have called *transdisciplinary critical discourse* is widely regarded as a position of the "left," since it seems most compatible with criticism of the political and cultural status quo ante. Whether this is so, however, is itself a contentious matter, and some proponents of antifoundationalism argue that it is mistaken to assume that "radical" epistemology and "radical" politics are part of a common enterprise. See Stanley Fish, *Doing What Comes Naturally: Change, Rhetoric, and the Practice of Theory in Literary and Legal Studies* (Durham, NC: Duke University Press, 1989), p. 350.

these concepts in order to treat them as variables in explanatory models has limited appeal for participants in the political and intellectual debates of the day.

Of course, not all sociologists go along with the scientistic style of research that dominates American sociology. Quantified and rationalized approaches to social phenomena are anathema to many sociologists, and the discipline is presently undergoing an intensification of its chronic crisis. As always, the crisis concerns whether sociology should continue to conduct itself as a late-blooming "infant" science or to take a more radically interpretive and humanistic approach. But even this debate tends to get caught up in archaic antinomies that no longer have a place in antifoundationalist discourse. Debates about micro versus macro orders of analytic scale, structure versus agency, science versus humanism, and quantitative versus qualitative methods tend to reiterate the familiar conceptual oppositions that many contemporary philosophers and literary scholars have endeavored to put aside. Somewhat late in the game, a growing number of sociologists have begun to appreciate postconventionalist, poststructuralist, or deconstructionist modes of writing, but their efforts too often amount to weak imitations of the longer-running exercises conducted in other fields. This is a particularly ironic development for sociology, a field that should be in the forefront of the "sociological turn" experienced in so many other disciplines.

For different reasons, ethnomethodology and the sociology of science are exceptions to what I just asserted about the irrelevance of professional sociology. Long before it became fashionable, ethnomethodologists took up the writings of Husserl, Heidegger, Merleau-Ponty, and Wittgenstein and developed a distinctive approach to discourse and practical reasoning, and more recently, sociologists of science have become embroiled in debates associated with "new wave" history and philosophy of science. The writings of Kuhn, Popper, Lakatos, Feyerabend, Polanyi, Hanson, Toulmin, and, more recently, Hacking, greatly influenced the current research programs in the sociology of science, and to a considerable extent, sociologists have contributed to transdisciplinary interests in scientific rhetoric and practical "skills" that have emerged in the science studies field.

Like other contributors to transdisciplinary critical discourse, ethnomethodologists and sociologists of scientific knowledge confront "an ancient tension between a notion of truth as something independent of local, partial perspectives and a notion of truth as whatever seems perspicuous and obvious to those embedded in some local, partial perspective."[5] For the most part, they opt for the latter – antifoundationalist – position by seeking to describe the "achievement" of social order and the "construction" of social

[5] Ibid., p. 5.

and scientific "facts." They explicitly renounce the use of transcendental standards of truth, rationality, and natural reality when seeking to describe and/or explain historical developments and contemporary practices.

Although ethnomethodologists and sociologists of science are often well informed about contemporary philosophical movements, their investigations tend to be more "empirical" (whatever might be meant by that term) than is usually the case for philosophical and humanistic scholarship. They conduct case studies of actions in particular social settings; they pay attention to detail; and they try to describe or explain observable (or at least reconstructible) events. Terms of the trade like *empirical observation* and *explanation* are problematic, given their association with empiricism and positivism, but it should be clear that ethnomethodologists and sociologists of science are especially attuned to "actual" situations of language use and practical action. Their studies enable a more differentiated understanding of language, science, and technology than can be gained by making sweeping generalizations about the nature and development of modernity or examining the published reflections of scientists and inventors.

With the "linguistic turn" in postwar philosophy and the renewal of interest in rhetoric and practical action, philosophers and other scholars have begun to appreciate that the traditional epistemological topics of rationality, practical reason, meaning, truth, and knowledge cannot be isolated from the immensely variable linguistic and practical circumstances in which reasons are given for actions, rules are invoked, meanings are explicated, and truth is demanded. Going beyond the ideal-typical investigations of earlier generations of pragmatist and ordinary language philosophers, contemporary scholars now are paying more attention to "actual" usage. For instance, contemporary philosophers of science are increasingly relying on historical and sociological investigations,[6] and some analytically inclined philosophers have turned to cognitive science and artificial intelligence for inspiration and guidance.[7]

In a development that is particularly relevant to my concerns, philosophers like Richard Rorty and Thomas McCarthy suggest that philosophical investigations should draw on ethnographic and related empirical studies of "language games." This is concisely summarized by McCarthy in a discussion of Rorty's "new pragmatism": "Explicating rationality and epistemic authority is not, then, a matter of coming up with transcendental arguments but of providing thick ethnographic accounts of knowledge-producing ac-

[6] See, for instance, Ian Hacking, *Representing and Intervening: Introductory Topics in the Philosophy of Science* (Cambridge University Press, 1983); Larry Laudan, *Progress and Its Problems: Towards a Theory of Scientific Growth* (Berkeley: University of California Press, 1977).

[7] See Paul Churchland, *Scientific Realism and the Plasticity of Mind* (Cambridge University Press, 1979).

tivities: 'if we understand the rules of a language-game, we understand all that there is to understand about why moves in that language-game are made.'"[8]

As McCarthy goes on to say, ethnomethodological studies offer an especially appropriate resource for antifoundationalist investigations of practical action and situated rule use.

Fragmentary programs and complex interweavings

My task in this book would be much easier if I could simply present coherent lessons from the literatures in ethnomethodology and sociology of science. Unfortunately, I cannot pretend to do this, and so I am compelled to carry out immanent critiques of both approaches while reconstructing them for expository purposes. Thus far, I have characterized these two fields as though each exemplified a unified approach to a subject matter. This is far from the case. Although both fields are small enough that most practitioners know, or at least know of, one another and although both include specialized journals and commonly recognized landmark writings, neither is integrated by a single set of epistemic commitments. Both ethnomethodology and the sociology of science include confusing arrays of research programs, and both fields harbor an entire range of epistemic commitments. Virtually all of the familiar divisions between formalist versus antifoundationalist, value-free versus politicized, and positivistic versus reflexive modes of inquiry appear in the disputatious literatures of ethnomethodology and the sociology of science, and virtually every familiar position in the philosophy of language, science, and action has been expounded at one time or another.

To compound the expository difficulties, both fields, and especially ethnomethodology, can be notoriously difficult to understand. This is especially the case for some of the best work in these fields. It is also very difficult to do ethnomethodology and the sociology of science in an innovative way. Numerous studies pass themselves off under the banners of ethnomethodology and the "new" sociology of science, without strongly exemplifying the radical initiatives in those areas. Consequently, I need to be selective when I characterize ethnomethodology and sociology of science. But more than that, I need to do a great deal of critical preparation before recommending the research in either or both of these fields to scholars in a transdisciplinary community.

My expository task is also made difficult by the complex interweavings among different programs in ethnomethodology and sociology of science. As

[8] Thomas McCarthy, "Private irony and public decency: Richard Rorty's new pragmatism," *Critical Inquiry* 16 (1990): 355–79, quotation on p. 359. Quotation from Richard Rorty, *Philosophy and the Mirror of Nature* (Princeton, NJ: Princeton University Press, 1979), p. 174.

I explain in Chapter 1, ethnomethodology was founded in the late 1950s by Harold Garfinkel, and shortly thereafter it became familiar as a phenomenologically inspired program for studying ordinary discourse and practical reasoning. From the outset, Garfinkel and his colleagues became infamous for their criticisms of established theoretical and methodological approaches to sociology. These criticisms, along with some of the conceptual themes developed by ethnomethodologists, influenced subsequent research and argumentation in the sociology of science.

In the 1970s, a group of British scholars broadened the scope of the sociology of knowledge and began to investigate the social production of knowledge in the "exact" sciences and mathematics. The early social studies of scientific knowledge were programmatic or historical in focus, but by the mid-1970s a few researchers hit on the idea of treating contemporary scientific laboratories as workplaces in which knowledge and facts were "constructed" or "manufactured," and they began to conduct what came to be known as *laboratory studies:* observational studies organized around some of the themes that had been raised earlier by ethnographic and ethnomethodological studies of other practical activities.

Roughly at the same time, and in an independent development, Garfinkel and a few of his students began to pay serious attention to the discourse and practical actions of laboratory scientists and mathematicians. Although there were, and continue to be, affinities between these studies and the larger body of studies in sociology of scientific knowledge, they differ in a number of important respects. To examine these differences can be very confusing, among other things, because Garfinkel and his students have developed an approach that differs in significant respects from other programs in ethnomethodology.

The term *ethnomethodology* has, to a large extent, taken on a life of its own, and it is often used casually to describe any of a variety of ethnographic or hermeneutic approaches to situated social practices. While recognizing that an attempt to distinguish an authentic "ethnomethodology" from various pretenders would be a tendentious exercise in hairsplitting and internecine rivalry, I think there is a need to clarify what the approach does or can promise. Although as I have suggested, ethnomethodology and sociology of science offer distinctive empirical – although not necessarily empiricist or foundationalist – approaches to epistemology's traditional topics, their radical potential has been undercut by recent developments in both fields. Just as their studies are beginning to be appreciated in the wider field of science studies, constructivist sociologists of science have become caught up in skeptical questions about their own research. This concern with what is sometimes called *reflexivity* has worked to the detriment of the naive energy that once inspired studies of "actual" scientific practices.

At the same time and especially in the United States, the older programs in functionalist and institutional sociology have coopted many of the radical initiatives raised by the constructivists. A similar fate has befallen ethnomethodology. At the very time when philosophers like McCarthy and literary critics like Stanley Fish mention ethnomethodology as an exemplary antifoundationalist approach to discourse and social practice, much of the research in the field has taken a decidedly foundationalist turn. The spin-off program of conversation analysis has become the most visible exemplar of ethnomethodology in the fields of sociology, linguistics, and communication studies. Conversation analysts have advanced increasingly formalist and foundationalist claims about the organization of language use, and many of them treat Garfinkel as a distant "father figure" whose radical initiatives are now mainly of historical interest. At the same time, as I discuss at length, Garfinkel's continuing program of studies offers a strong alternative to the formalist and foundationalist approach advocated by many of the more influential conversation analysts.

Given these complications, I do not want to construct an overview of ethnomethodology and the sociology of science, in the sense of presenting a comprehensive taxonomy of the different styles of research represented in the two fields. Rather, my endeavor is far more tendentious and destabilizing. I argue that studies in ethnomethodology and sociology of science not only offer critical purchase on topics in epistemology and social theory but also provide leverage for an immanent critique of the modes of explanation and analysis that are employed in both fields. The sociology of science offers critical leverage against some of the scientistic tendencies expressed in many ethnomethodological and conversational analytic studies. At the same time, ethnomethodological studies offer what I believe is a more sophisticated understanding of language use and practical action than is found in constructivist sociology of science. Consequently, although I recommend ethnomethodology and the sociology of science as research fields that offer empirical approaches to epistemology's traditional topics, I devote a great deal of critical attention to questions about just how these fields can more effectively address those topics.

It also has become clear that there is no one-way street between empirical studies of practical actions and philosophical approaches to discourse and practical reasoning. I cannot simply insist that ethnomethodology and the sociology of science provide empirical foundations for discussions on epistemological issues. Nor can I simply attribute developments in these fields to a priori philosophical commitments. It is certainly the case that a great deal of (often dubious) philosophy is advanced under the banner of empirical sociology, but philosophers and humanities scholars are no less likely to advance dubious claims about society, language, technology, and science when addressing the condition of "modern" or "postmodern" knowledge.[9] The systematic differences between the ways that ethnomethodologists and sociologists of science take up epistemology's topics demonstrate that there is no unequivocal standard for what counts as "empirical" sociological research. Consequently, rather than ending familiar philosophical debates about meaning, rationality, objectivity, and the like, the programmatic claims and "empirical" research strategies in ethnomethodology and sociology of science position themselves within those debates.

Ethnomethodologists and sociologists of science tend to draw on phenomenology and Wittgenstein's later writings, but even though their philosophical commitments remain significant, their research is not philosophical in any established sense. Although there is no clear-cut basis for separating the sense and adequacy of the empirical claims advanced by these two research programs from various discursive accounts of language and knowledge advanced by philosophers, I argue that they offer treatments of epistemology's topics that are neither philosophical nor sociological in the usual sense.

The plan of this book

This book provides the theoretical policies for a set of empirical studies and exercises that I intend to publish in later work. It is a review of research in ethnomethodology and the sociology of science that sets up a critical dialogue within and between the two fields. The first three chapters focus mainly on developments in sociology. Chapter 1 discusses the "invention" of ethnomethodology and reviews some of the themes and developments associated with the research program. Chapter 2 traces the development of a "new" sociology of knowledge that attempted to broaden the application of Mannheim's "non-evaluative total conception of ideology" and to displace Merton's functionalist program for studying scientific norms and institutions. Chapter 3 presents a critical discussion of the more prominent programs in the new sociology of scientific knowledge: the "strong program" in the sociology of science, the "empirical relativist" program, the ethnographic "laboratory studies," and others.

[9] For example, see Heidegger's essay, "The question concerning technology," pp. 3–35, in Martin Heidegger, *The Question Concerning Technology and Other Essays*, trans. William Lovitt (New York: Harper & Row, 1977). Heidegger offers some illuminating conceptual rubrics, but his pronouncements are launched from such abstract heights that they beg a more differentiated examination of the history of science and technology. Lyotard's much celebrated "report" on the postmodern condition is another conspicuous example. Although Lyotard does draw on the literature in the social studies of science, his claims are extraordinarily sweeping and unsubstantiated. See Jean-François Lyotard, *The Post-Modern Condition: A Report on Knowledge*, trans. G. Bennington and B. Massumi (Minneapolis: University of Minnesota Press, 1984).

The next three chapters broaden the scope of the discussion by examining some of the problems associated with the empirical approaches to language, practical action, science, and technology discussed in the previous chapters. Although many ethnomethodologists and sociologists of science assert that their studies are empirical and that they no longer need to address "philosophical" considerations concerning them, I believe that we cannot so easily put aside the chronic problems associated with skepticism, scientism, and linguistic representation. These problems often are peremptorily "solved" by programmatic claims to the effect that an accumulation of empirical findings justifies calling an end to "metatheoretical" debate. Although I do not claim that definite solutions to such problems can be discovered by more careful study of the philosophical literature, I do contend that many ethnomethodologists and sociologists of scientific knowledge hold dubious and self-contradictory preconceptions of language, science, and practical action. As a sociologist I am in no position to advance a philosophy of science that corrects such "deficiencies," but I hope to establish that many of the topics of epistemology can be addressed in an interesting and informative way by examining contemporary scientific practices. My recommendation is not to adopt a "new and improved" set of assumptions about language, practical action, science, and knowledge but to suggest how these and other epistemic matters should be *topicalized* for empirical investigation. This recommendation, of course, can itself be criticized for making undefended assumptions or for setting up an infinite regress, but I argue that epistemic matters can be reviewed without falling into the aporias of an endlessly skeptical "reflexivity."

Chapter 4 discusses ethnomethodology's (and, to a lesser extent, the sociology of scientific knowledge's) debt to phenomenology and existential philosophy. After a brief discussion of Husserl's phenomenological explication of the mathematization of nature, the chapter lays out an ethnomethodological conception of the "local production" of technical actions. The latter part of the chapter then criticizes the way that phenomenological research (particularly that of Alfred Schutz) has been incorporated into "protoethnomethodological" studies that draw a distinction between "scientific" analysis and "everyday" knowledge.

Chapter 5 examines the significance for research in ethnomethodology and the sociology of scientific knowledge of Wittgenstein's later investigations of language and mathematics. The chapter begins with a discussion of how a skeptical interpretation of Wittgenstein's argument about rules in arithmetic has become an established tenet in the sociology of science. I then look at some of the criticisms of rule skepticism in post-Wittgensteinian philosophy while arguing that Wittgenstein's writings problematize the aims of an explanatory sociology of knowledge just as much as they undermine foundationalist philosophy. I finish the chapter by suggesting how ethnomethodology offers a way out of the paradoxes of a relativist or skeptical sociology of knowledge.

Chapter 6 describes and criticizes the program in "molecular sociology" that became established in the field of conversation analysis (CA). CA was once closely affiliated with ethnomethodology, and it is often considered to be ethnomethodology's most successful empirical program. I believe, however, that CA's descriptive program has taken a formalist and foundationalist path that differs profoundly from the orientation to practical actions that is taken in ethnomethodological studies of science. By critically expounding on these differences, in reference to the analytic language and communal research strategies in CA, the chapter introduces a proposal for a "postanalytic ethnomethodology" that is developed in Chapter 7.

Chapter 7 addresses what I believe to be a common problem faced by ethnomethodology and sociology of scientific knowledge: how to analyze particular settings of social practice without trading on the terms by which members make partisan claims and conduct their disputes. I contend that there is no escape from this problem, but that the problem arises out of a misunderstanding. The idea that there is a general problem in the first place implies the possibility of such an escape, and if we recognize that the possibility of escape is an illusion, the problem vanishes. I suggest that ethnomethodological studies of science provide a way to examine epistemic activities without buying into dualistic oppositions between scientism or subjectivism.

In later work, I intend to build on the program outlined in this volume and to present a series of studies and exercises that demonstrate how a postanalytic ethnomethodology can *respecify* selected topics in philosophy and history of science. These topics include observation, representation, measurement, discovery, and explanation. By respecifying them, I hope to treat these familiar epistemological topics as terms that gloss over immensely varied practical phenomena.[10] The aim of such respecification is to provide a set of detailed and vivid cases for describing the locally organized production of epistemic language games, thus enriching our understanding of the complex fields of activity called science.

[10] My initiatives in this regard are taken from an unpublished source informally known as Garfinkel's "blue book": Harold Garfinkel, Eric Livingston, Michael Lynch, Douglas Macbeth, and Albert B. Robillard, "Respecifying the natural sciences as discovering sciences of practical action, I & II: doing so ethnographically by administering a schedule of contingencies in discussions with laboratory scientists and by hanging around their laboratories," unpublished manuscript, Department of Sociology, University of California at Los Angeles, 1989. For published sources that include some of the arguments from the "blue book,"see Garfinkel, "Respecification: evidence for locally produced, naturally accountable phenomena"; and Harold Garfinkel and D. Lawrence Wieder, "Evidence for locally produced, naturally accountable phenomena of order*, logic, reason, meaning, method, etc., in and as of the essentially unavoidable and irremediable haecceity of immortal ordinary society: IV two incommensurable, asymmetrically alternate technologies of social analysis," pp. 175–206, in G. Watson and R. Seiler, eds., *Text in Context: Contributions to Ethnomethodology* (London: Sage, 1992).

CHAPTER 1

Ethnomethodology

Ethnomethodology can be described briefly as *a way to investigate the genealogical relationship between social practices and accounts of those practices.* It generally is considered to be a subfield of sociology, although it also is represented in communication studies, science studies, anthropology, and philosophy of the social sciences. Its connection to twentieth-century traditions in social theory and sociological methodology is both deep and ambivalent. It is deep because ethnomethodology offers a way to address a set of themes that have unquestioned pride of place in theories of social structure and social action. They include action, order, rationality, meaning, and structure, among others. These themes are also prominent in methodological debates about the relationship between commonsense reasoning and scientific analysis. The connection is ambivalent because ethnomethodology's orientation to these foundational topics puts it at odds with most of the established theories and methodologies in the social sciences.[1]

Ethnomethodology's relationship to sociology is difficult to describe and comprehend. Many ethnomethodologists work in sociology departments, so in that sense they are sociologists, but one of their most infuriating research policies has been to place "professional" sociological methods for generating knowledge of social structure alongside the "lay" know-how that is substantively part of the society that sociologists study. As a matter of research policy (if not personal conduct), ethnomethodologists treat the "family concerns" of professional sociology with studious indifference.[2] By treating lay and professional methods as part of the same domain of study, they distance themselves from the disciplinary form of life in which they and their sociologist colleagues conduct their professional affairs. The infuriated

[1] These issues are addressed at length in Graham Button, ed., *Ethnomethodology and the Human Sciences* (Cambridge University Press, 1991).

[2] The policy of ethnomethodological indifference is discussed in Harold Garfinkel and Harvey Sacks, "On formal structures of practical action," pp. 337–66, in J. C. McKinney and E. A. Tiryakian, eds., *Theoretical Sociology: Perspectives and Developments* (New York: Appleton-Century-Crofts, 1970), repr. in H. Garfinkel, ed., *Ethnomethodological Studies of Work* (London: Routledge & Kegan Paul, 1986), pp. 160–93. Also see Benetta Jules-Rosette, "Conversation avec Harold Garfinkel," *Sociétés: revue des sciences humaines et sociales* 1 (1985): 35–39. I elaborate on the policy of ethnomethodological indifference in Chapter 4.

reaction by the colleagues down the hall can be no less understandable and no less well founded than the high rage expressed by family members toward the students who performed an ethnomethodological "breaching experiment" by pretending to be strangers in their family households: "Why must you always create friction in our family harmony?" "I don't want any more of *that* out of you, and if you can't treat your mother decently, you'd better move out!" "We're not rats, you know."[3]

What would motivate such a disconcerting and apparently disrespectful "attitude" toward the profession?[4] Different answers can be given in different cases, but for many of us, it has to do with the combination of an intense interest in the topics that sociologists address and a profound disappointment with what happens to those topics when they are subjected to professional sociological analysis. This disappointment is due not only to the widely acknowledged "looseness" of sociology's analytic procedures or the inconclusiveness of the predictions based on them. In my judgment, it has more to do with the way that sociological perspectives and methods have been designed to give unified treatments of an entire roster of topics: families, religions, riots, gender relations, race and ethnicity, class systems, and the like. Of course, plenty of room is left for the empirical study of these topics, but the dominant trend in late-twentieth-century sociology has been to subsume the entire roster of social phenomena under an overarching conceptual framework that defines a set of analytic categories that apply at the "level" of society as a whole.[5] Consequently, familiar facets of family life, religious experience, economic activity, and daily life generally become thinned out when they are treated as cases to be analyzed by referring to a stock of sociological variables and using statistical procedures to determine sources of variation from norms. Sociology also includes so-called qualitative and micro approaches, but often the tendency is to treat these as preliminary or degraded modes of analysis, and the descriptive orientation in such studies is often dominated by an effort to show how the observable

[3] Harold Garfinkel, *Studies in Ethnomethodology* (Englewood Cliffs, NJ: Prentice-Hall, 1967), pp. 48–49.

[4] For rhetorical purposes, I am presenting the story as though all ethnomethodologists evinced the same (extreme) attitude toward their colleagues. Actually, persons who identify themselves as ethnomethodologists have worked out various accommodations to their disciplinary situations. These can be likened to the different "stances" that inmates in an asylum take toward the institution and its staff: openly rebelling, going along with, or making the best of the situation. See Erving Goffman, *Asylums* (Garden City, NY: Doubleday, 1961).

[5] This approach derived from Talcott Parsons's theories of social action and social system. The dominance of Parsons's "structural functionalist" program began to wane in the late 1960s, and many sociologists today treat it as defunct. To a large extent, however, the Parsonian conception of science and basic orientation to social order remains very much alive in the glossaries of concepts and in treatments of theory and method in contemporary sociology textbooks.

details of the cases studied "reflect" the behind-the-scenes operation of abstract social "forces." There is a clear and understandable reason for sociology's dominant theoretical and methodological orientation: It instantiates an *image* of a unified science inherited from logical–empiricist philosophy of science, and it conveys a *sense* of a progressive and authoritative approach that incorporates the essence of natural science inquiry.

Debates about whether sociology should be scientific have gone on for at least a century, and those who opposed scientism in sociology lost the battle long ago, even though their position continued to survive as a minority view. However, ethnomethodologists opened a new front in this classic battle by closely studying the relationship between sociological practice and everyday language, and by so doing they provided a kind of empirical rebuttal to sociology's scientism that could no longer be dismissed as "merely" a philosophical argument. Since the early 1970s still another battlefront was opened up when constructivist sociologists of science, and a few ethnomethodologists, turned their attention to the "actual" practices in the natural sciences and found (or at least claimed) that the logical–empiricist view of scientific inquiry was an idealized and substantially mistaken version of scientific practice. These studies were far from definitive, and many of them remained respectful of established sociological traditions, but they raised new kinds of trouble for analytic sociology's conception of science. First, they "problematized" many of the assumptions about scientific theory and method that are entrenched in sociology's pedagogy, and second, they challenged the tendency in sociology to address "social aspects" that were analytically distinct from the "concrete" and "internal" details of the practices studied.

For now, I shall defer a more extended discussion of how the sociology of scientific knowledge confronted established sociological conceptions of scientific theory and method. In the rest of this chapter, I present a brief history of ethnomethodology, in order to introduce some of the distinctive research policies that inform the discussions in later chapters. As mentioned in the Introduction, my overall purpose is to set up a critical convergence between the treatments of scientific practice and ordinary language use developed by ethnomethodologists and sociologists of science. This, in turn, will lead to a set of recommendations for "respecifying" some of the central topics of social theory and epistemology.

Garfinkel's invention of ethnomethodology

Unlike many other developments in sociology, ethnomethodology can be traced to a definite origin. Harold Garfinkel is universally acknowledged as the "founding father" of the field, although he occasionally has joked that he has sired "a company of bastards." He coined the term *ethnomethodology* in the mid-1950s, but it did not become familiar until the mid-1960s when several of his and his students' works were published. During a symposium held in the late 1960s, Garfinkel recounted the story of how he came up with the term ethnomethodology while he was preparing a series of reports for a multidisciplinary study of jury deliberations at the University of Chicago:

> I was interested in such things as jurors' uses of some kind of knowledge of the way in which the organized affairs of the society operated – knowledge that they drew on easily, that they required of each other. At the same time that they required it of each other, they did not seem to require this knowledge of each other in the manner of a check-out. They were not acting in their affairs as jurors as if they were scientists in the recognizable sense of scientists. However, they were concerned with such things as adequate accounts, adequate description, and adequate evidence. They wanted not to be "common-sensical" when they used notions of "common sensicality." They wanted to be legal. They would talk of being legal. At the same time, they wanted to be fair. If you pressed them to provide you with what they understood to be legal, then they would immediately become deferential and say, "Oh, well, I'm not a lawyer. I can't be really expected to know what's legal and tell you what's legal. You're a lawyer after all." Thus, you have this interesting acceptance, so to speak, of these magnificent methodological things, if you permit me to talk that way, like "fact" and "fancy" and "opinion" and "my opinion" and "your opinion" and "what we're entitled to say" and "what the evidence shows" and "what can be demonstrated" and "what actually he said" as compared with "what only you think he said" or "what he seemed to have said." You have these notions of evidence and demonstration and of matters of relevance, of true and false, of public and private, of methodic procedure, and the rest. At the same time the whole thing was handled by all those concerned as part of the same setting in which they were used by the members, by these jurors, to get the work of deliberations done. That work for them was deadly serious.[6]

As Garfinkel elaborates, the jurors conducted themselves predominantly as practical reasoners with no professional credentials or technical expertise for collecting and assessing evidence, performing methodic demonstrations, or making judgments about matters of fact and opinion. Nevertheless, in their own way they addressed familiar "methodological" concerns during

[6] In Richard J. Hill and Kathleen Stones Crittenden, *Proceedings of the Purdue Symposium on Ethnomethodology* (Purdue, IN: Institute for the Study of Social Change, Department of Sociology, Purdue University, 1968), pp. 6–7. This excerpt was reprinted in Harold Garfinkel, "On the origins of the term 'ethnomethodology,'" pp. 15–18, in Roy Turner, ed., *Ethnomethodology* (Harmondsworth: Penguin Books, 1974). Also see John Heritage, *Garfinkel and Ethnomethodology* (Oxford: Polity Press, 1984), p. 45.

their deliberations, and the way they did so was substantively related to the verdict they negotiated. By analogy with the various "ethnoscience" approaches in social anthropology, Garfinkel proposed that "ethnomethodology" would be a way to study these largely untutored methods of practical reasoning.

While pursuing a Ph.D. in Harvard's Department of Social Relations in the late 1940s, Garfinkel had been exposed to some of the early efforts by anthropologists to develop "ethnoscience" approaches. A few years later, while working on the Chicago jury project, he read a review of research in those fields and hit on the idea of developing an "ethnomethodology." The ethnosciences – ethnobotany, ethnomedicine, ethnophysics, and so forth – were designed to elicit culturally specific taxonomies of plants, animals, medicines, color terms, and other semantic domains and to map them against a backdrop of relevant scientific knowledge.[7] So, for instance, ethnobotany is the study of "native" classifications of plants, that is, culturally specific systems of names for differentiating types of plants, including the systematic axes and hierarchical relationships between plant categories. After eliciting these classifications from native informants, the ethnobotanist compares them with the taxonomies developed by contemporary botanists. The differences between native and scientific taxonomies can then be interpretively related to characteristic patterns of native custom, ceremonial practice, and kinship organization. By analogy, ethnomethodology would be the study of the ordinary "methods" through which persons conduct their practical affairs. The relationship between ethnomethodology and the other ethnosciences should, however, be treated with some caution, since there are some striking differences:[8]

1. Garfinkel characterized the jurors' methodology in terms of broad epistemic distinctions.[9] The jurors' methods for assessing the veracity of

[7] The ethnosciences retained some of Durkheim and Mauss's interest in native classifications. See Emile Durkheim and Marcel Mauss, *Primitive Classification*, trans. J. W. Swain (London, 1915). Claude Lévi-Strauss's more general discussion of the "sciences of the concrete" draws on ethnoscience research in *The Savage Mind* (Chicago: University of Chicago Press, 1966), chap. 1. Later, some of these approaches became integrated into cognitive anthropology. See Charles Frake, "The diagnosis of disease among the Subanum of Mindanao," *American Anthropologist* 63 (1961): 113–32; Harold Conklin, "Hanunóo color categories," *Southwestern Journal of Anthropology* 11 (1955): 339–44; William C. Sturtevant, "Studies in ethnoscience," *American Anthropologist* 66 (1966): 99–131; Stephen Taylor, ed., *Cognitive Anthropology* (New York: Holt, Rinehart and Winston, 1969).

[8] See Harvey Sacks's remarks on the difference between ethnomethodology and ethnoscience in Hill and Crittenden, eds., *The Purdue Symposium on Ethnomethodology*, pp. 12–13.

[9] What Garfinkel developed can be described as an "epistemic sociology." See Jeff Coulter, *Mind in Action* (Oxford: Polity Press, 1989), pp. 9 ff. The idea of "epistemic sociology" runs the risk of overintellectualizing the juror's work, but it does enable us to see that Garfinkel was respecifying the central topics in epistemology by identifying them as commonplace discursive and practical activities. This theme is picked up in Garfinkel's and his students' studies of work in the sciences (see Chapter 7).

testimony and correctly interpreting the evidence did not closely compare with technical research procedures in jurisprudence, nor did they even dimly resemble standard designs for experimental or other scientific methods. The jurors' deliberations over "facts," "reasons," "evidence," and so on were expressed in a natural language shared with professional jurists and scientists, but the jurors' methodic practices did not closely parallel the methodological techniques in the natural or human sciences. Garfinkel's ethnomethodology was thus not, strictly speaking, an ethnoscience, since it was not precisely based on the model of scientific methodology.[10] Instead, he treated the jurors' commonsense methods as phenomena in their own right. In this, he was influenced by Alfred Schutz's efforts to explicate the phenomenology of the social world.[11]

2. Garfinkel was not proposing to develop *taxonomies* of ordinary methods, nor was he trying to delimit a semantic domain common to particular scientific fields and native cultures. The phenomenon for ethnomethodology was not the system of *names* that jurors used to refer to methodological matters; instead, it was how they accomplished their "methodological" determinations over the course of their deliberations. For Garfinkel, such "methods" include the entire range of lay and professional practices through which social order is produced. By conceiving of these methods as subject matter, Garfinkel was proposing an encompassing approach to the study of social actions and not a study of a particular domain of native classifications (of plants, animals, colors, kin relations, etc.).

3. The "natives" on the jury were not members of an exotic culture; they were variously educated, English-speaking members of the same society that Garfinkel inhabited. For him to study the terms and procedures according to which jurors conducted their deliberations did not require a tutorial in a specialized language, nor did it require anything like the sort of ethnography of distinct horticultural or medicinal arts that might inform a study of ethnobotany or ethnomedicine. The jurors' practical reasoning was intuitively transparent, so much so, in fact, that it was easily disregarded by the other researchers in the Chicago jury project. The researchers wanted most of all to get "behind" the apparent surface of the deliberations to find out what really motivated the jurors. Garfinkel related an anecdote about the jury project that concisely illustrates this:

> In 1954 Fred Strodtbeck was hired by the University of Chicago Law School to analyze tape recordings of jury deliberations obtained from a bugged jury room. Edward Shils was on the committee that hired him. When Strodtbeck

[10] In some respects ethnomusicology would be more comparable because no single genre of modern music provides the measure of native musical practices.

[11] See Alfred Schutz, *Collected Papers*, vols. 1 and 2 (The Hague: Nijhoff, 1962, 1964). I give a more elaborate account of Garfinkel's relation to phenomenology in Chapter 4.

> proposed to a law school faculty that they administer Bales Interactional Process Analysis categories, Shils complained: "By using Bales Interaction Process Analysis I'm sure we'll learn what about a jury's deliberations makes them a small group. But we want to know what about their deliberations makes them a jury." ... Strodtbeck replied and Shils agreed: Shils was asking the wrong question![12]

Strodtbeck's reply that "Shils was asking the wrong question" alerts us to a dilemma for sociological studies of practical actions that remains contentious to this day. Bales Interactional Analysis is a method of "content analysis" that was once widely used in social psychology.[13] When using the procedure, the analyst observes an interactional event (either live or on tape), and while doing so he or she "codes" the utterances by selecting among an inventory of categories, for example, deciding that a particular speaker's utterance "gives support" to another party in the conversation. Strodtbeck recommended Robert Bales's procedure as a way to reduce the hours of tape-recorded deliberations to a manageable, statistically analyzable data base. Shils's complaint acknowledged that Bales's coding scheme would enable the researchers to construct the sort of data that conventional sociological methods were designed to "handle" but that by using it they would fail to address a more contextually specific set of "methods" that the jurors used. Had Shils and Strodtbeck taken the complaint seriously, they would have been left with no methodological basis beside their own commonsense, prescientific understandings of the "raw data" recorded on their tapes. Nevertheless, Garfinkel took Shils's complaint very seriously, and by doing so he courted the difficulties that Strodtbeck's reply avoided.

4. No developed natural or social science provided a comparative basis for defining and assessing the jurors' methods. Ethnobotanists can often presume a relatively complete scientific inventory of the plants in the region studied, or short of that, they can begin to produce one by conducting laboratory analyses of specimens supplied by native informants. With the scientific taxonomy in hand, the botanist can then consult the native inventory of plant categories to find out whether it includes names for each scientifically validated species. By doing so, the botanist can discover whether the natives assign different names to morphologically different male and female plants of the same species and whether their taxonomy includes distinctions or entire axes that the scientific classification system lacks. In cases in which natives use an herb for medicinal purposes or they prohibit the

[12] Harold Garfinkel, Michael Lynch, and Eric Livingston, "The work of a discovering science construed with materials from the optically-discovered pulsar," *Philosophy of the Social Sciences* 11 (1981): 131–58, quotation from p. 133.

[13] Robert Bales, *Interactional Process Analysis: A Method for the Study of Small Groups* (Reading, MA: Addison-Wesley, 1951).

eating of a plant, laboratory analysis can be used to assess the "actual" biochemical composition and physiological effects of the plant substances. In contrast, under a research policy that Garfinkel later termed "ethnomethodological indifference," he and his students made no assumption that a particular set of methodological prescriptions – whether drawn from sociology, a particular natural science, or formal logic – can act as a standard for defining a rationality operating "beneath" native methods. This policy articulated a decisive break with the established Weberian and Parsonian theories of action that dominated sociology at the time. Although Max Weber's action theory is quite complicated, and it prescribes a legitimate societal place for different types of nonrational as well as rational action,[14] it proposes a descriptive method that uses an idealized "standpoint" of an omniscient scientific observer:

> When we adopt the kind of scientific procedure which involves the construction of *types*, we can investigate and make fully comprehensible all those irrational, affectively determined, patterns of meaning which influence action, by representing them as "deviations" from a pure type of the action as it would be if it proceeded in a rationally purposive way. For example, in explaining a panic on the stock exchange, it is first convenient to decide how the individuals concerned would have acted if they had *not* been influenced by irrational emotional impulses; then these irrational elements can be brought in to the explanation as "disturbances." Similarly, when dealing with a political or military enterprise, it is first convenient to decide how the action would have proceeded if all the circumstances and all the intentions of those involved had been known, and if the means adopted had been chosen in a fully rationally purposive way, on the basis of empirical evidence which seems to us valid. Only then does it become possible to give a causal explanation of the deviations from this course in terms of irrational factors.[15]

The first-person plural pronouns in this passage identify the narrator's voice with the hypothetical scientific observer's viewpoint. In order to "make fully comprehensible" actions in complex historical fields, "our" descriptions would have to have extraordinary scope and specificity. A sociology constructed along these lines would aspire to be a science of all sciences, capable of rendering complete judgments about singular events in countless circumstances. Just imagine what it would take to "decide how the

[14] See Stephen Kalberg, "Max Weber's types of rationality: cornerstones for the analysis of rationalization processes in history," *American Journal of Sociology* 85 (1980): 1145–79.

[15] Max Weber, "The nature of social action," pp. 7–32, in W. G. Runciman, ed., *Weber: Selections in Translation* (Cambridge University Press, 1978), quotation from p. 9. Also see Weber, *Economy and Society*, ed. and trans. G. Roth and C. Wittich (Berkeley and Los Angeles: University of California Press, 1978), p. 6; and Jürgen Habermas, *The Theory of Communicative Action*, vol. 1, trans. Thomas McCarthy (Boston: Beacon Press, 1981), pp. 102–3.

action would have proceeded" in the battle of Waterloo "if all the circumstances and all the intentions of those involved had been known."[16] It can fairly be said that Weber's "we" is inserted in the grammatical place occupied by a omniscient deity in more explicitly theological discourse.[17]

Although Talcott Parsons distinguished his voluntaristic theory of action from positivistic and utilitarian versions of rationality, his theory of social action similarly retained the judgmental position of an idealized scientific observer. Like Weber, Parsons acknowledged that actions could be governed by socially accepted, legitimate standards other than those verified by empirical science. As he defined it, "intrinsically rational" action adhered to scientifically verifiable standards, but this form of action was only a subclass of a much broader domain of normatively guided social action. Parsons never let go of the theoretical leverage provided by an *idealization* of a scientific observer's knowledge of the conditions and choices relevant to actual situations. Parsons used the *imaginable possibility* of obtaining verifiable empirical knowledge of the situation as a standard for defining and distinguishing among the various subjective elements and normative standards composing a conceptual framework for action.[18]

Garfinkel completed his Ph.D. dissertation under Parsons's guidance,[19] and his earlier studies were heavily influenced by Parsons's action theory, but his account of jurors' reasoning entirely recast the relation of the scientific observer to the social actions being analyzed. In contrast with long-standing traditions in sociology, economics, and psychology, Garfinkel advocated neither a man-the-scientist model of rational action[20] nor a virtual "scientific" standpoint from which to assess the rationality of any practice. What was distinctive about his agenda was *not* that he wanted to study ordinary methods of practical reasoning but that he disavowed the privilege of an academic or administrative science. The traditions of social study he rejected also investigate commonsense knowledge and practical reasoning, typically by using formal models and prepackaged analytic techniques, but Garfinkel decided to make a topic of commonsense knowledge of social

[16] These complexities can be appreciated by reading John Keegan, *The Face of Battle: A Study of Agincourt, Waterloo and the Somme* (New York: Viking, 1976).

[17] James Edwards, *The Authority of Language: Heidegger, Wittgenstein, and the Threat of Philosophical Nihilism* (Tampa: University of South Florida Press, 1990).

[18] See Talcott Parsons, *The Structure of Social Action,* vol. 1 (New York: McGraw-Hill, 1937). A useful discussion of Parsons's theory of action and the phenomenological criticisms of it is presented in Heritage, *Garfinkel and Ethnomethodology,* pp. 7–36.

[19] Harold Garfinkel, "The perception of the other: a study in social order" (Ph.D. diss. Harvard University, 1952).

[20] Examples of "man-the-scientist" models of action and reasoning are presented in several of the psychological and cognitive science studies in Denis J. Hilton, ed., *Contemporary Science and Natural Explanation: Commonsense Conceptions of Causality* (New York: New York University Press, 1988).

structures without first setting up a scientific counterpoint to that knowledge.

As I elaborate in Chapters 4, 6, and 7, ethnomethodology's disavowal of classic studies of action, reasoning, and social structure can be extremely puzzling. Garfinkel's proposals seem to divest ethnomethodology of any special entitlement to knowledge about the practices it studies, and it can be baffling to consider what an ethnomethodological pedagogy or an ethnomethodological demonstration of research findings might look like. The entire nexus of methodological concepts and preliminary arguments for establishing a study's evidential claims and conclusions would seem to dissolve in a confused relationship to the fields of commonplace methods described in the research. If methods are to be situated on the side of *what* is described, what can ensure the truth, relevance, and adequacy of any description? These concerns have often been raised about, and against, ethnomethodology, and later I address them at length. But for the present, let me simply mark them as chronic puzzles and complaints about the consequences of Garfinkel's invention.

Early initiatives

Although Garfinkel came up with the term ethnomethodology in the mid-1950s and some of the studies and research policies that later came to be associated with the field were already present in his Ph.D. dissertation,[21] it was not until the mid-1960s that ethnomethodology became a familiar part of sociology.[22] A few of Garfinkel's ethnomethodological studies were presented at conferences and published in the late 1950s and early 1960s.[23]

[21] Garfinkel, "The perception of the other." Talcott Parsons was Garfinkel's dissertation adviser, and the dissertation attempts to develop Parsons's theory of action by infusing it with phenomenological insights largely drawn from Schutz's writings. While at Harvard, Garfinkel was exposed to Aron Gurwitsch's teachings in phenomenology, and he also read some of Husserl's original works. At the time, he did not decisively break with Parsons's program, and it can be argued that the rupture was not complete until the publication of Garfinkel's *Studies in Ethnomethodology*. However, evidence of Garfinkel's "deep disquietude" with the Parsonian system is evident in his dissertation and in various unpublished manuscripts Garfinkel circulated in the late 1950s and early 1960s comparing Parsons's and Schutz's theoretical positions.

[22] An account of some of the early seminars in the 1950s and 1960s is provided in a rather strange account of the history of ethnomethodology: See Pierce Flynn, *The Ethnomethodological Movement* (Berlin: Mouton de Gruyter, 1991), pp. 33 ff.

[23] Some of Garfinkel's early papers include "Aspects of common-sense knowledge of social structure," *Transactions of the Fourth World Congress of Sociology* 4 (1959): 51–65; "The rational properties of scientific and commonsense activities," *Behavioral Science* 5 (1960): 72–83; and "Studies of the routine ground of everyday activities," *Social Problems* 11 (1964): 225–50. The latter two papers were reprinted in *Studies in Ethnomethodology*. As the first reference indicates, Garfinkel presented drafts of some of his papers at conferences in the late 1950s. Somewhat earlier he published a paper in which "ethnomethodology" was not mentioned, although with hindsight it might as well have been. See his "Conditions of successful degradation ceremonies," *American Journal of Sociology* 61 (1956): 240–44.

Perhaps the most significant of these was a paper published in 1963 entitled "A Conception of, and Experiments with 'Trust' as a Condition of Stable Concerted Actions."[24] A year later, his student and colleague Aaron Cicourel published an influential study, *Method and Measurement in Sociology*.[25] Garfinkel's and Cicourel's studies referenced each other, although not without some tension. In the preface to his book, Cicourel acknowledges his debt to Garfinkel's exposition of Schutz's work, but he adds that his book "may depart significantly from [Garfinkel's] own ideas about the same or similar topics. I have not had the benefit of his criticisms, but have sought to footnote his ideas contained in published and unpublished works within the limits of not being given permission to quote from them directly."[26] Although his book was and is still widely regarded as an example of ethnomethodology, the ambivalent relationship between Cicourel's program and ethnomethodology was present from the outset. As I explain in Chapter 4, Cicourel's book might better be regarded as a "protoethnomethodological" treatment of methodological problems in sociology. Cicourel did not claim to be doing an ethnomethodological study, and he drew on a range of sources in sociolinguistics, psychology, and philosophy.[27] With hindsight, it is also possible to say that Garfinkel's "trust" paper was a "protoethnomethodological" treatment of the various orders of rules that can be discerned through the systematic disruption of games and ordinary social routines.

Garfinkel's "trust" paper and Cicourel's book established ethnomethodology as (1) a "method" for discerning the taken-for-granted background assumptions, tacit knowledge, behavioral norms, and standard expectancies through which participants constitute ordinary social scenes and routine interactions and (2) a perspective from which to launch relentless and unsparing investigations of the tacit research practices used in "conventional" social science. Such ethnomethodological investigations were easily read as critiques. As Cicourel and others articulated them, these critiques were based on the observation that survey analysis, social–psychological experimentation, ethnography, interviewing, and other typical research techniques necessarily rely on commonsense reasoning and ordinary interactional practices. This observation applied to an entire range of methodic procedures for gathering, coding, and interpreting data. Although such research techniques require

[24] In O. J. Harvey, ed., *Motivation and Social Interaction* (New York: Ronald Press, 1963), pp. 187–238.

[25] Aaron Cicourel, *Method and Measurement in Sociology* (New York: Free Press, 1964).

[26] Ibid., pp. iv–v.

[27] In later studies, Cicourel increasingly referred to the cognitive sciences, and he preferred to identify his approach with the title "cognitive sociology." See Aaron Cicourel, *Cognitive Sociology* (Harmondsworth: Penguin Books, 1973).

researchers and subjects to understand and interpret vernacular discourse, social science researchers do not specifically examine these prescientific procedures of ordinary reasoning and social interaction.[28]

For instance, survey analysts count on the fact that the standardized items in their questionnaire instruments will be intelligible to their respondents, and they try to design the surveys to ensure that different respondents will treat each question in a comparable way. Similarly, interview schedules are designed so that a series of questions will elicit stable and meaningful attributes and attitudes. Cicourel argued that such methods were problematic, and unless they were grounded in a more sophisticated understanding of the logic of social interaction, the situated determination of meaning, the workings of memory, and many other related matters, they would remain dubious. Following Garfinkel, he suggested that standard sociological methods presuppose the unproblematic accomplishment of the very social phenomena that deserve further sociological study. As might be expected, this critique touched off confused, hostile, and resentful reactions in the professional sociological community.

The central text and its policies

Garfinkel's book, *Studies in Ethnomethodology,* published in 1967, introduced the basic policies and objectives of ethnomethodology, and it included several papers that Garfinkel had written over a span of several years prior to its publication. Because some of the chapters were written long before the book was published, they do not exemplify the book's program and policies with the same clarity as do the sections written later. Some of the studies, particularly the chapter "The Rational Properties of Scientific and Common Sense Activities,"[29] are heavily indebted to Alfred Schutz's and Felix Kaufmann's methodological writings.[30] As I discuss in Chapter 4, Garfinkel's earlier studies made more explicit use of Schutz's theory-centric view of science, and they adopted Schutz's distinction between scientific and commonsense rationality. Different chapters in *Studies* also exemplify vari-

[28] An argument could be made that social psychologists did indeed examine such prescientific procedures – for example, Solomon Asch, *Social Psychology* (Englewood Cliffs, NJ: Prentice-Hall, 1952). However, although such experimental studies were often insightful, they used a contrast between the experimenter's scientific knowledge and the subject's perception and cognition to elucidate the asymmetry between the two. In that respect they differed from ethnomethodology's nonironic interest in commonsense knowledge, and especially from its *procedural* emphasis on such knowledge.

[29] *Studies in Ethnomethodology,* pp. 262–83. This article originally appeared in *Behavioral Science* 5 (1960): 72–83.

[30] See Alfred Schutz, "The problem of rationality in the social world," *Economica* 10 (1943): 130–49, repr. in Schutz, *Collected Papers,* vol. 2, pp. 64–90; and Felix Kaufmann, *Methodology of the Social Sciences* (New York: Oxford University Press, 1941).

ous phases of Garfinkel's investigations of the practical and organizational production of "conventional" sociological methods.[31] Despite the book's unevenness and although many readers complained of its opaque jargon and obscure polemics, it became and rightly remains the central ethnomethodological text. It introduced many of the conceptual themes, demonstrative exercises, and critical lines of argument that are still regarded as ethnomethodology's distinctive contribution to sociology and the philosophy of the social sciences.

Garfinkel's prolegomenon was written as a practical document[32] rather than a statement of theoretical doctrines. Its explicit aim was to announce a program of studies and to present some of the research in hand. Garfinkel deliberately eschewed methodological principles, and instead outlined a few "maxims" and "policies" for ethnomethodological investigations. He presented these policies as throwaway items rather than names for generic concepts, and did not use them to define an abstract system of organizational principles and analytical elements for explaining the constitution of society. Nevertheless, despite all of its terminological and rhetorical precautions, Garfinkel's text was and still is treated as a theoretical, or even metaphysical, statement. Moreover, ethnomethodologists are often held accountable for doctrinaire readings and criticisms of the text's policies.[33] There is no point in bemoaning such misreadings, since an argument could be made that they are exactly what is responsible for the ambiguous prominence that Garfinkel and ethnomethodology have enjoyed in sociology. As a creature of textbooks, curricula, polemical debates, theoretical lineages, and academic politics, ethnomethodology has been sustained through communal misreadings of its central text: a virtual consensus constituted by deep misunderstandings of a common set of slogans.[34] In a way, ethnomethodology's fate should not

[31] Compare, for instance, chaps. 6 and 7 of *Studies in Ethnomethodology:* "Good organizational reasons for 'bad' clinic records" (pp. 186–207), an incisive account of the tacit organizational reactions to that study's interventions; and "Methodological adequacy in the quantitative study of selection criteria and selection practices in psychiatric outpatient clinics" (pp. 208–61), a review of some of the procedures and problems arising in a study of clinic activities.

[32] In the 1970s and early 1980s Garfinkel entitled various collections of his and his students' unpublished papers and manuscripts as *A Manual for the Study of Naturally Organized Ordinary Activities.*

[33] For a recent example, see David Bloor, "Left- and right- Wittgensteinians," in Andrew Pickering, ed., *Science as Practice and Culture* (Chicago: University of Chicago Press, 1991), pp. 266–82. Bloor translates Garfinkel's text into two fundamental "doctrines" and then asserts that they are contradictory.

[34] For a suggestive elaboration of this picture of a "knowledge community" see Peter Galison, "The trading zone: coordination between experiment and theory in the modern laboratory," paper presented at International Workshop on the Place of Knowledge, Tel Aviv and Jerusalem, May 15–18, 1989. This is a general discussion of the phenomenon of how academic communication can be sustained through systematic creation of a zone of mutual misunderstanding.

be surprising in light of what its practitioners say about the constitution of the social world.

In the next section I briefly review a few of the themes most frequently mentioned in expositions of ethnomethodology. Although there are many other options to choose from, I comment on only three of them: accountability, reflexivity, and indexicality. These themes implicate one another, and each of them indexes an entire swarm of related issues. Garfinkel tends not to state these terms with the *ity* suffix, although expositors invariably do so in the interest of turning them into "key concepts."

Accountability

In the preface of his text, Garfinkel (p. vii) introduces one of his distinctive shibboleths and defines it with a hyphenated phrase: "Ethnomethodological studies analyze everyday activities as members' methods for making those same activities visibly-rational-and-reportable-for-all-practical-purposes, i.e., 'accountable,' as organizations of commonplace everyday activities." The term *accountable* has become an established part of ethnomethodology's unusual vocabulary, and it is often translated into a more economical phrase, "observable-and-reportable," or simply "observable-reportable." When written with the appropriate suffix and (mis)understood as a stable theoretical concept, accountability coimplicates ethnomethodology's research policies and the phenomena it studies. Perhaps an appreciation of this concept can be conveyed by decomposing it into a set of proposals:[35]

1. Social activities are *orderly.* In significant aspects they are nonrandom, recurrent, repeated, anonymous, meaningful, and coherent.
2. This orderliness is *observable.* The orderliness of social activities is public; its production can be witnessed and is intelligible rather than being an exclusively private affair.
3. This observable orderliness is *ordinary.* That is, the ordered features of social practices are banal, easily and necessarily witnessed by anybody who participates competently in those practices.
4. This ordinarily observable orderliness is *oriented.*[36] Participants in orderly social activities orient to the sense of one another's activities, and while doing so they contribute to the temporal develop-

[35] This list is a distilled, simplified, and alliterative rendering of Garfinkel's various recitations of ethnomethodological policies in *Studies in Ethnomethodology* and in more recent lectures and public presentations.

[36] What exactly might be meant by "oriented" or "orientation" is easily misunderstood and subject to different technical specifications. I discuss the matter further in Chapter 6, but meanwhile it should be sufficient to say that "orientation" does not imply a psychological disposition or intention but a gestural or verbal display described using the second- or third-person grammatical case.

ment of those activities. A pedestrian's mere glance can *display* her projectable orientation to "crossing the street," and this display is available to approaching drivers and is thereby constitutively embedded in the socially organized traffic of a public street scene.[37]

5. This orientedly ordinary observable orderliness is *rational.* Orderly social activities make sense to those who know how to produce and appreciate them. Such activities are analyzable and predictable; often they are no less predictable than a sunrise, and indeed, they can be geared to a sunrise and a sunrise to them.[38]
6. This rationally oriented ordinary observable orderliness is *describable.* Masters of the relevant natural language can talk about the order of their activities, and they can talk *in and as* the order of their activities. Consequently, sociological description is an endogenous feature of the fields of action that professional sociologists investigate.[39]

"Accountable" social activities are produced in connection with the possibility of their description in a natural language. In a variety of ways, members can be held to account for what they are doing. They can be asked to keep records, to show that they have followed instructions, to justify their actions in terms of a set of rules or guidelines, and to inform others about what to do and where to go. Such accountability is implied in the *instructable reproducibility* of social actions: practical efforts to instruct and inform members about methods for reproducing and recognizing a "same" action on different occasions. Moreover, the accountable *display* of social order is not produced by a cognitive schema, a set of beliefs, or a society in the mind. Instead, it is identical with the concerted order of driving in traffic, the recognizable and routine orders of moves in a game, and the visible order of service provided by the evident line-up of bodies in a queue.[40]

Reflexivity

One of the more puzzling themes in Garfinkel's text is the matter of the "'reflexive' or 'incarnate' character of accounting practices and accounts"

[37] See David Sudnow, "Temporal parameters of interpersonal observation," pp. 259–79, in D. Sudnow, ed., *Studies in Social Interaction* (New York: Free Press, 1972).

[38] I am not suggesting that orderly social actions can "make the sun come up" (aside from any literal debt that this expression owes to the Ptolemaic tradition) but that a work schedule, for example, or a ceremony can contextually specify a sunrise's anticipatory significance, phase structure, sequential place in an order of activities, and so forth.

[39] See Harvey Sacks's early paper, "Sociological description," *Berkeley Journal of Sociology* 8 (1963): 1–16.

[40] These examples and the way they explicate the issue of accountability were furnished by Garfinkel's lectures and unpublished writings after the publication of *Studies in Ethnomethodology.*

(p. 1). Strikingly different implications can be drawn from Garfinkel's and other sociologists' proposals regarding reflexivity. For the time being, I will give a preliminary characterization. *Reflexivity* is implicated in the phenomenon of accountability. If, as mentioned earlier, sociological descriptions are endogenous to the fields of action that professional sociologists investigate, then such descriptions are *reflexive* to the settings in which they originate. Even if they inadequately represent "society" or some part of "society," such accounts contribute to the discourse and actions in particular social scenes. For example, as we noted, Garfinkel observed that jurors in their own fashion examined various evidential documents and testimonies. Jurors referred to the evidence, and they ruminated about the import of that evidence for the case at hand. They speculated about how the society outside the courtroom could have produced the events under dispute. What they determined was reflexive to their ways of determining it, and their descriptions and evidential arguments were reflexively embedded in their deliberations. Moreover, the jurors themselves treated the reports and testimonies presented by the litigants as variously plausible descriptions that expressed or reflected the litigants' purposes, aims, motives, entitlements, obligations, and social statuses.

To speak of the reflexive character of such ordinary sociological reasoning is not the same as saying that jurors and other ordinary members of the society are, in fact, masters of the discipline called sociology. Nor does it have anything to do with the commonplace complaint that sociology is "mere" common sense dressed up in pseudoscientific jargon. From the standpoint of professional sociology, ordinary sociological descriptions are uninteresting or faulty, or at the very least in need of scientific verification. On the one hand, common knowledge is dismissed as trivial, since "everybody knows" how to describe commonplace actions and social events before pursuing higher (or even lower) education. On the other hand, sociologists often stress that commonsense beliefs can be prejudiced by cultural ideologies and perspectival limitations. Various programs of sociological research attempt to supersede common sense with a more comprehensive understanding of sources of historical and cultural variation and/or with a rigorous application of inferential statistics.[41] Garfinkel did not deny the rhetorical, practical, and informative value of the contrast between sociological knowl-

[41] To appreciate how ubiquitous and appealing this view of common sense can be, see Ian Robertson's popular introductory textbook, *Sociology*, 2nd ed. (New York: Worth, 1983). In the introduction of his book Robertson tries to convince students of the value of sociology by presenting a list of true/false questions. The questions are written to evoke "commonsense" prejudices about the categories of people who receive welfare, the statistical distributions of murders across gender categories, and other matters of common opinion and politicized judgment. Each of the commonsense opinions is contrasted with "scientific" knowledge derived from sociological research (or, in many cases, government statistical reports). Publishers' advertisements for a revised edition of Robertson's text highlighted the fact that the number of items in this popular list had been increased from fifteen to twenty.

edge and commonsense belief, but he did argue that an exclusive preoccupation with surpassing or correcting "mere" common sense was guaranteed to obscure ethnomethodology's phenomenon of the local and accountable production of social order.

Garfinkel's proposals about reflexive phenomena tie together ethnomethodology's substantive interest in ordinary social actions and its often critical stance toward the discipline of sociology. However, it is not entirely clear that an interest in reflexivity necessarily implies methodological skepticism. Ethnomethodologists have been able both to assert and to deny the criticalness of their relation to "conventional" sociology. On the one hand, they have argued that they study reflexive aspects of society that other programs in sociology overlook, and thus they propose to expand sociology's domain. If it turns out that even the most banal conversation is an immensely rich social achievement, so much the better for sociology. On the other hand, ethnomethodologists have also made the more disputatious claim that social science researchers conduct interviews, run experiments with human subjects, or interpret subjects' self-reported survey responses by systematically relying on commonsense methods that they cannot justify by reference to general precepts of scientific method.

If sociologists who aimed to construct general models or other representations of real-world social structures were to accept such criticisms, they would be left in a quandary. They could try to improve the validity and reliability of their methods by taking account of ethnomethodological findings about biases and distortions arising from researchers' and subjects' use of vernacular language and commonsense judgments. But this would present them with an endless and hopeless task, since ethnomethodology is of little use in detecting and eliminating biases. Alternatively, sociologists could revise their views about what it means to do science. If they were to do so, they would find the sociology of science to be a valuable source of help, since studies in that field often demonstrate that formal accounts of scientific method do not describe what scientists actually do. Consequently, many of the methodological criticisms of sociology could be shown to apply to other disciplines whose scientific status is not in doubt. However, when assumptions about the efficacy and universality of formal scientific methods provide a primary basis for legitimating sociology, it is less easy to put those assumptions aside. This possibility not only haunts "conventional" sociology, as I argue, but it also haunts many of the research programs in ethnomethodology and sociology of science.

Indexicality

Indexicality is the most obvious throwaway term in Garfinkel's text. It is not a distinct concept, since it is simply another way of speaking of the entire picture of social order that Garfinkel presents under such rubrics as "reflexivity," "accountability," and the "local production" of social order. Although indexicality became a hallmark of the "ethnomethodological perspective," Garfinkel eventually discarded the term, implying that for him it was simply an optional item in the analytic vocabulary. Moreover, it seems likely that he introduced the term precisely as a throwaway item in an argument. This can be appreciated by reading the discussions of indexical expressions in Chapter 1 of *Studies in Ethnomethodology* and in Garfinkel and Harvey Sacks's paper, "On Formal Structures of Practical Actions."[42] In both cases, the category of "indexical expressions" initially seems to include particular types of words or idiomatic expressions, but it eventually becomes a way of speaking about the entire field of language use that ethnomethodologists investigate. Indexicality is a ticket that allows entry into the ethnomethodological theater, and it is torn up as soon as one crosses the threshold.

Garfinkel borrowed the term *indexical expressions* from Y. Bar-Hillel.[43] In the early 1950s Bar-Hillel participated in one of the earliest machine translation projects, and like many others who attempted to develop computerized methods for translating texts from one language to another, he discovered persistent and unanticipated complications. He found that it was not enough to encode a dictionary of equivalent terms and a set of rules for syntactic transformations, since both the input and output texts would still need to be worked over by competent speakers of the particular languages. The input text needed to be prepared for the computer's operations, and the

[42] Harold Garfinkel and Harvey Sacks, "On formal structures of practical actions." As the coauthorship of the paper indicates, Garfinkel's account of indexical expressions was developed in collaboration with Harvey Sacks's ethnomethodologically inspired investigations of tape-recorded conversation. In many of his transcribed lectures, Sacks exemplifies and elaborates on the phenomenon of indexical expressions. See especially Harvey Sacks, "Omnirelevant devices; settinged activities; indicator terms," transcribed lecture (February 16, 1967), pp. 515–22, in G. Jefferson, ed., *Lectures on Conversation,* vol. 1, (Oxford: Blackwell, 1992). I discuss Sacks's program of conversation analysis in Chapter 6. For an interesting analysis of the argument in Garfinkel and Sacks's paper, see Paul Filmer, "Garfinkel's gloss: a diachronically dialectical, essential reflexivity of accounts," *Writing Sociology* 1 (1976): 69–84.

[43] Y. Bar-Hillel, "Indexical expressions," *Mind* 63 (1954): 359–79. A less obvious source of Garfinkel's interest in "indexicality" was his exposure to Calvin Mooers's "zatocoding" system for cataloging bibliographic sources for libraries of small engineering firms in the Boston–Cambridge area. See Calvin N. Mooers, "Zatocoding applied to mechanical organization of knowledge," *Aslib Proceedings* 8 (1956): 2–32. Garfinkel took an interest in how the Mooersian catalog could provide a practical solution to the "big topics of context and practical action" by specifying key items in an index that, by using a mechanical device, enabled a stack of sources relevant to an engineering project at hand to be retrieved from a card catalog. Endless combinations of sources could thus be organized around the changing uses of "indexical" items. (Garfinkel, personal communication.) Also see Garfinkel et al., "Respecifying the natural sciences as discovering sciences of practical action," p. 138, n. 23.

many mistakes and grammatical oddities in the output text needed to be repaired before it could be regarded as an adequate translation. One of the persistent sources of trouble that Bar-Hillel identified was a broad class of terms he called *indexical expressions.* These included many of the most commonly used words in the English language: pronouns such as *he, she,* and *it,* deictic expressions *(here, this, over there),* auxiliary verbs *(have, be, can),* anaphoric usages (terms whose meanings vary when they are placed in different clauses), and other less well defined tokens and idiomatic expressions. Indexical expressions presented a problem for Bar-Hillel's enterprise, because their dictionary equivalents could not be specified in advance and their sense varied with the occasion of use. Moreover, the "contexts" implicated by such terms were themselves variable, as they included the placement of particular words in phrase structures, different aspects of time and place implicated by utterances, and a wide range of presuppositions about typical and singular features of the scenes in which these expressions were actually or imaginably uttered. Consequently, it was difficult, if not impossible, to write a set of rules for assigning meanings to such terms in "context," since the particular "contextual" orders implicated by any indexical expression were themselves unstable.

Garfinkel understood that Bar-Hillel's problems with indexical expressions were far from unique and that they were a perspicuous instance of a far more general phenomenon. Indeed, the "problem" with indexicals is as ancient as the history of philosophy.[44] Whenever logicians or philosophers try to affix truth values to particular formal statements or to give stable definitions to terms, they invariably must contend with the fact that when a statement contains indexical expressions, its relevance, referential sense, appropriateness, and correctness will vary whenever it is used by different speakers, on different occasions, and in different texts.[45] In order to remedy this problem, philosophers attempted in various ways to replace indexicals with spatiotemporal references, proper names, technical terms and notations, and "objective expressions."[46] So, for instance, to assess the truth

[44] Garfinkel and Sacks, "On formal structures of practical actions," pp. 347–48, mention the *Dissoi logoi,* a fragment from approximately 300 B.C., as an example of an ancient study that observes that the truth of the expression "I am an initiate" can vary with the speaker and the time of its utterance. Garfinkel and Sacks list many of the philosophers and logicians who have raised the topic of indexical expressions (or various close relatives like "indicator terms," "deixis," and "occasional expressions"). These philosophers include Husserl, Russell, Goodman, and Wittgenstein. For further discussion, see J. Coulter, "Logic: ethnomethodology and the logic of language," pp. 20–49, in Button, ed., *Ethnomethodology and the Human Sciences;* and Heritage, *Garfinkel and Ethnomethodology,* p. 142.

[45] These features of vernacular language, which thwart the effort to construct natural languages, were often viewed as "defects" associated with human imperfection. The logician's task of inventing a notational system that transcended these limitations was likened to an ascension to a more perfect state of rationality, above the plane of mundane human existence.

[46] Proper names do not by themselves "remedy" indexicality; indeed, on many occasions they

value of an utterance like "The water is hot enough, now," an analyst would try to translate it into an "objective" or context-free statement like "At 16:53, Eastern Standard Time, the temperature of the H_2O reads one hundred degrees Celsius."[47] Garfinkel points out that such a "programmatic" substitution of objective expressions for indexicals is, at best, provisional and satisfied as a matter "of practical social management."[48] In saying this he was clearly influenced by Wittgenstein's critique of the classical philosophy of language,[49] although he extended that critique and transformed it into a preliminary description of practices in the natural and social sciences.

For ethnomethodology, indexical expressions and indexical actions constitute an entire *field* of phenomena to be investigated. To understand how this could be so, consider a simple case of the deictic reference *here* used as an indicator term in the question "What are you doing here?"[50] We might imagine that any indicative use of the term *here* could be translated into a

are used indexically. Take, for example, the following instance of a proper name used during a phone call: ML is speaking to MS about a New Year's celebration to be held later that evening at MS's home. MS is married to and resides with a person named "Jeff": ML: "Is Jeff coming?" MS: "Uhm, yeah?" A moment of confusion ensues, as MS does not immediately grasp that the "Jeff" in this reference is a mutual friend and not her husband. That a misunderstanding occurs does not necessarily demonstrate that MS would need to know ML's "intention" in order to discern to whom he is referring, nor is the problem with the reference due to linguistic grammar alone. Rather, the reference uses an unmentioned set of detailed understandings and presumptions about ML's and MS's mutual friends and aquaintances, their living arrangements, and festive occasions. ML supposes (and supposes that MS will recognize) that Jeff the spouse will already "be there" and so will not be "coming" to the party, so that the name "Jeff" will be a precise enough reference without having to mention a surname or social security number.

[47] The contrast between objective and indexical expressions is misleading if taken too literally. Although it has argumentative significance, it does not implicate a difference between "self-referential" and "objective" statements. To understand why this is so, consult Gerard Genette, *Narrative Discourse: An Essay in Method* (Ithaca, NY: Cornell University Press, 1980), p. 212. Genette also uses a "water boils . . ." example and contrasts it with another form of statement exemplified by "For a long time I used to go to bed early." The latter expression "can be interpreted only with respect to the person who utters it and the situation in which he utters it. *I* is identifiable only with reference to that person, and the completed past of the 'action' told is completed only in relation to the moment of utterance." But as Genette goes on to say, "I am not certain that the present tense in 'Water boils at one-hundred degrees' (iterative narrative) is as atemporal as it seems," since it too can be sensibly uttered by particular speakers on appropriate occasions. He argues that the contrast nevertheless has "operative value."

[48] Garfinkel, *Studies in Ethnomethodology*, p. 6.

[49] Ludwig Wittgenstein, *Philosophical Investigations*, trans. G. E. M. Anscombe (Oxford: Blackwell Publisher, 1958).

[50] This example of an "indicator term" is discussed in Sacks, "Omnirelevant devices; settinged activities; indicator terms." Also see Wittgenstein, *Philosophical Investigations*, sec. 8 ff., for a discussion of how the expressions "this" and "there" become problematic when treated as signifiers or referential terms.

proper name for the place the speaker intends, but such an effort would encounter the problem of deciding just which name should correspond to a particular use of the term. In any particular case, does *here* refer to a geographical place, an address, a social occasion like a meeting or celebration, or all of the above? As Harvey Sacks points out, the problem is more complicated than selecting the proper referent from such a list. Using examples from a tape-recorded group therapy session, he demonstrates that indicator terms do not simply stand proxy for names, "since each formulation of 'here' may well be consequential, i.e., if 'here' is say, 'the group therapy session' there might be good reasons for wanting to say 'here,' e.g., . . . what are you doing here,' rather than saying 'What are you doing in group therapy.' "[51] The term is thus not standing in place of an objective expression but has a distinctive use in its own right. Moreover, Sacks contends that far from always being ambiguous or problematic, indicator terms have "stable" uses in conversation. Speakers ordinarily use indicator terms effectively and intelligibly without having to establish (ostensively or otherwise) what they stand for. In other words, rather than simply standing in the way of logical investigations, indexical expressions have "rational" properties in their own right. And as Garfinkel argues, the "demonstrably rational properties of indexical expressions and indexical actions is an ongoing achievement of the organized activities of everyday life."[52]

Unlike Bar-Hillel, Garfinkel and Sacks greatly expand the relevance of "indexicality" beyond the analysis of specific classes of words. They use the contrast between objective and indexical expressions as a placeholder in their argumènt, without implying a grammatical distinction between the two classes of expressions. If the replacement of "indexical" by "objective" expressions is programmatically unsatisfiable and yet is achieved as a matter of "practical social management," then there is no context-free way to distinguish between the two classes of expressions.

For instance, the example I gave earlier, "At 16:53, Eastern Standard Time, the temperature of the H_2O reads one hundred degrees Celsius" may count as an objective (or adequately objective) expression for some purposes, or it may be faulted for not specifying the locale or time more precisely or not stating the barometric pressure. Alternatively, this phrase may be faulted for being a stilted and not very helpful way of answering the question "Is the water hot enough yet?" The candidate "objective expression" can thus be subject to many of the complaints that have been lodged about the more commonly mentioned types of indexical expressions. One conclusion to be drawn from this is that indexical expressions

[51] Sacks, "Omnirelevant devices," pp. 518–19.
[52] Garfinkel, *Studies in Ethnomethodology*, p. 34.

are ultimately "irreparable." To argue this perhaps can carry some weight in philosophical debates, but it would be misleading to treat ethnomethodology's interest in indexicality as a basis for a general skeptical position. Once it is agreed that all utterances and activities are indexical, then it no longer makes sense to suppose that a system of context-free and standardized meanings can apply to all occasions of natural language use.[53] Less obviously, however, it no longer makes sense to treat the *unrealizable possibility* of such a context-free system as a general backdrop for analyzing situated practices.

It no longer clarifies anything to say that every possible utterance, statement, or representation is indexical. When, for instance, Garfinkel and Sacks mention that Emile Durkheim's fundamental rule of method – "The objective reality of social facts is sociology's fundamental principle" – is an example of an "indexical expression" for members of the American Sociological Association, this only seems to challenge one of sociology's sacred cows. Garfinkel and Sacks add that Durkheim's expression can be used on different occasions as a definition of professional sociologists' activities, "as their slogan, their task, aim, achievement, brag, sales-pitch, justification, discovery, social phenomena, or research constraint."[54] Because a similar thing can be said about the statement of a law in physics for an association of physicists, it hardly counts against sociology's scientistic ambitions to point out that Durkheim's maxim is indexical. Nor does Garfinkel's reference to the "practical social management" of indexical expressions devalue the local relevance and local adequacy of the "objective expressions" that replace them. This is because the distinction in principle between objective and indexical expressions no longer has much value when the topic of study shifts to the "demonstrably rational properties of indexical expressions." At this point, what becomes prominent is not that all expressions are indexical but that members manage to make adequate sense and adequate reference with the linguistic and other devices at hand. The question for ethnomethodology is, How do they do that?

Two programs of study in ethnomethodology

The arguments about indexical expressions served a valuable purpose for ethnomethodology, since they introduced an approach to the study of language use that was very different from the dominant perspectives in sociology, analytic philosophy, and linguistics. With hindsight, it is possible to say

[53] For a discussion of how Garfinkel and Sacks critically transform the Russelian project of implementing the Leibnizian goal of formulating an ideal language, see Jeff Coulter, "Logic: ethnomethodology and the logic of language."

[54] Garfinkel and Sacks, "On formal structures of practical actions," p. 339.

that Garfinkel's and Sacks's proposals gave rise to at least two related but distinct developments in ethnomethodological research.

Ethnomethodological studies of work

The first of these developments took its point of departure from several of the ethnographic investigations presented in *Studies in Ethnomethodology.* These studies described how practitioners in various settings made their activities, along with their "settinged" features, objectively accountable for all practical purposes. They included Garfinkel's descriptions of how staff in a medical clinic managed and manipulated case files and other records of their activities, how coroners composed adequate and defensible accounts of their investigations in a practical case-by-case fashion, how a transsexual (or "intersexed person") struggled to produce (her) unquestionably objective and matter-of-fact membership in the category "female," and how participants in various ordinary communicative activities managed to achieve an unequivocal and unmistakable sense of their activities. These studies treated the practical management of "the unsatisfied programmatic distinction between and substitutability of objective for indexical expressions" as a substantive phenomenon, more than a point of criticism to be leveled against analytic philosophy and social science.

In the decade following the publication of his central text, Garfinkel and several of his students and colleagues turned their attention to the natural sciences and mathematics. This was not an entirely new development for ethnomethodology, since many of the earlier discussions and criticisms of social science methods played off comparisons with the natural sciences. But by the mid-1970s, the program had come full circle: Whereas "ethnomethodology" initially pointed to ordinary methods that had little to do with the precepts of scientific methodology, Garfinkel and his colleagues now turned their attention to the ordinary, day-to-day production of scientific methods. In part, this new set of studies was an extension of the overall project of investigating the practical management of the unsatisfied programmatic distinction between indexical and objective expressions. As Garfinkel stated in his book, "Research practitioners' studies of practical activities of a science, whatever their science, afford them endless occasions to deal rigorously with indexical expressions."[55] Consequently, there was no reason to exempt the most "rigorous" fields of mathematics and natural science from studies of the programmatic substitution of objective for indexical expressions. Indeed, such studies promised to respecify what "rigor" could mean, given the assumption that indexical expressions were no

[55] Garfinkel, *Studies in Ethnomethodology,* p. 6.

less obstinate nuisances for mathematicians and natural scientists than they were for machine translation programmers, social science data analysts, and logicians.[56] Ethnomethodological studies of science did more than simply develop earlier initiatives. As I discuss later in more detail, they provided a critical perspective on some of the continuing programs in ethnomethodology.

Conversation analysis

The second development, which came to be called *conversational analysis* or, in more recent usage, *conversation analysis,* gradually became an autonomous program of study.[57] Current research in that field may or may not have much to do with ethnomethodology, but it is clear that there was once an intimate relationship between the two programs. In the early 1960s, Harvey Sacks, who at the time was pursuing a Ph.D. at the University of California at Berkeley, began to exploit the capacities of audiotape recording to study "naturally occurring" or "spontaneously produced" social activities. Tape recordings of phone calls, group therapy sessions, dinner conversations, and other ordinary exchanges, provided the "data" that Sacks and his colleagues used for investigating richly detailed ethnomethodological phenomena. Sacks was influenced by Garfinkel's ethnomethodological initiatives, Erving Goffman's studies of face-to-face interaction,[58] and, especially, extensive discussions and collaborative projects with Emanuel Schegloff, Gail Jefferson, and David Sudnow.[59] As I mentioned, Garfinkel and Sacks jointly wrote a paper in which they discussed the phenomenon of indexical expressions.[60] In addition, many of Sacks's transcribed lectures at the University of California at Irvine in the late 1960s and early 1970s elucidated what Garfinkel characterized as the "demonstrably rational properties" of indexical expressions. In many instances, such as in the case of his analysis of the deictic expression *here* discussed earlier, Sacks was able

[56] For an exemplary treatment of mathematical "rigor," see Eric Livingston, *The Ethnomethodological Foundations of Mathematics* (London: Routledge & Kegan Paul, 1986).

[57] In Chapter 6 I go into detail about some of the differences between conversation analysis and ethnomethodology.

[58] See Emanuel Schegloff, "An introduction/memoir for Harvey Sacks – Lectures 1964–1965," in Gail Jefferson, ed., *Harvey Sacks–Lectures1964–1965,* a special issue of *Human Studies* 12 (1989): 185–209. Schegloff downplays Goffman's influence on Sacks, even though Sacks and Schegloff studied with Goffman at Berkeley, and in another paper, "Goffman and the analysis of conversation," pp. 89-135, in P. Drew and A. Wootton, eds., *Erving Goffman: Perspectives on the Interaction Order* (Oxford: Polity Press, 1988), Schegloff argues that Goffman used Sacks's and his colleagues' conversation-analytic studies without adequately appreciating them or contributing to their development.

[59] Sacks, Schegloff, and Sudnow were graduate students at Berkeley in the early to mid-1960s. Gail Jefferson was one of Sacks's first students at UCLA and UC, Irvine, and she collaborated with him on some of his most important projects.

[60] Garfinkel and Sacks, "On formal structures of practical actions."

to show that analytic philosophers, linguists, and others who treat language as being composed of inherently meaningful tokens or statements miss the extent to which speakers achieve sensible and precise communication by placing apparently vague lexical particulars within locally coordinated sequences of activities.

As I show in Chapter 6, Sacks hoped to develop a science of practical social actions, and he and his colleagues developed a formal approach to "talk-in-interaction." The current form of their approach differs from Garfinkel's projects in a number of respects. Conversation analysts investigate the demonstrably rational properties of indexical expressions by describing recurrent sequential actions in conversation and specifying formal rules for generating their organizational features. Their aim is to develop a grammar for conversation that describes how different speakers coordinate their actions to produce coherent sequences of two-party or multiparty talk. Garfinkel's continuing program of studies investigates the uses of grammar. Although this is not necessarily incompatible with conversation analysis, the difference is especially clear in the program of science studies pursued by Garfinkel and his students. The original aim in those studies was not to construct a formal science of practical actions but to examine how formalizations are developed and used in and as local courses of practical actions. As I argue in Chapter 7, ethnomethodological studies of the sciences are necessarily reticent about their own scientific status. Although ethnomethodologists do not aspire to produce exercises in "metascience" and they are not "antiscientific" in orientation, they are necessarily indifferent to the illusory security and preliminary justifications provided by a scientific program.

Criticisms of ethnomethodology

An indication of the seriousness of ethnomethodology's challenge to sociology was the fact that it drew vehement denunciations by some of the more prominent sociologists in North America and Britain. The most vitriolic of these were given in the decade following the publication of *Studies in Ethnomethodology.* In recent years the hostilities between ethnomethodology and sociology have subsided, perhaps because sociology has become an increasingly fragmented field and is thus more difficult to police. More important, many ethnomethodologists have deliberately tried to link their studies to established themes, theoretical perspectives, and methodological strategies in sociology. Not least, there are fewer unabashed ethnomethodologists than there once were, as many of the most promising students left the field in response to the shrinking job markets of the 1970s and 1980s.

It is worth reviewing these criticisms, partly because I expect that many readers will agree with some of them, but also because they help clarify the challenge that ethnomethodology presents to classic sociological and science studies approaches. I will start with the most hostile criticisms and finish with the most sympathetic.

Matters of style and professional conduct

Perhaps the most widely known criticism was made by Lewis Coser in his 1975 presidential address to members of the American Sociological Association.[61] Coser's criticism was heavy-handed, extremely uncharitable, and profoundly ignorant of the philosophical background and research initiatives of ethnomethodology. Nevertheless, he repeated many of the complaints sociologists have made in more casual discussions, and his version was significant because of the ceremonial occasion at which it was presented. Coser made strange bedfellows of ethnomethodology and quantitative analysis, as he argued that both approaches are so preoccupied with method that they lose sight of the substantive history and constitution of entire societies.[62] He dismissed the subject matter of ethnomethodology as obviously trivial, chiding Garfinkel and his colleagues for wasting their own and their readers' time by conducting elaborate studies and writing prolix descriptions of what anybody already knows, for example, "methods" for crossing the street or starting a conversation. Moreover, he complained that ethnomethodologists violated conventional academic standards by forming a "cult," and he lamented the fact that members of this "cult" circulated unpublished drafts of research papers to one another, without first submitting them to peer review. In his view, ethnomethodology was sustained by the collective delusion that Garfinkel's incomprehensible writing must be saying something profound, and Coser charged that ethnomethodologists were simply making academic headway by studying what any clear-sighted person should recognize as uninteresting and trivial matters.

A similar offensive was launched by Earnest Gellner, a veteran anthropologist and philosopher of social science, who informed his readers that the "ethnos" are affected by a peculiarly Californian form of irrationality and that among other things they perform like rock stars in front of admiring

[61] Lewis Coser, "ASA presidential address: two methods in search of a substance," *American Sociological Review* 40 (1975): 691–700.

[62] Coser's remarks borrowed the tenor, if not the cogency, of C. Wright Mills's attack on "abstracted empiricism" in chap. 3 of Mills's *The Sociological Imagination* (New York: Oxford University Press, 1959), pp. 50–75.

audiences of "ethno-chicks."[63] Like Coser, he imputed an unprofessional prose style as well as life-style to the ethnomethodologists, and also like Coser he advanced an ad hoc explanation to account for the irrational appeal of this band of renegades. For him, ethnomethodology's extreme overemphasis on subjectivity promoted the popular themes of the 1960s youth culture. In his reading, "two facts at any rate emerge quite clearly from *Studies in Ethnomethodology:* it is pre-occupied with the inner meanings, to the actors, of their actions; and secondly, it places the study of those inner meanings within the sociological tradition, of which it sees itself as a continuation."[64] Aside from his distaste for the cultural origins he imputes to ethnomethodology, theoretically Gellner sees no reason to abandon the view that "it is culture, or language, which provides the ready-made material potential for giving an account of this act or that, and on individual occasions, men simply draw upon this available wealth of characterizations . . . individuals do not use any methods at all to make things accountable – they just fall back on available accounts, without further ado."[65]

Gellner correctly recognizes that Garfinkel attacks the view he avows – the use of social scientific "models of man" to turn practical actions into unreflective, rule-governed behaviors of a "cultural dope"[66] – but he entirely miscasts Garfinkel's position when he equates "accounts" and "methods" with "inner meanings" and "subjective" determination. This charge is so far off base that it is not worth taking seriously, but what is serious about Gellner's and Coser's attacks is their disdainful effort to "explain away" ethnomethodology by defining it as irrational. Although their explanations of ethnomethodology are not well informed, they are typical of a genre: They stipulate the social or cultural functions of various manifestations of "tribal"

[63] Earnest Gellner, "Ethnomethodology: the re-enchantment industry or the California way of subjectivity," *Philosophy of the Social Sciences*, 5(1975):431–50. Gellner says that based on his experience of having attended an ethnomethodology conference, "it was noticeable, and I think significant that the quality and quantity of ethno-chicks surpassed by far those of chicks in any other movement which I have ever observed – even Far Out Left Chicks, not to mention ordinary anthropo-chicks, socio-chicks or (dreadful thought) philosophy chicks." This precise and rigorous observation is stated on the same page as the following complaint, "Let us face it, they do not write well, and their stylistic failings spring from these very features – careless neologism, a slapdash indifferences to precision and rigour in exposition, an eager willingness to say more and to say it again rather than refining what one has already said, and so forth" (p. 435).

[64] Ibid., p. 432.

[65] Ibid., p. 433.

[66] Garfinkel, *Studies in Ethnomethodology*, p. 68, defines "cultural dope" as "the man-in-the-sociologist's-society who produces the stable features of the society by acting in compliance with preestablished and legitimate alternatives of action that the common culture provides."

life while assuming an epistemic privilege that rules out of relevance any objection that might arise from within the life of the tribe.[67] Accordingly, ethnomethodology's "jargon" serves to prevent outsiders from criticizing a group's commitments; its "ritual" enactments help participants achieve a sense of common mission; and its "deviant values" serve to enhance subcultural solidarity. In this case, persons committed to ethnomethodology are able to experience firsthand what it is like to be somebody else's "cultural dope," and such experience only helps inspire further distaste for Coser's and Gellner's explanatory enterprises.[68] Although I suppose that Coser and Gellner would view this as the reaction of a brainwashed member of the tribe, their impressionistic and inaccurate "observations" of ethnomethodology lead me to wonder about their analytic accounts of other tribes, religious groups, occupations, gangs, and collectivities who live further beyond the pale of the academic profession.

Matters of scale and context

A more reasoned mode of criticism pertains to the scale of inquiry. In its most common form, this criticism begins with the following objection: "It is all very well to conduct close studies of sales encounters, family dinner conversations, shoptalk among co-workers in an office, jury deliberations, and other routine and intricate orders of social interaction, but how can you understand these if you do not also take account of the broader social, economic, and historical contexts in which the events take place?" In the sociology of science, a related question is often raised about observational studies of scientists and technicians working in laboratories: "How can you understand scientific practices if you do not look beyond the walls of the laboratory to take into account sources of funding and supplies, public values supporting science, and the competitive dynamics working at the level of entire disciplines?" These are sensible questions, and they are often raised in a friendly way, but they cover over a rather deep difference between ethnomethodology and sociology.

Context is a word that does heavy duty in sociological discourse. In the simplest terms, it means a relevant set of "surrounding" factors that influence the particular actions or events of interest. Participants in sociological disputes often promote their favored views of context: "You haven't taken account of the historical context!" "But what about class?" "And what about

[67] See W. W. Sharrock and R. J. Anderson, "Magic, witchcraft, and the materialist mentality," *Human Studies* 8 (1985): 357–75.

[68] The published replies to Coser were far too polite: Don Zimmerman, "A reply to Professor Coser," *American Sociologist* 11 (1976): 413; and H. Mehan and H. Wood, "De-secting ethnomethodology: a reply to Lewis A. Coser's presidential address to the American Sociological Association," *American Sociologist* 11 (1976): 13–21.

gender, and race?" "But these events are taking place in court, what about the legal context?" "You're forgetting the power dimension." "You haven't focused closely enough on the immediate circumstances." Making a complete, or at least an adequate, sociological description of any event might thus seem to involve an endless task.[69] Indeed, this is what is known as the "etcetera problem." In an early paper, Sacks summarized the problem as follows:

> Consider the problem of comparing proposed descriptions. The feature of any description that it will not only be incomplete but that (a) it could be indefinitely extended and (b) the extension cannot be handled by a formula for extrapolation, implies that any description can be read as far from complete, or as close to complete, as any others. From simply reading two descriptions of variant length, style, etc., one could conclude that while one is more elaborate the other is more terse, while one is more extensive the other more intensive, etc.
>
> How could one, then, by simply reading a variety of descriptions, decide which had a better correspondence, i.e., which was "more sociological"? Obviously, the accreditation of the authors provides no reasonable solution. Nor does the appending of a methods section, for it is given in the recognition of the etcetera problem, that if application of "the same methods" does not produce "the same description" this does not reflect on either (a) the actual methods used, or (b) the reporting of the methods. It is obviously no solution to use "the author's purpose" or for that matter the reader's purpose in reading the paper to decide adequacy of description. That merely shifts the question of using correspondence to establish adequacy from (a) correspondence between description and intended object to (b) correspondence between purpose, description and intended object. We still face a problem of reconciliation.[70]

As noted earlier in my discussion of indexicality, ethnomethodologists also make "contextualist" arguments, but with an important difference: Instead of viewing context as a variably configured array of "factors" that surround any given event and determine its meaning and significance, they treat context and event together. When our inquiries begin with observations of singular events in context (i.e., of a laboratory technician performing a specific run of an experimental procedure or of a particular joke told in a conversation among friends), the very terms we use to identify what is going on – that is, the way we characterize the events, participants, and actions – already imply the relevance of context. Where ethnomethodology takes off

[69] One sometimes can get the impression from sociological arguments that there is no distinction between describing an object in an intelligible way and claiming that the object described is "nothing but" what the description says about it.

[70] Sacks, "Sociological description," pp. 12–13.

at this point is, first, noticing that "members" commonly have no trouble with seeing, and seeing at a glance, "what's going on" in the situations in which they and others act. They are undeterred by the skepticism expressed by the etcetera problem. Having noticed this, ethnomethodologists then try to describe how members manage to produce and recognize contextually relevant structures of social action. Although such inquiries do not directly answer such questions as "How is the technician affected by the fact that she's not given credit for the lab's discoveries?" they do turn our attention to the reflexive way in which the identities of persons, actions, things, and "contexts" become relevantly and recognizably part of an unfolding "text" (or, better, "contexture") of practical details. Such an inquiry no longer tries to explain those details by "connecting" them to a corresponding set of contextual "factors"; instead, it attempts to describe the primordial sensibility of linguistic and embodied actions.

Criticisms about the scale of ethnomethodology's subject matter often convey an image of "society" as a big thing that contains the actions and events we witness in daily life. Because the big thing is considered more massive and stable than any of the little events that take place within it, it is given explanatory priority. The sociologist's task is thus to construct a map of society that supplies a set of coordinates for locating and identifying local events. The complication raised by the etcetera problem is that there are no scenic overlooks or orbiting satellites from which to get a clear glimpse of "society as a whole," nor is the everyday life-world the sort of thing that we can begin to understand by transposing its objective shape onto a more convenient set of coordinates.[71] The relevant axes, coordinates, and dimensions of social space seem to have no finite limit; they vary far more freely and radically than do the conventions used for constructing geographers', geologists', and astronomers' maps. A common solution proposed in sociology is to end debate on such "metatheoretical" matters by accepting one or another conceptual framework and getting on with the empirical work. This strategy might be called *fictive consensus*. The problem is, whose framework should we elect? Worse, why should we think any particular "framework" is appropriate in the first place?

Questions of power and emancipation

A subset of the criticisms of ethnomethodology's unduly restrictive approach arise from the tradition(s) of historical materialism. These criticisms are, in

[71] David Bogen observes that it is a curious use of grammar to liken the social "world" to something like a planet. See David Bogen, "Beyond the limits of *Mundane Reason*," *Human Studies* 13 (1990): 405–16, which is a review of M. Pollner's *Mundane Reason: Reality in Everyday and Sociological Discourse* (Cambridge University Press, 1987).

a sense, more interesting that those generated by latter-day adherents to functionalism. Perhaps because of their efforts to infuse historical materialism with selective aspects of existential philosophy, the more eclectic proponents of critical theory, Marxist hermeneutics, and left-structuralism like Jürgen Habermas, Anthony Giddens, and Pierre Bourdieu are better prepared to understand ethnomethodology's language and interests.[72] Since ethnomethodology is readily construed as a study of social praxis, it seems relevant to the long-standing problem of how to bridge the gap between Marxist theory and everyday experience in late-capitalist society. Inevitably, ethnomethodology disappoints such hopes, but not uninterestingly.

The problem in a nutshell is that ethnomethodology is not clearly aligned with an "emancipatory" politics or, for that matter, with any transparent political agenda. On the one hand, it has sometimes been argued that the approach is "conservative" because ethnomethodologists rarely talk about power or coercion, and superficially understood, the approach seems to suggest that enterprising actors freely create the world(s) in which they act. On the other hand, ethnomethodology promotes an avowedly "radical" agenda, albeit of an "epistemological" variety, and both ethnomethodologists and politically radical sociologists launch attacks against more "conventional" approaches. The most serious problem that Habermas, Giddens, Bourdieu, and others find with ethnomethodology is that it disavows structural determinism. Although it might be hoped that ethnomethodological research would document the systematic "distortions" of communication, and the routinized "reproduction" of historically structured relational asymmetries, the best-known ethnomethodologists seem to work hardest at demonstrating circumstantial relativity and describing local contingencies that confound a priori configurations of status and power at the very point of their presumptive application.[73]

Habermas, Giddens, and Bourdieu each take ethnomethodology seriously, but each tries to transcend its limitations by retaining elements of rationalism, objectivism, and foundationalism. Likewise, each also construes ethnomethodology as a "theoretical position" that sits squarely to one side of the well-worn oppositions between agency and structure, *Verstehende* and causal determination, and constructivism and objectivism. Bourdieu, for instance, treats ethnomethodology as though it were a kind of phenomenological anthropology that tries to recover the "native experience and the

[72] See, for instance, Anthony Giddens, *New Rules of Sociological Method: A Positive Critique of Interpretive Sociologies* (London: Hutchinson, 1978), pp. 33–44; Jürgen Habermas, *The Theory of Communicative Action,* vol. 1, pp. 102–41; Pierre Bourdieu, *Outline of a Theory of Practice,* trans. Richard Nice (Cambridge University Press, 1977), pp. 1–29.

[73] An excellent example of this is E. A. Schegloff, "Between micro and macro: contexts and other connections," pp. 204–34, in J. Alexander, B. Giesen, R. Munch, and N. Smelser, eds., *The Micro–Macro Link* (Berkeley and Los Angeles: University of California Press, 1987).

native theory of experience" endogenous to a culture. Bourdieu counterposes this position to Claude Lévi-Strauss's "objective analysis" of the gift exchange, in which the "observer's totalizing apprehension" defines a material economy that the native denies and "misrecognizes." Rather than opting for either side, Bourdieu proposes a dialectical reconciliation, through which the stylized practice of the gift exchange suppresses the analytically discerned economic equivalence of gift and return-gift. Native strategies for differentiating gifts and deferring gift giving until the "appropriate" occasion arises systematically bury the economic relevancy of the exchange. (It would be an insult or gaffe simply to return the same gift, or a gift of equivalent value, without "hiding" the economic nature of the exchange behind a ritual screen.) For Bourdieu, the objective "mechanism" of the exchange still hovers over the ritual process, but its salience to any single gift is hidden by the apparently spontaneous or ceremonial circumstances of the presentation. Although somewhat more nuanced, this solution is not markedly different from earlier attempts to salvage structural determinacy. The observer gets the last word, and he imputes self-deception to the natives who do not acknowledge or recognize the analytic explanation.[74]

Although Giddens gives an appreciative view of ethnomethodology's research policies, he too treats ethnomethodology primarily as a theoretical position that fails to answer to the need to provide a strong basis for confronting a larger set of structures. His theory of *structuration* (a term Bourdieu also uses) subsumes ethnomethodology's account of "agency" within a dialectical relationship to stable, institutionally structured, and historically founded systems of dominance. This theory outlines how institutionalized social interaction "reproduces" structurally patterned hierarchies of inequality and power as though from the "ground" up. Giddens's themes have been incorporated in interactionist studies of educational decision making, classroom conduct, medical examinations, and legal interrogations, so in a way his version has become less a criticism of ethnomethodology than a strong suggestion of how ethnomethodologists can make their studies theoretically significant. A problem in this case is that the theory is almost too easily documented by studies of the appropriate settings, and a residue of "surplus details" simply appears to drop out of relevance when the theme of "structuration" is secured to the documentary instances. And as I elaborate in Chapter 6, the ethnomethodological studies that follow this line often do so by giving a foundationalist reading of conversational analytic research, that is, by using a rule-based version of conversational actions in order to gain critical leverage for examining "institutionalized talk."[75]

[74] Bourdieu, *Outline of a Theory of Practice*, p. 46.

[75] See Schegloff, "Between micro and macro, contexts and other connections," for an immanent critique of such uses of conversation analytic research.

Habermas also construes ethnomethodology as a *theory* that opposes any effort to transcend the immanent understandings that are implied in the production of communicative actions. Unlike Bourdieu, Habermas does not oppose this position dialectically to an objectivistic analysis; instead, he insists that communicative actions necessarily contain immanent validity claims. For Habermas, the *"rational infrastructure of actions oriented to reaching understanding"* is not a structure discovered through empirical research; it is implicated by the very attempt to interpret and describe social activities.[76] Standards of truth, truthfulness, and intelligibility apply a priori to substantive utterances and to professional descriptions of them, and because of the combination of their abstractness and intrinsic relevance, they can be used to criticize empirical departures from those standards. The trouble with ethnomethodology as Habermas sees it is that it does not accord special analytic attention to validity claims:

> Garfinkel treats as *mere phenomena* the validity claims, on whose intersubjective recognition every communicatively achieved agreement does indeed rest – however occasional, feeble, and fragmentary consensus formation may be. He does not distinguish between a valid consensus for which participants could if necessary provide reasons, and an agreement without validity – that is, one that is established *de facto* on the basis of the threat of sanctions, rhetorical onslaught, calculation, desperation, or resignation.[77]

As Habermas construes it, ethnomethodology does not acknowledge the "standards of validity" that an observer must employ when interpreting the utterances studied: "If he does not credit himself with such an extramundane position, he cannot claim a theoretical status for his statements."[78] The solution to this "dilemma" is as follows:

> The social-scientific interpreter, in the role of an at least virtual participant, must in principle orient himself to the *same* validity claims to which those immediately involved also orient themselves; for this reason, and to this extent, he can start from the always implicitly shared, immanent rationality of speech, take seriously the rationality claimed by the participants for their utterances, and at the same time critically examine it. In thematizing what the participants merely presuppose and assuming a reflective attitude to the interpretandum, one does not place oneself *outside* the communication context under investigation; one deepens and radicalizes it in a way that is in principle open to *all* participants.[79]

Although this is a very inspiring view, it makes communicative practice appear to be a docile matrix for exercising a theoretical will. This theoretical

[76] Habermas, *The Theory of Communicative Action,* p. 106, emphasis in original.
[77] Ibid., pp. 128–29.
[78] Ibid., p. 130.
[79] Ibid.

will comes armed with a set of categorical distinctions and a logical framework that, we can be assured, will always find a way to make actual discourse "fit" the mold, in however "fragmentary," "occasional," and "feeble" a way. Habermas's analytic strategy requires that we translate utterances into "statements" that take a yes/no position with respect to a priori validity claims.[80] Garfinkel's interest in "mere phenomena" thus is bypassed in order to reinstate an idealized simulacrum of a "rational" discourse.

Ethnomethodologists are not oblivious to politics, and like others they are capable of discussing and taking strong positions on contentious matters of the day. For the most part, however, they do not try to use their investigations as instruments for advancing one or another popular cause, remedial program, or normative policy. Nor do they endeavor to use their studies to lend "scientific" authority to their own political commitments.[81] This has nothing to do with a personal indifference to such matters. The desire for an authoritative critique of power can be overwhelming and understandable, but all too often it encourages a principled (and sometimes an unprincipled) effort to pursue the unrealized dreams of transcendental analysis. It seems that no other alternative can be acceptable given the stakes of the game. Consequently, the overwhelming need to ascend to a more comprehensive, objectively based, and normatively grounded position from which to oppose the powerful forces of oppression tends to be realized by turning the field of study into a docile projection of a theoretical will.

Questions of meaning and self-reflection

Several recent critiques have been made by persons who profess sympathy for the "radical" epistemic policies of ethnomethodology but who are disappointed by how the field has developed. A recent (if late) upsurge of interest among sociologists in "postmodern" approaches inspired by deconstructionist literary theory has given impetus to efforts to "outradicalize" ethnomethodology. Part of the complaint is that the early orientation to indexicality and the later development of sequential analysis of conversation give ethnomethodology no apparent theory of referential "meaning." As Paul

[80] A more extensive critique of Habermas's theory is provided by David Bogen in "A Reappraisal of Habermas's *Theory of Communicative Action* in light of detailed investigations of social praxis," *Journal for the Theory of Social Behaviour* 19 (1989): 47–77.

[81] There is disagreement about this, as might be expected. See, for instance, Alec McHoul, "Language and the sociology of mind: a critical introduction to the work of Jeff Coulter," *Journal of Pragmatics* 12 (1988): 229–86; M. Lynch and D. Bogen, "Social critique and the logic of description: a response to McHoul," *Journal of Pragmatics* 14 (1990): 505–21; A. McHoul, "Critique and description: an analysis of Bogen and Lynch," *Journal of Pragmatics* 14 (1990): 523–32; Lena Jayyusi, "Values and moral judgement: communicative praxis as a moral order," in Button, ed., *Ethnomethodology and the Human Sciences*, pp. 227–51.

Atkinson argues in a review article, ethnomethodological studies tend to reduce questions of meaning to rather dry explications of sequential order.[82] He and others also accuse ethnomethodology of having taken an objectivistic path that seemingly forgets the more "reflexive" orientation to "interpretive practices" that was once more prominent in ethnomethodology.

Although there may be a point to such complaints, the proposed solutions tend to conjure up a familiar roster of ghostly entities – meanings, intentions, goals, consciousness, and the like – associated with correspondence theories of reference and mentalistic explanations of meaning. The aim of trying to come to terms with what we are saying and doing when we speak of intentions, thinking, meaning, knowing, and so forth seems to become lost when these terms are once again given a foundational role in explanations of practical action. It is not that ethnomethodologists want to banish all references to "mind" and "interpretation" when investigating communicative actions but that they are struggling to avoid the Cartesian oppositions between word/world, sign/referent, signifier/signified, thought/object, and so on, which classic theories of meaning call into play and that so often bolster "mythological" accounts of mental entities and forces.[83]

Although inspired by avowedly antifoundationalist approaches to textual analysis, many of the critiques are indebted to Enlightenment conceptions of self-reflexive consciousness. This is particularly clear in Melvin Pollner's plea for reviving a "radically reflexive" ethnomethodology.[84] Pollner notes correctly that for ethnomethodologists the term *reflexivity* describes how the sense of a question, indicative gesture, or silence in conversation is "achieved" as part of the setting in which it occurs. Conceived in this way, the "incarnate" or "reflexive" achievement of sense is an endogenous property of the fields of social action that ethnomethodologists study. Pollner adds that this version of reflexivity can, but often does not, enable more self-reflexive appreciation of the concept:

> *Referential* reflexivity conceives of all analysis – ethnomethodology included – as instances of constitutive processes. . . . Not only are members deemed to be involved in endogenous constitution of accountable settings but so too are analysts without exception. Thus, ethnomethodology is referentially reflexive to the extent that it appreciates its own analyses as constitu-

[82] Paul Atkinson, "Ethnomethodology, a critical review," *Annual Review of Sociology* 14 (1988): 441–65

[83] The "mythology" does not imply that we have no business speaking of thoughts, mind, intentions, and so forth but, rather, that our academic accounts of these matters tend to presume an inappropriate picture. "A myth is, of course, not a fairy story. It is the representation of facts belonging to one category in the idioms appropriate to another. To explode a myth is accordingly not to deny the facts but to re-allocate them" (Gilbert Ryle, *The Concept of Mind* (Chicago: University of Chicago Press, Chicago, 1949), p. 8).

[84] Melvin Pollner, "'Left' of ethnomethodology," *American Sociological Review* 56 (1991): 370–80.

> tive of endogenous accomplishments. Referentially reflexive appreciation of constitution is *radicalized* when the appreciator is included within the scope of reflexivity: that is, when the very formulation of reflexivity – as well as every other feature of analysis – is appreciated as an endogenous achievement.[85]

Simply put, this radicalization extends the hermeneutic circle that encompasses acts-in-context to include the act of describing that very relationship. *Radical reflexivity* is thus a kind of "reflective" examination of the researcher's relation to particular "reflexive" operations in the social field investigated. This recommendation may appear to promote a more complete comprehension of reflexivity than would derive from a description of a meeting, conversation, or written text presented from a third-person vantage point. Conversely, it can be argued that a first-person reflexive account (or some other stylistic way of "referentially" highlighting or mentioning an observer's, interpreter's, or narrator's relation to the scene described) offers no general advantage over a third-person account and further that it invites a regress of further "reflections" on previous "reflections."[86]

The problem is that Pollner links ethnomethodology's version of reflexivity with a traditional concept of self-reflection that can, and I would argue should, be distinguished from it.[87] The "incarnate" reflexivity of accounts that Garfinkel introduces is unavoidable; it has no antonym; and it has to do with contextual placement and background understandings.[88] If, for instance, a silence in a conversation is "heard" by the analyst, and presumably by the participants, as an awkward pause in response to an invitation, the analytic "meaning" of this silence will be reflexively constituted in and through its sequential occurrence after the invitation, along with any embodied expressions of hesitancy or doubt that accompany that silence. In contrast, the "referential reflexivity" that Pollner advocates is avoidable; in his terms, actors and analysts can "evade, avoid, or finesse radical reflection in the

[85] Ibid., p. 372.

[86] Pollner recommends (" 'Left' of ethnomethodology," p. 374, n. 3) the kind of "reflexive" orientation that Steve Woolgar, Michael Mulkay, Malcolm Ashmore, and others experimentally explore by devising "new literary forms" that disrupt and call attention to the limits of narrative reportage. See Steve Woolgar, ed., *Knowledge and Reflexivity: New Frontiers in the Sociology of Knowledge* (London: Sage, 1988).

[87] See Marek Czyzewski, "Reflexivity of actors and reflexivity of accounts," in *Theory, Culture, and Society*, 11 (1994): 161–68.

[88] When "reflexivity" is treated as a part of a methodological program to recover culturally specific practices or orientations by situating one's own analysis in accountable affairs and scenes, one can have more or less of it. Such reflexivity is not a matter of the depth or explicitness of an interpreter's self-reflectiveness, however, as it is a way to modify an investigator's initial conceptions of observation, data, and findings in order to incorporate systematic features of exotic methods. See Benetta Jules-Rosette, "The veil of objectivity: prophecy, divination and social inquiry," *American Anthropologist* 80 (1978): 549–70.

course of practical activities."[89] Thus, according to his account, it is possible to be more or less "reflexive" by mentioning or not mentioning "one's own" constitutive relationship to the field described. This kind of reflexivity is a matter of formulating what one is doing – of "reflecting" and using "metalanguage" – and it is associated with skeptical concerns about the referential correspondence between statements and what they describe. It also reflects back on an abstract individual and cognitive "source." In contrast with a reflexive coordination of acts witnessed in a public domain and described in a common language, referential reflexivity is a matter of explicitly interpreting, reflecting on, and saying out loud. Both the interpretive acts that constitute a sense of an objective social structure and the reflections that attempt to "appreciate" their constitutive role are grounded in an analytic consciousness.

I would not want to dismiss the virtues and occasional appropriateness of "self-reflection" – that is, pausing to consider what one has just said, thinking out loud, admitting self-doubt, wondering whether others see things in the same way that you do, confessing bias, and so on. However, such conventionally "self-reflective" actions do not refer systematically to constitutive acts. They are acts in and of themselves: acts of confessing, prefacing, hesitating, acknowledging, wondering, qualifying, and so forth. Their sense is "achieved" in particular sequential environments, and they are often accompanied by characteristic poses, expressions, and reactions. Like other discursive acts, their sense, relevance, and appropriateness is reflexively tied to the pragmatic and relational circumstances in which they are uttered and/ or written.

Pollner's version of reflexivity appeals deeply and articulately to a radical constructivist struggle against objectivism. He is right to complain that much of the research in ethnomethodology exhibits empiricist and scientistic tendencies. At times it can seem as though the research is animated by little more than a confidence that the findings will be significant because they are "analytic." As Pollner argues, the waning hostility in recent years between ethnomethodology and "conventional" sociology is comforting to some but disturbing to those of us who figure that there is little point to an "ethnomethodology" that does not challenge the foundationalist view of theory and method that still prevails in professional sociology. But it is one thing to propose an antifoundationalist approach and another to follow through on it. Antifoundationalism is not synonymous with antiobjectivism, and Pollner, like many who oppose objectivism, ultimately replaces one abstract foundation with another. In place of an independent "mundane world" he installs the "work of worlding": acts emanating from a subject that

[89] Pollner, "'Left' of ethnomethodology," p. 374.

produce a world, acts the subject then "forgets" by presuming the independence of that world.[90]

Numerous other constructivist and ethnomethodologically informed treatments make similar arguments when "interpretive work," "ethnomethods," "representations," "persuasion," and "rhetoric" account for the appearance of a stable, consensual, objective "reality."[91] In place of thoughts or ideas in older antiobjectivist traditions, these studies install social, textual, interactional, and rhetorical practices and devices. What they have in common is a preoccupation with a referential or representational picture of language: They hold "reality" separate from language and then stress the foundational role of linguistic acts in achieving a semblance of reality.[92] By criticizing these studies, I do not mean to defend realism or objectivism but, rather, to question the representational picture of language that frames the classic argument.

Conclusion

Garfinkel once said that sociologists can "have none of" ethnomethodology if they are to hope to preserve a comprehensive theory of society. This is not to say that ethnomethodology has no academic place or pretensions, that it is atheoretical, or that its mode of investigation would better be situated in corporate industry or the Central Intelligence Agency. It is doubtful that ethnomethodology could sustain itself separate from the academic professions, because its program is critically bound to traditional analytic investigations of practical action and natural language use. Although, as I argue, ethnomethodology should not be construed as an "analytic" discipline, lay and professional analytic practices provide ethnomethodology with its subject matter. In a sense, ethnomethodology is a parasite of the host discipline of sociology, but unlike a parasite that reduces its host to a lifeless husk, ethnomethodology tries to reinvigorate the lifeless renderings produced by formal analysis by describing the "life" from which they originate.

[90] For a more extensive critical account of Pollner's approach, see Bogen, "Beyond the limits of *Mundane Reason*."

[91] For a critique of such accounts, see Graham Button and Wes Sharrock, "A disagreement over agreement and consensus in constructionist sociology," *Journal for the Theory of Social Behavior* 23:1–25. Also see David Bogen and Michael Lynch, "Do we need a general theory of social problems?" pp. 213–37 in G. Miller and J. Holstein, eds., *Reconsidering Social Constructionism* (Hawthorne, NY: Aldine de Gruyter, 1993).

[92] Button and Sharrock, "A disagreement over agreement and consensus in constructionist sociology," p. 12.

CHAPTER 2

The demise of the "old" sociology of science

In the early 1970s, Barry Barnes, David Bloor, Michael Mulkay, David Edge, Harry Collins, and other British sociologists confronted the structural–functionalist sociology of science developed by Robert Merton and his followers and assembled a loosely federated array of constructivist, relativist, and discourse-analytic programs. Since then, related variants of a "new" sociology of science proliferated on the Continent, in Australia, and in North America. Merton's program is still very prominent in American sociology, as his latter-day disciples have withstood the challenge of the new sociology of science by selectively assimilating some of the initiatives coming from Britain and the Continent.

Although proponents of the new sociology of science drew on a variety of sources, they were influenced by ethnomethodology's critical treatment of "constructive analysis" in the social sciences. Like ethnomethodologists they focused on informal day-to-day practices, but their constructivist interpretations mainly applied to the activities of natural scientists and not social scientists. In many cases the arguments and explanations generated by sociologists of scientific knowledge relied on scientistic versions of sociological method that ethnomethodologists had previously criticized. This apparent incongruity between a skeptical view of natural science theories, methods, and findings and a positive view of sociological analysis has not gone unnoticed by both critics and proponents of the new sociology of science, and it has recently become a focus of much consternation and debate. By reviewing and criticizing developments in the sociology of science in this and the next chapter, I attempt to clarify the problems that led to these debates. I do this in order to identify an epistemic "trading zone" of cognate issues and rival claims between the ethnomethodology and sociology of scientific knowledge[1] that I hope will illuminate some of the abiding issues and chronic debates in the sociology, history, and philosophy of science.

[1] The metaphor of a "trading zone" was taken from Peter Galison, "The trading zone: coordination between experiment and theory in the modern laboratory," paper presented at the International Workshop on the Place of Knowledge, Tel Aviv and Jerusalem, May 15–18, 1989. Territorial metaphors like "margin" and "zone" can misleadingly suggest discretely

The critique of the "old" sociology of science

The story of the how the sociology of scientific knowledge emerged in the 1970s has been recited on many occasions.[2] Indeed, the repetition and circulation of the story were instrumental to that emergence. What is new about the programs in sociology of science developing in Britain and elsewhere over the past two decades is their aim to investigate, and sometimes to explain, "the very content and nature of scientific knowledge."[3] Thomas Kuhn's landmark study, *The Structure of Scientific Revolutions,*[4] is widely acknowledged as the most significant source for a "sociological turn" in the history and philosophy of science,[5] but as the proponents of sociology of scientific knowledge admit, they went well beyond Kuhn's explicit suggestions on how the sociology of science might support and use his work.[6] They also built on some key revisions of the established philosophy of social science[7]

bounded fields, like two island communities linked by a distinct communicative channel. In this case, a more accurate picture would be of proximate neighborhoods in a city, where the boundaries between them are hopelessly gerrymandered. The alleged boundaries nevertheless are highly relevant to the disputes that flare up among the various gangs and factions in both neighborhoods.

[2] See, for instance, Barry Barnes, *Scientific Knowledge and Sociological Theory* (London: Routledge & Kegan Paul, 1974), and *Interests and the Growth of Knowledge* (London: Routledge & Kegan Paul, 1977); David Bloor, *Knowledge and Social Imagery* (London: Routledge & Kegan Paul, 1976; 2nd ed., Chicago: University of Chicago Press, 1991); Michael Mulkay, *Science and the Sociology of Knowledge* (London: Allen & Unwin, 1979); H. M. Collins, "The seven sexes: a study in the sociology of a phenomenon, or the replication of experiments in physics," *Sociology* 9 (1975): 205–24; Steven Shapin, "History of science and its sociological reconstructions," *History of Science* 20 (1982):157–211; Karin Knorr-Cetina and Michael Mulkay, "Introduction: emerging principles in social studies of science," in K. Knorr-Cetina and M. Mulkay, eds., *Science Observed: Perspectives on the Social Study of Science* (London: Sage, 1983); Bruno Latour, *Science in Action* (Cambridge, MA: Harvard University Press, 1987); and Steve Woolgar, *Science: The Very Idea* (Chichester: Ellis Horwood; and London: Tavistock, 1988).

[3] Bloor, *Knowledge and Social Imagery,* p. 1.

[4] (Chicago: University of Chicago Press, 1962); see 2nd ed., with "Postscript" (Chicago: University of Chicago Press, 1970).

[5] Ludwik Fleck, *Genesis and Development of a Scientific Fact* (Chicago: University of Chicago Press, 1979), gives an "insider's" account of the development of the Wasserman test. The text was originally published in 1935, and it is sometimes regarded as an important precursor to Kuhn's *The Structure of Scientific Revolutions* and for the recent developments in the sociology of scientific knowledge. Fleck's work did not circulate in the sociology of science community until after its republication in 1979. Some of Kuhn's other predecessors could also be cited, not the least of which would be Wittgenstein.

[6] In the "Postscript" of the 1970 ed. of *The Structure of Scientific Revolutions* (p. 176, n. 5), Kuhn cites a number of quantitative studies on "invisible colleges" by Nicholas Mullins, Diana Crane, Warren Hagstrom, Derek de Solla Price, and Donald de B. Beaver. These studies represent an approach to the "mapping" of scientific communities by using citation networks and related bibiometric indices, which largely omit any reference to the "contents" of the disciplines studied. For an exposition of Kuhn's work from the point of view of the sociology of scientific knowledge, see Barry Barnes, *T. S. Kuhn and Social Science* (London: Macmillan, 1982).

[7] See Peter Winch, *The Idea of a Social Science* (London: Routledge & Kegan Paul, 1958).

and sociology of knowledge[8] by invoking Ludwig Wittgenstein's later writings. In addition, they developed constructivist interpretations of Garfinkel's ethnomethodological studies and Cicourel's critiques of social science methods, which they applied in their investigations of routine "constructive" activities in the natural sciences.[9]

The sociology of scientific knowledge also developed against the negative precedent set by Robert Merton's well-established "paradigms" for the sociology of knowledge and sociology of science.[10] Merton incorporated Karl Mannheim's sociology of knowledge in a modified structural–functionalist framework that he and his colleagues applied to the study of scientific institutions and scientific change.[11] The critiques of the Mertonian program by enthusiasts for a new sociology of knowledge may have been overdrawn at times, but they served to announce the development of a rival program centered in Britain rather than the United States.[12]

Mannheim's approach to the sociology of knowledge was also criticized directly, but in a more limited way.[13] The criticisms of Mannheim and Merton were linked, since Merton's reading of Mannheim was very influential in American sociology.[14] But because the criticisms differed in some important respects, I discuss them separately.

[8] Peter Berger and Thomas Luckmann, *The Social Construction of Reality* (New York: Doubleday, 1966).

[9] The two key texts were Harold Garfinkel, *Studies in Ethnomethodology* (Englewood Cliffs, NJ: Prentice-Hall, 1967); and Aaron Cicourel, *Method and Measurement in Sociology* (New York: Free Press, 1964).

[10] See Robert K. Merton, "Science and technology in a democratic order," *Journal of Legal and Political Science* 1 (1942): 115–26; "A paradigm for the sociology of knowledge," pp. 7–40 in Merton, *The Sociology of Science: Theoretical and Empirical Investigations,* ed. with introduction by Norman W. Storer (Chicago: University of Chicago Press, 1973). Merton did not use the term *paradigm* in Kuhn's sense. For him it meant a general outline or model for sociologists to follow.

[11] See Robert K. Merton, *Social Theory and Social Structure,* enlarged ed. (New York: Free Press, 1968; originally published in 1949), chaps. 2 and 3, for his revisions of older functionalist programs, his program of "middle-range" theory, and his distinction between "manifest" and "latent" functions.

[12] Some of the earlier critiques of Merton's program were by M. J. Mulkay, "Some aspects of cultural growth in the natural sciences," *Social Research* 36 (1969):22–52; and "Norms and ideology in science," *Social Science Information* 15 (1976): 637–56; B. Barnes and R. G. A. Dolby, "The scientific ethos: a deviant viewpoint," *European Journal of Sociology* 11 (1970): 3–25; and Ian Mitroff, *The Subjective Side of Science* (New York: Elsevier, 1974). The last source is by an American writer, whose work was an early exception to the British–American division. In the preface of his collection of essays, *The Essential Tension: Selected Studies in Scientific Tradition and Change* (Chicago: University of Chicago Press, 1977), pp. xxi–xxii, Kuhn defends Merton against his British critics.

[13] Karl Mannheim, *Ideology and Utopia,* trans. Louis Wirth and Edward Shils (New York: Harvest Books, 1936); *Essays on the Sociology of Knowledge,* trans. and ed. Paul Kecskemeti (New York: Oxford University Press, 1952). A critical revision of Mannheim's approach is presented in David Bloor, "Wittgenstein and Mannheim on the sociology of mathematics," *Studies in the History and Philosophy of Science* 4 (1973): 173–91.

[14] See Robert K. Merton, "Karl Mannheim and the Sociology of Knowledge," chap. 15 of *Social Theory and Social Structure;* and Merton, "A paradigm for the sociology of knowledge."

The "correction" of Mannheim

The critique of Mannheim by Bloor, Barnes, and others was more of a *correction* and an *expansion* of Mannheim's program for the sociology of knowledge than it was an attack on it. Mannheim was not the founder of the sociology of knowledge *(Wissensoziologie),* but he is generally regarded as its most significant progenitor.[15] Before emigrating from Germany in the early 1930s, Mannheim wrote a series of essays that developed and applied a distinctive sociological approach to the production of knowledge. He proposed that the sociology of knowledge resulted from a historical transformation of an older and more particularistic concept of ideology. According to this argument, the concept of ideology emerged from a continuing tradition of political discourse in which it was used as a rhetorical weapon for "unmasking" a political opponent's arguments by associating them with narrowly personal and partisan interests.

Mannheim traces this rhetorical form to Napoleon's denunciation of the philosophes for being impractical "ideologists."[16] This denunciation linked an older sense of "ideology" as a "theory of ideas" with a more modern deprecation of the practical validity of mere ideas. Marx's polemics against the young Hegelians' efforts to combat "ideas" with other "ideas" similarly valorize practical action, but Marx and Engels go well beyond this form of argument when they explain "ideology" in reference to class position. Mannheim's treatment of ideology broadens Marx's form of explanation and inverts the Napoleonic denunciation of the "free" intelligentsia. Whereas Napoleon complained about the remoteness of academic philosophy from economic and political action, Mannheim treated that detachment as a practical condition for value freedom. Although he credited Marx with transforming the critique of ideology into a sociological analysis of the class conditions underlying "false consciousness," he significantly broadened and transformed Marx's treatment to the point of developing what easily could be regarded as an anti-Marxist approach.[17]

Although Marx extended the left-Hegelian critique of religion to include the state and its ruling ideology, he never abandoned the assumption that

[15] Mannheim's contemporary, Max Scheler, preceded him by developing an approach to *Wissensoziologie.* Scheler transformed the Marxist analysis of ideology into a more abstract and less politicized standpoint for explaining how social conditions give rise to "relatively natural views of the world." See Max Scheler, *Problems in the Sociology of Knowledge* (London: Routledge & Kegan Paul, 1980). For a general reader on the sociology of knowledge, see J. E. Curtis and J. W. Petras, eds., *The Sociology of Knowledge* (New York: Praeger, 1970).

[16] Mannheim, *Ideology and Utopia,* pp. 71 ff.

[17] Ibid., pp. 75 ff. For the Marxist conception of ideology, see Karl Marx and Friederich Engels, *The German Ideology, Parts I and II* (New York: International Publishers, 1947).

historical materialism provided a scientifically correct basis for exposing the ideological distortions promulgated by the ruling elites and their intelligentsia. Although Marx also traced the perspective of historical materialism to a particular class origin – the splinter group of the bourgeois intellectuals who took up the cause of the proletariat – he proposed that the (non)position of the proletariat in a classless society would create the social conditions for a universal "ideology" that was not distorted by narrow class interests. Marx was, of course, mainly interested in fostering a revolutionary transformation of the rapidly industrializing capitalist societies of mid-nineteenth-century Britain and Western Europe, and both his critique of capitalism and the grounds he proposed for it were positioned within those social conditions. In contrast, Mannheim tried to develop a general theory of the relationship between social conditions and ideology. He agreed substantially with the Marxist effort to place "ideas" on a sociohistorical base, but he questioned the scientific status of historical materialism, and he sought to generalize and de-politicize the analysis of relevant existential conditions.[18]

Unlike Marx, Mannheim had the benefit of hindsight. Like many of his contemporaries, he had occasion to doubt a Marxist or any other claim to a universally "correct" analysis of the conditions in postwar German society. Moreover, he had the dubious privilege of living in the shadow of the fractious academic debates involving advocates of natural scientific versus hermeneutic approaches to the human sciences.[19] This *Methodenstreit* absorbed many of Mannheim's contemporaries and predecessors, including Wilhelm Dilthey, Heinrich Rickert, Max Scheler, Max Weber, and Georg Simmel, all of whom took different positions in the debate. Mannheim's solution to the problems raised by the academic and political strife in Germany was to treat the very existence and structure of the factional squabbles as a historical condition for the development of a nonpartisan conception of ideology. He held that participants in these interminable debates would be able to realize that their own arguments were no less ideological than those of their opponents, and consequently they would be able to see that the existential determination of ideas held across the board. Such a realization could, of course, support a relativistic or nihilistic view of politics and epistemology, but this was not what Mannheim recommended. Instead, he wanted to dissociate ideological analysis from the forms of ad hominem argument that are commonly used to "explain away" or debunk

[18] Mannheim's broadened concept of "existential conditions" may have been influenced by his having read Lukacs's *History of Class Consciousness* (London: Merlin Press, 1971), and Heidegger's *Being and Time*, trans. John Macquarrie and Edward Robinson (New York: Harper & Row, 1962).

[19] For a discussion of the influence of these debates on Mannheim's sociology of knowledge, see Susan J. Hekman, *Hermeneutics and the Sociology of Knowledge* (Notre Dame, IN: University of Notre Dame Press, 1986).

particular knowledge claims by linking them to idiosyncratic biases and particularistic social affiliations.

Mannheim distinguished the sociology of knowledge from a "relativist" position by saying that relativism retains an absolutist standard of evaluation when it confuses the insight that "all knowledge is relative to the knower's situation" with the conclusion that "all knowledge-claims must be doubted." Presuming to doubt all knowledge is no less absolutist than presuming that there must be a ground for all true knowledge. So instead of advocating relativism, Mannheim argued for a "relationist" concept of knowledge. Rather than opting for a radically individualist conception of knowledge, he suggested that particular ideas are situated in historical and social circumstances. Such ideas might not accord with Western standards of purposive rationality, but this does not discount their adequacy in terms of the relevant epistemic community's "sphere" of categorical judgments and validity claims. Accordingly, "relational" knowledge – knowledge cultivated in a living community of understandings – could be dynamic without necessarily being arbitrary. Like Weber, Mannheim tried to construct an integrative theory that situated ideas in the very constitution of social order.[20] He attempted neither to reduce ideas to economic interests nor to detach them from historically specific existential conditions. Moreover, like Weber, he tried to preserve an idealized "scientific" vantage point for explaining the existential determination of knowledge:

> The non-evaluative general total conception of ideology is to be found primarily in those historical investigations, where, provisionally and for the sake of the simplification of the problem, no judgments are pronounced as to the correctness of the ideas to be treated. . . . The task of a study of ideology, which tries to be free from value-judgments, is to understand the narrowness of each individual point of view and the interplay between these distinctive attitudes in the total social process.

While trying to establish such a position, Mannheim struggled with a dilemma that he never resolved to the satisfaction of his critics.[21] Given the general sociology of knowledge he advocated, it would have been contradictory for him to assume a transcendental position from which to specify a nonevaluative total conception of the relations between all other systems of ideas and their respective existential conditions. Such a position could only have been secured by exempting the sociology of knowledge from its own substantive explanatory program. Nevertheless, Mannheim needed the assurance that somehow the sociology of knowledge could explain the social

[20] Mannheim expresses this by saying: "Actually, epistemology is as intimately enmeshed in the social process as is the totality of our thinking" (*Ideology and Utopia*, p. 79).

[21] See A. Von Schelting's review of *Ideology and Utopia*, in *American Sociological Review* 1 (1936): 664–74.

conditioning of other knowledge systems without being explained away in reference to its own limited circumstances.

Again, he turned the peculiar existential conditions of the sociology of knowledge into a decisive methodological advantage. He suggested that the ideological fragmentation and relative autonomy of the German academic community promoted a detached and differentiated understanding of the ideological field.[22] For Mannheim, the "unanchored, *relatively* classless stratum"[23] of the intelligentsia was a vanguard of a sort, but rather than representing the interests of the proletariat, this vanguard could ascend to a disinterested position removed from partisan politics.[24] Consequently, the sociology of knowledge would represent a further stage in the historical advance of the ideology concept, in which a nonevaluative conception of ideology would provide a basis for criticizing absolutist conceptions of knowledge. Mannheim did not claim that the practical supports for the "objectivity" of the sociology of knowledge guaranteed the truth of its analyses. He argued instead that it would be absurd to use the standards of mathematics and the "exact" sciences to evaluate the ultimate truth of historically situated knowledges. And because the practical validity of the sociology of knowledge was grounded in its historical situation, he contended that it too should not be held to the standards of an exact science.

This apparently modest proposal implied a threefold distinction among scientific, social scientific, and ordinary systems of knowledge:

1. At least some of the knowledge produced in mathematics and the exact sciences seems to be nonrelational. Although the knowledge produced in these disciplines can be traced to specific historical

[22] Mannheim, *Ideology and Utopia,* p. 85. For a review of this argument, see Hekman, *Hermeneutics and the Sociology of Knowledge,* pp. 52 ff. Mannheim's position on how the very conditions of modernity confer analytic advantages for understanding and synthesizing diverse epistemic cultures recalls Durkheim's discussion of the epistemic advantages of societal differentiation in *The Division of Labor in Society* (Glencoe, IL: Free Press, 1964).

[23] Mannheim, *Ideology and Utopia,* p. 155.

[24] The position is akin to Weber's remarks in "Science as a vocation," pp. 129–56, in H. Gerth and C. Wright Mills, eds. and trans., *From Max Weber* (New York: Oxford University Press, 1946), about the way that academic life *can* and *should* provide an existential foundation for a relatively value neutral analytic attitude. As Weber made explicit, the "privilege" of this position is a very rare one, but not because of any rarefied epistemological foundation. Instead, it rests on a sincere exploitation of the social advantages and leisure of the academic life. Weber recognized that the academy was far from immune from economic hustling and status contests, but C. Wright Mills (*The Sociological Imagination* [New York: Oxford University Press, 1959]) presented a more jaundiced view of the academy when he took stock of American sociology in the 1950s and found his colleagues caught up in theoretic posturing and scientistic scamming. Mills did not entirely dismiss the Weber–Mannheim position, however. Instead, he recommended a more individualistic stance that takes advantage of the academic's legitimate irresponsibility and advocated a detachment from the vulgar life of the academy in favor of diffusely populist causes.

origins, the content of this knowledge (or at least some of it) no longer bears the imprint of history.

2. The academic intelligentsia produce relational knowledge, but their institutional and historical situation enables a degree of value freedom. The "perspective" of the sociology of knowledge develops out of this situation. It does not transcend its historical and social origins, but as a matter of policy and practical situation, it is practically and fallibly more comprehensive and nonpartisan than are the systems of knowledge it seeks to explain.
3. Religious, moral, and political ideologies are practically grounded in a communal setting of beliefs and practices. The content of knowledge in such systems and the criteria for evaluating the validity of such knowledge are essentially situated.

It is important to keep in mind here that Mannheim made these distinctions in relation to a methodological demand rather than an epistemological or ontological commitment.[25] For him, the problem was one of empirically demonstrating how historical and social conditions affect the validity and content of ideas:

> Are the existential factors in the social process merely of peripheral significance, are they to be regarded merely as conditioning the origin or factual development of ideas (i.e., are they of merely genetic relevance), or do they penetrate into the "perspective" of concrete particular assertions? . . . The historical and social genesis of an idea would only be irrelevant to its ultimate validity if the temporal and social conditions of its emergence had no effect on its content and form. If this were the case, any two periods in the history of human knowledge would only be distinguished from one another by the fact that in the earlier period certain things were still unknown and certain errors still existed which, through later knowledge, were completely corrected. This simple relationship between an earlier incomplete and a later complete period of knowledge may to a large extent be appropriate for the exact sciences (although indeed to-day the notion of the stability of the categorical structure of the exact sciences is, compared with the logic of classical physics, considerably shaken).[26]

Mannheim's allusion in the last line of this passage to the theory of relativity indicates that he did not view the development and results in mathematics and the exact sciences as essentially or eternally beyond the

[25] For a discussion of Mannheim's efforts to distinguish the sociological grounds of the sociology of knowledge from epistemological criteria, see Nico Stehr, "The magic triangle: in defense of a general sociology of knowledge," *Philosophy of the Social Sciences* 11 (1981): 225–29.

[26] Mannheim, *Ideology and Utopia*, p. 271.

scope of the sociology of knowledge. Rather, he argued that the historical stability and consensual use of a statement like "two times two equals four" make it impossible to show how the content of the statement reflects the particular social position of its users.[27] The form of the statement "gives no clue as to when, where, and by whom it was formulated," unlike an artistic work whose composition can give art historians many clues for assigning it to a particular artist or genre of art, associating it with historically relative stylistic conventions, and explicating the relevant artistic community's presuppositions about the nature of the artistic subject. Similarly, a social science text or argument typically gives many clues that enable it to be traced to a "school" or "perspective" like radical behaviorism, Jungian psychoanalysis, French structuralism, or classical economics.

Mannheim's alleged exemption of mathematics and the exact sciences from the purview of sociology of knowledge became a major point of attack in proposals for a "strong program" in the sociology of scientific knowledge.[28] Some writers simply labeled this exemption as a "mistake"[29] in Mannheim's program, but such a label overlooks the importance of that exemption. What Mannheim was trying to establish with his contrast was not an exemption for natural science so much as a legitimation of practically and historically situated knowledge. Because he avowed that the sociology of knowledge could itself aspire only to a strong form of relational knowledge, he was attempting to legitimate his own mode of investigation.[30]

The relevant exemption Mannheim sought was for the sociology of knowledge, since he tried to exempt its validity claims from the stringent epistemological standards he attributed to mathematics and exact science. He did not dispute the applicability of such standards to some areas of mathematics and natural science, but again, he was mainly interested in justifying the apparently weaker claims of the sociology of knowledge. The main problem, as he formulated it, was one of demonstration:

> The existential determination of thought may be regarded as a demonstrated fact in those realms of thought in which we can show (a) that the process of

[27] Ibid., p. 272; also see p. 79. Stephen Turner observes that contrary to what is assumed in many criticisms, Mannheim's exemption of the truths of arithmetic from sociological explanation was not made "on the ground of a criterion of 'rationality'." See Turner, "Interpretive charity, Durkheim, and the 'strong programme' in the sociology of science," *Philosophy of the Social Sciences* 11 (1981): 231, n. 3.

[28] David Bloor, "Wittgenstein and Mannheim on the sociology of mathematics," *Studies in History and Philosophy of Science* 4 (1973): 173–91; Barnes, *Scientific Knowledge and Sociological Theory*, pp. 147–48.

[29] Steve Woolgar, *Science: The Very Idea*, p. 23.

[30] See Hekman, *Hermeneutics and the Sociology of Knowledge*, p. 58. Hekman argues that Mannheim makes contradictory statements about the relationship between scientific and relational knowledge and that he does not take as clear a foundationalist position as many expositors say he does.

> knowing does not actually develop historically in accordance with immanent laws, that it does not follow only from the "nature of things" or from "pure logical possibilities," and that it is not driven by an "inner dialectic." On the contrary, the emergence and crystallization of actual thought is influenced in many decisive points by extra-theoretical factors of the most diverse sort. These may be called, in contradistinction to purely theoretical factors, existential factors. This existential determination of thought will also have to be regarded as a fact (b) if the influence of these existential factors on the concrete content of knowledge is of more than mere peripheral importance, if they are relevant not only to the genesis of ideas, but penetrate into their forms and content and if, furthermore, they decisively determine its scope and the intensity of our experience and observation, i.e. that which we formerly referred to as the "perspective" of the subject.[31]

In his influential discussion of Mannheim's sociology of mathematics, David Bloor quotes part of the preceding passage and challenges Mannheim's association of "social causes" with "extra-theoretical factors." Bloor raises the question, "But where does this leave behaviour conducted in accordance with the inner logic of a theory?"[32] He goes on to contend that the strong program in the sociology of knowledge can answer this question, and he offers an innovative set of proposals for extending Mannheim's program to cover mathematical and scientific knowledge. But at the same time, Bloor's reading of Mannheim creates some confusion, pertaining to the way he interprets Mannheim to be advancing a "realist" or a "platonist" ontology of mathematics.[33]

In my reading of the preceding quotation, Mannheim is not subscribing to an absolutist position on the inherent nature of mathematical objects any more than he is endorsing a Hegelian conception of the "inner dialectic" of ideas. Instead, he is discussing the requirements for demonstrating the "existential determination of thought" against the claims of various absolutist and transcendental philosophies. The quotation marks he places around "nature of things," "pure logical possibilities," and "inner dialectic" signify that he is treating these phrases as familiar idioms. These idioms are drawn from some of the arguments that the sociology of knowledge confronts when it attempts to demonstrate the social determination of knowledge. Rather than endorsing, for example, philosophical realism, logical determinism, or dialectical reason, Mannheim is acknowledging that the sociology of knowledge faces obstinate arguments generated from within, or on behalf of, the systems of knowledge it tries to explain. Such arguments are not easily

[31] Mannheim, *Ideology and Utopia,* p. 267, quoted in Bloor, "Wittgenstein and Mannheim on the sociology of mathematics," p. 179.
[32] Bloor, "Wittgenstein and Mannheim," p. 179.
[33] Ibid., p. 176.

displaced, and Mannheim recommends a methodical procedure for accomplishing their displacement in particular cases. This procedure has two basic steps:

1. A use of historical comparison for showing that an "immanent theory" cannot entirely explain the contents and historical development of the system of knowledge in which it is situated. This procedure is used to demonstrate that such a theory cannot unequivocally and exhaustively attribute the present state of its knowledge to "the nature of things," "pure logical possibilities," or an "inner dialectic."
2. A specification of the social conditions (the local historical milieu, class interests and group "mentalities," rhetorical strategies, etc.) that influenced the development and content of the given state of knowledge.

Because Mannheim strongly opposed transcendental and absolutist philosophies, it might seem that he would dismiss the very possibility that knowledge could ever "develop historically in accordance with immanent laws." Nevertheless, he was unable to find a way to demonstrate that an expression like 2 x 2 = 4 could be explained by extratheoretical "existential" factors. Mannheim did not stipulate a sweeping exclusion of science from his explanatory program, since his method surely could apply to such cases as Darwin's theory of evolution. It is possible to show (1) that the extent to which the theory correctly interprets the "fossil record" was originally, and is still, contested in significant respects[34] and (2) that the theory emerged at a particular time and place, served particularistic social interests, was widely contested both within science and in more public arenas, and is still a subject of ideological controversy.

Although adherents to the theory may successfully resist various efforts to dispute their position, the fact that the theory remains contested and that

[34] To point to the absence of a necessary documentary grounding does not dismiss the "documentary method of interpretation" through which the theory is hermeneutically "grounded." See K. Mannheim, "On the interpretation of *Weltanschauung,*" pp. 33–83, in *Essays on the Sociology of Knowledge.* Garfinkel investigates the "documentary method of interpretation" (*Studies in Ethnomethodology,* pp. 76–103) by devising a mock psychiatric counseling session in which the "counselor" answers a series of yes–no questions in a random fashion. Unwitting subjects struggled to assimilate the series of answers within the developing text supplied by their questions. They were able to do so by successively transforming the sense of what they were asking about in accordance with the series of yes and no answers. Although this exercise did not employ Mannheim's historical method of demonstration, it did rely on the demonstrable difference between the "inherent" basis of the subjects' interpretation and the temporally developing sense they made of that (also developing) "basis." In the case of Darwin's theory, disputes about the authority of the fossil record and how to read it do not discount the consistency of the theory and its documentary base, but they do provide leverage for relationist analyses.

understandings of its meaning and application can vary even within the confines of specialized scientific fields provides enough leverage for Mannheim's relationist program of explanation.[35] Partisan battles can be quite fierce on evolutionary matters, and thus it seems at least possible that members of the "free intelligentsia" can remain detached enough to begin to elaborate the various religious, political, regional, and class correlates to the various argumentative positions in those disputes.[36] But how can $2 \times 2 = 4$ be contested by anyone who understood the formula?

Bloor addresses this question by invoking Wittgenstein's various writings on mathematics.[37] Bloor does not search for evidence of controversy over the truth of mathematical propositions; instead, he transforms Mannheim's concern with the historical conditions of a proposition's validity to a more basic question about the conditions that support the very meaning and intelligibility of a statement. Bloor reads Wittgenstein to be offering the beginnings of a "social theory of knowledge" that can explain "correct" as well as "incorrect" mathematical expressions and operations.[38] Although Wittgenstein did not offer an explanatory theory and made very limited use of empirical social science, Bloor uses his writings to support a broadened conception of the sociology of knowledge. Mannheim and Wittgenstein made no apparent use of each other's work, but in the spirit of Bloor's argument it would be easy to set up an imaginary dialogue between Mannheim and Wittgenstein:

[35] Bloor and other advocates of the strong program characterize Mannheim's approach to mathematics and exact science as a "sociology of error." This means that Mannheim's explanatory program comes into play only after it can be demonstrated that a particular "idea" strayed from the relevant discipline's immanent ability to account for it. This characterization is accurate only for the particular sense of "error" as an unexplained residual in a causal theory. It is not as though Mannheim is saying that the sociology of knowledge can explain only "false" beliefs but that it explains what an "immanent" theory cannot fully account for.

[36] Such detachment is not so easily attained, especially when their commitments to science and secular education tend to associate members of the intelligentsia with one side of a dispute and not the other. For instance, in her sociological study of the recent legal disputes about the teaching of evolution, Dorothy Nelkin sides rather decisively with the evolutionists, and in fact she gave testimony as an expert witness for the plaintiffs in *McLean v. Arkansas Board of Education,* a case heard before a U.S. district court in Arkansas in 1982. Nelkin was not necessarily committed to a nonpartisan examination of the controversy, but such a commitment is at least imaginable. See D. Nelkin, *The Creation Controversy: Science or Scripture in the Schools* (New York: Norton, 1982), p. 146, n. 5.

[37] See especially Ludwig Wittgenstein, *Remarks on the Foundations of Mathematics,* rev. ed., ed. and trans. G. E. M. Anscombe (Cambridge, MA: MIT Press, 1983).

[38] See Chapters 3 and 5. Also see Bloor's and my exchange in a series of three papers: M. Lynch, "Extending Wittgenstein: the pivotal move from epistemology to the sociology of science," pp. 215–65; D. Bloor, "Left- and right-Wittgensteinians," pp. 266–82; and M. Lynch, "From the 'will to theory' to the discursive collage: a reply to Bloor's 'Left- and right-Wittgensteinians,'" pp. 283–300, in Andrew Pickering, ed., *Science as Practice and Culture* (Chicago: University of Chicago Press, 1992).

Mannheim: Even a god could not formulate a proposition on historical subjects like 2 × 2 = 4, for what is intelligible in history can be formulated only with reference to problems and conceptual constructions which themselves arise in the flux of historical experience.[39]

Wittgenstein: "2 × 2 = 4" is a true proposition of arithmetic – not "on particular occasions" nor "always" – but the spoken or written sentence "2 × 2 = 4" in Chinese might have a different meaning or be out and out nonsense, and from this is seen that it is only in use that the proposition has its sense.[40]

Wittgenstein's remark can be read to suggest that 2 × 2 = 4 is something like an indexical expression, in which the meaning of the statement depends on the circumstances of its use. Using the hypothetical example of an exotic culture, Wittgenstein implies that even if speakers of Chinese understood the expression 2 × 2 = 4, they might apply it differently to their own system of number use. Wittgenstein is not making a historical argument, but it would be very easy to develop one along the lines of his example. By citing examples of historical societies that did not use the numeral 2 or the concept of multiplication as we do, it could be argued that 2 × 2 = 4 is not a universally valid or intelligible expression.[41]

Bloor cites a related example when he observes that Babylonian mathematics did not include the concept of zero. He argues that this provides "evidence for the idea that mathematical notions are cultural products."[42] Wittgenstein, according to Bloor, demonstrates a way for mere mortals to attribute to historical subjects what Mannheim says "even a god" could not. Although Wittgenstein does not contest the validity of 2 × 2 = 4 as a proposition in what we call arithmetic, he shows how its intelligibility is inseparable from its linguistic–cultural use. By replacing Wittgenstein's "imaginary ethnography" with actual historical and anthropological examples, Bloor suggests a way to strengthen Mannheim's program in the sociology of knowledge to cover even the most basic mathematical propositions.[43]

[39] Mannheim, *Ideology and Utopia*, p. 79.

[40] Wittgenstein, *On Certainty*, ed. G. E. M. Anscombe and G. H. von Wright (Oxford: Blackwell Publisher, 1969), sec. 10.

[41] The quotation from Wittgenstein can also be read to suggest that if a group does not use the numeral 2 or the symbol x as we do, we might very well doubt that they are doing anything comparable to arithmetic. I pursue this line of argument in Chapter 5.

[42] Bloor, "Wittgenstein and Mannheim on the sociology of knowledge," p. 187. He cites O. Neugebaure, *The Exact Sciences in Antiquity* (Princeton, NJ: Princeton University Press, 1952).

[43] See D. Bloor, *Wittgenstein: A Social Theory of Knowledge* (New York: Columbia University Press, 1983), for a more elaborate treatment of this issue.

Wittgenstein's bearing on Mannheim's program can also be extended to experimental science. Wittgenstein touches on something very close to Mannheim's concerns when he mentions the "world-picture" *(Weltbild,* a close cognate of Mannheim's *Weltanschauung)* implicated by Lavoisier's chemistry experiments:

> Lavoisier makes experiments with substances in his laboratory, and now he concludes that this and that takes place when there is burning. He does not say that it might happen otherwise another time. He has got hold of a definite world-picture – not, of course, one that he invented: he learned it as a child. I say world-picture and not hypothesis, because it is the matter of course foundation for his research and as such also goes unmentioned.[44]

Again, Wittgenstein appears to be offering a way to extend Mannheim's conceptual framework to cover the procedures in the exact sciences. If Lavoisier's experiments are set within the world picture he learned as a child, it makes sense to say that his "socialization" constituted an existential condition for his taken-for-granted "tacit knowledge."[45]

This example is roughly akin to Kuhn's discussion of Lavoisier's "discovery" of oxygen in the late eighteenth century.[46] Kuhn mentions that Joseph Priestley and Lavoisier both held "legitimate" claims to the discovery, since both succeeded in isolating what later was called "oxygen" by heating red oxide of mercury.[47] Kuhn argues that Priestley's sample was not "pure" and, more important, that Priestley did not recognize that he had isolated a distinct species of gas, and so he could not reasonably be credited with discovering oxygen: "If holding impure oxygen in one's hands is to discover it, that had been done by everyone who ever bottled atmospheric air."[48] When he performed his experiment in 1775, Priestley regarded the highly combustible gas that he had isolated as "common air with less than its usual quantity of phlogiston."[49]

[44] Wittgenstein, *On Certainty,* sec. 167.

[45] See Michael Polanyi, *The Tacit Dimension* (New York: Doubleday, 1966).

[46] Kuhn, *The Structure of Scientific Revolutions,* pp. 53 ff., and elsewhere. Kuhn (p. 45) acknowledges Wittgenstein's *Philosophical Investigations,* although not in connection with his discussion of Lavoisier and Priestley's experiments, and I have no idea whether he had read the preceding passage from *On Certainty.*

[47] Kuhn (ibid, p. 53) also mentions a third claimant to the "discovery," C. W. Scheele, but he dismisses this claim as having been publicly made too late to count on the historical record. Augustine Brannigan (*The Social Basis of Scientific Discoveries* [Cambridge University Press, 1981], pp. 20 ff.) disputes Kuhn's method for assigning the discovery to Lavoisier and thus adjudicating a dispute over the discovery by reference to its outcome. An entirely different line of attack on Kuhn's example is provided by Philip Kitcher ("Theories, theorists and theoretical change," *Philosophical Review* 87 [1978]: 519–47). Kitcher takes issue with the relativistic implications of Kuhn's example and contends that any historical understanding of how the concept "oxygen" eventually came to replace "dephlogisticated air" must take into account the referential adequacy of the respective terms; whereas "oxygen" refers to something in the world, "phlogiston" does not.

[48] Kuhn, *The Structure of Scientific Revolutions,* p. 54.

[49] Ibid., p. 53.

A few years later, Lavoisier concluded after performing a series of similar experiments that he had isolated one of the two main constituents of the atmosphere. His interpretation implicated a theoretical picture entirely different from Priestley's, since he treated the combustible product of the experiment not as "air" with the phlogiston removed but as a purified constituent of air. Moreover, as Kuhn argues, the displacement of phlogiston theory brought into play a complex of definitions and explanatory concepts for demonstrating the causes of combustion and investigating the chemical makeup of substances. In Kuhn's terms, one paradigm (which now seems more compatible with modern chemistry) displaced an older world picture. Kuhn's discussion enables us to interpret Wittgenstein's remark in a more precise way, since now it can be said that following Lavoisier's discovery of oxygen, chemists were socialized to become members of a "normal science" community sharing a conceptual framework, a set of ostensive definitions, and an array of established experimental devices and practices. For subsequent generations of chemists, the makeup of air and the explanation of combustion were not hypotheses because (to paraphrase the preceding quotation from Wittgenstein) "they are accepted as a matter-of-course foundation for their research and as such they go unmentioned." The knowledge cultivated in such a stable disciplinary community would be no less "relational" than the political and religious beliefs that Mannheim tried to explain with his sociology of knowledge.

Kuhn's historiographic method enabled proponents of the strong program to argue that developments in the exact sciences did not show the "simple relationship between an earlier incomplete and a later complete period of knowledge" that Mannheim suggested. On those occasions when one paradigm replaced another, such as during the Copernican revolution or with the rise of quantum theory, the changes not only involved particular theoretical corrections within a stable system of accepted knowledge, but they also included procedures for verifying facts and defining what counted as relevant tests or demonstrations.

As in the case of Lavoisier's discovery of oxygen, it is only by means of retrospective judgments that historians are able to articulate the objective grounds for accepting one world picture over another. Priestley used phlogiston theory to account for the experimental facts that Lavoisier and his followers placed in the organizational framework of a different conceptual and practical gestalt.[50] Although the outcome of the controversy may be explained retrospectively in terms of "immanent laws" that are now accepted in physical chemistry, these laws were not definitively articulated at the time of the controversy. The laws of modern physical chemistry are inseparable from a commitment to the system that replaced phlogiston theory, and so they do

[50] See Paul Feyerabend, *Against Method* (London: New Left Books, 1975) for further arguments and examples.

not provide an impartial basis for explaining the outcome of the controversy. Consequently, there is a sense in which the history of science can support Mannheim's contention that "the process of knowing does not actually develop historically in accordance with immanent laws," since these laws explain only those events that gave rise to them through the retrospective illusions of "Whig" historiography.[51]

The attack on Merton's self-exemplifying sociology of science

In the 1930s, Robert Merton studied with Talcott Parsons at Harvard, and he later developed his own variant of the structural–functional approach in sociological theory.[52] Merton's theoretical approach was eclectic and hegemonic, as it expanded Parsons's framework to subsume the broadest possible range of sociological researches. Merton's theory of the "middle range" endeavored to bridge the gap between Parsons's highly abstract theory and more concrete modes of empirical research on social institutions and social attitudes.[53] Merton also elaborated an interpretive scheme for taking account of "dysfunctional" as well as "functional" aspects of institutions and "latent" as well as "manifest" functions.[54] Merton and his associates captured the institutional center of American sociology and covered the whole of the discipline with a thin and flexible theoretical gloss. Consequently, it became impossible to critique Merton's approach without court-

[51] For an early critique of this sort of historiography of science, see Joseph Agassi, *Towards an Historiography of Science* (The Hague: Mouton, 1963).

[52] As mentioned in Chapter 1, Garfinkel studied under Parsons in the 1940s. Despite their common connection to Parsons and despite the fact that both are regarded as major contributors to contemporary approaches to sociological theory and the sociology of science, Merton and Garfinkel had little to do with each other. Merton, to my knowledge, never mentioned Garfinkel or ethnomethodology in his voluminous writings on sociological theory and sociology of science. And although Garfinkel developed ethnomethodology partly as a reaction to Parsonian structural functionalism, he ignored Merton's elaboration and transformation of Parsons's approach. Their paths diverged very early in the game: Merton became the major spokesman for the "center" of American sociology, and Garfinkel conducted a radical campaign at the margin of the discipline. Whereas Merton tried to build on the broadest possible base in existing sociology, Garfinkel attacked the "curious absurdities" in the classic tradition and turned to phenomenology and Wittgenstein for inspiration.

[53] For an account of the way that Merton and his colleagues helped install the dominant theoretical–empirical program in mid-century American sociology, see Stephen Turner and Jonathan Turner, *The Impossible Science: An Institutional Analysis of American Sociology* (London: Sage, 1990).

[54] Functionalism had often been criticized as an inherently conservative perspective that implicitly justifies social institutions by focusing on their "functional" aspects. Merton attempted to defuse this argument by developing the concept of "dysfunction" (an institution or practice that destabilizes a social order) and insisting that functionalists specify the reference point for any alleged function or dysfunction by indicating which of the groups in the society it serves or *dis*serves. Merton took apparent delight in demonstrating an ability to translate passages from Marx into functionalist idioms (cf. "Paradigm for sociology of knowledge," p. 35, and *Social Theory and Social Structure,* enlarged ed., pp. 99–100).

ing the accusation that one was (1) stipulating artificial limits to "Mertonianism" and/or (2) violating some of the conditions for maintaining good standing in the academic profession of sociology.[55] Given the hazards associated with both accusations, it is understandable that the more successful of Merton's critics were stationed overseas in Britain.[56]

Merton's contributions to the sociology of scientific knowledge began with his doctoral dissertation, "Science, Technology and Society in Seventeenth-Century England."[57] In the subsequent half-century, this durable and prolific scholar contributed to numerous areas of sociology. Especially after he moved to Columbia University in the 1950s, Merton and his students formed a virtual cartel that dominated the sociology of science, as well as a number of other subfields, until the 1970s.[58] The Mertonian sociologists produced an array of studies, including historical studies of the development of science, grand conceptual typologies of the scientific ethos," and more "micro" approaches to scientific organizations and communication networks. Although the Mertonian program was functionalist in its orientation, in the 1950s and early 1960s this meant little more than that it was part of the

[55] The rhetorical force with which Merton responded to some of his critics can be appreciated by reading his essay "The ambivalence of scientists: a postscript," pp. 56–64 in Merton, *Sociological Ambivalence and Other Essays* (New York: Free Press, 1976), esp. pp. 59–60. Perhaps the high-water mark in the dominance of functionalism in American sociology was Kingsley Davis's presidential address to the members of the American Sociological Association in which he argued that functional analysis was a "myth." Davis presented this as a defense of functionalism against its critics, reasoning that functionalism was not a distinct theoretical framework but a mode of argument that every sociologist used. Consequently, whether viewed as a perspective, school, argumentative style, or academic gang, functionalism proved hard to pin down, and by the same token it proved resilient in the face of criticism. See Kingsley Davis, "The myth of functional analysis as a social method in sociology and anthropology," *American Sociological Review* 24 (1959): 757–72.

[56] A kinder and gentler version of why the criticisms took hold in Britain is supplied by Arnold Thackray's essay "Measurement in the historiography of science," pp. 11–31, in Y. Elkana, J. Lederberg, R. K. Merton, A. Thackray, and H. Zuckerman, *Toward a Metric of Science: The Advent of Science Indicators* (New York: Wiley, 1978). After chronicling the successes and rapid progress shown by Mertonian sociology of science, Thackray (p. 21) observes: "Not altogether surprisingly, this sustained attention to the 'internal sociology' of science has not passed without comment. The critics have usually been residents in Europe – most often in Britain – and hence remote from the advantages and the limitations shared by a group of practitioners enjoying a common paradigm."

[57] Shortly after its completion, the dissertation was published in *Osiris: Studies on the History and Philosophy of Science* (Bruges: St. Catherine's Press, 1938; new ed., New York: Harper & Row, 1970). For a series of articles discussing the Merton thesis, see the special issue of *Isis* 79 (1988): 571–623.

[58] Some of Merton's more prominent students and colleagues were Bernard Barber, Jonathan and Stephen Cole, Norman Storer, Nicholas Mullins, Diana Crane, Lowell Hargens, and Harriet Zuckerman. Their research linked up with Derek de Solla Price's and Joseph Ben-David's studies of scientific institutions and publication patterns. More recent students, Thomas Gieryn and Susan Cozzens, have integrated the Mertonian approach with the more recent approaches in the sociology of scientific knowledge.

mainstream of sociology. Only in retrospect did the program seem limited in its conception of how social factors were related to science.

The British attack on Merton focused on two related aspects of his approach: (1) his distinction between "external" and "internal" explanations of scientific progress and (2) his account of the autonomy and integrity of science.

Internal and external explanations. Merton's argument in "Science, Technology and Society in Seventeenth-Century England" closely followed Weber's essay "The Protestant Ethic and the Spirit of Capitalism."[59] In parallel with Weber's thesis on the relationship between Calvinist doctrines and entrepreneurial activity, Merton argued that the worldly ascetic values associated with northern European Protestantism motivated many of the founders and patrons of the Royal Society. Although the Protestant clergy were often hostile to science, Merton argued that Puritan values stimulated an esteem for secular achievement, especially when such achievements seemed indifferent to personal motives for profit and pleasure. Consequently, scientific innovations were highly valued, since they were justified as disinterested contributions to humankind that testified to the intricacy of God's plan.[60] Although some historians read Merton to be giving an "externalist" argument in which religious factors explain the content of Newton's or Boyle's discoveries,[61] Merton qualified his argument much along the lines of Weber's famous "switchman" analogy.

Weber used the image of a "switchman" at a railroad yard to suggest how the Protestant ethic provided a catalyst, although not a determinant, for the development of capitalist industry. The switchman does not determine the layout of the track or the momentum of the train, just as the Calvinist emphasis on "worldly asceticism" did not determine the historical preconditions for the rise of capitalism or the competitive dynamic that later sustained the progressive rationalization of industry. Instead, Puritanism was a catalyst for motivating entrepreneurial activity and spurring economic development along a historical track that it might otherwise not have taken. Merton makes a similar argument about the influence of Puritan doctrines on a related field of practical endeavor, scientific innovation.

[59] Max Weber, *The Protestant Ethic and the Spirit of Capitalism,* trans. Talcott Parsons (London: Allen & Unwin, 1930). Parsons, *The Structure of Social Action,* vol. 1 (New York: McGraw-Hill, 1937) p. 511, cites Merton's study as a source of "facts, which confirm Weber's position" on the tendency for Protestantism to be affiliated with innovative occupations.

[60] Merton qualifies his argument sufficiently to protect it against the counterexample of Italian science. For Merton, because the historical development of science was not "caused" by religion but only facilitated or inhibited by religious factors, his explanation can account for the development of Italian science despite the occasional interference by the church.

[61] See A. R. Hall, "Merton revisited," *History of Science* 2 (1963): 1–16.

Like Weber, he argued in favor of the "role of ideas in directing action into *particular* channels."[62]

In a recent essay, Steven Shapin defends Merton's thesis against various historians' criticisms by citing Merton's provisos about the absence of religious influence on the "internal history of science."[63] Shapin points out that although Merton contended that the values of Protestantism were motivationally and rhetorically significant for the activities of the Royal Society, he was careful to say that religious values did not cause or sanction particular discoveries and methodological innovations. Although Shapin says this to defend Merton's thesis, he does so with ironic intent. Given Shapin's commitment to the strong program, the fact that Merton expressed no intention to "adduce social factors to explain the form or content of scientific knowledge or scientific method" should be understood as an account of the disadvantage of Merton's program for the sociology of science.[64]

As we mentioned earlier, Bloor and Barnes devised methods for demonstrating how "internal" developments in science could be explained by "social factors." While doing so, they also redefined what was meant in the first place by a "causal explanation." They extended the concept of "cause" to fit the very sort of argument Merton made in his thesis. In his discussion of causal determinacy, Barnes gives the example of an explanation for a road accident in which an ice patch is cited as the cause of the accident. He points out that this "implies neither that 'whenever ice-patches occur there is an accident,' nor that 'there is never an accident unless there is an ice-patch.' "[65] The causal factor of interest is specified against a background of normal conditions, and the explanation implies that the event would not have occurred (or would have occurred differently) if this factor had been absent or different.

Barnes's way of conceptualizing causality applies to Weberian "switchman" explanations no less than to more familiar forms of mechanistic explanation.[66] So for instance, if a switchman erroneously shunts a train onto the wrong track, his "human error" can be cited as the cause of a resulting collision. An explanation might focus on the switchman's drunken state, his inadequate training, or his misinterpretation of some ambiguous information

[62] Merton, "The Puritan spur to science," in *The Sociology of Science,* p. 237. This is a reprint of "Motive forces of the new science," chap. 5, pp. 80–111 of Merton, *Science, Technology and Society in Seventeenth-Century England* (New York: Howard Fertig, 1970).

[63] Steven Shapin, "Understanding the Merton thesis," *Isis* 299 (1988): 594.

[64] Shapin makes this clear in a later essay, "Discipline and bounding: the history and sociology of science as seen through the externalism–internalism debate," pp. 203–37 in *Proceedings of Conference on Critical Problems and Research in the History of Science and History of Technology,* Madison, WI, October 30–November 3, 1991.

[65] Barnes, *Scientific Knowledge and Social Theory,* p. 71. He draws the example from A. MacIntyre, "The antecedents of action," in B. Williams and A. Montefiore, eds., *British Analytical Philosophy* (London: Routledge & Kegan Paul, 1963).

[66] Barnes (*Scientific Knowledge and Social Theory,* pp. 73–74) goes beyond MacIntyre's ("The antecedents of action") recommendations in this regard.

he was given about the situation. For Barnes and Bloor, causal explanation can apply not only to the switchman's alleged "error" in light of the disaster but also to his routine and unproblematic behavior. Accordingly, a train's normal passage through the switchyard can be said to have been "caused" or "determined" by the switchman's allegedly correct actions, since those actions could have been otherwise. In this case a social explanation might mention the switchman's trained capacity to understand the situation and act appropriately.

The application of this version of causality to Merton's thesis seems at first simply to broaden the definition of what counts as a causal explanation. In accordance with Barnes's account of causality, Merton's thesis explains the social determination of scientific knowledge, but such an "external" account does nothing to diminish the status of the scientific knowledge it explains. When Barnes suggests that routine or unproblematic scientific practices can be explained socially, he does not mean that, for example, religious factors would explain anything more than what Merton ascribes to them. Instead, Barnes cites other factors inherent in the "subculture" of a scientific field to explain an immanent progression of routine innovations.[67] He describes the apparently autonomous development of a mature or normal scientific discipline by mentioning such factors as subcultural socialization, and scientists' use and extension of analogies and semantic categories deriving from wider fields of discourse. Merton and his followers had not paid much attention to such factors, but to do so is not necessarily incompatible with their overall approach.

For the strong program, a distinction between internal and external aspects of a scientific discipline is important for determining what kind of social explanation of scientific development is appropriate, but not because of any permanent epistemological demarcation between science and nonscience.[68] Proponents of the strong program do not entirely discard the internal–external dichotomy, but they do believe that the alleged boundary between science and nonscience is a historically contingent rhetorical achievement.[69]

The consequences of this move are twofold. First, Merton's account of the origins of science is now redefined as a causal explanation, since science is

[67] Barnes, *Scientific Knowledge and Social Theory,* pp. 86 ff.

[68] In his ironic defense of Merton, Shapin ("Understanding the Merton thesis," p. 594) suggests that Merton may have had no small part in establishing the distinction: "It is a plausible hypothesis that our present-day language of 'internal' and 'external' factors, as well as the validation of an overwhelmingly 'internalist' historiography of scientific ideas, actually originated with Merton and the circle of scholars with whom he studied and worked in the 1930s." Shapin later ("Discipline and bounding") credited Bernard Barber for having established the distinction.

[69] This argument was developed in an article by Thomas Gieryn, a former student of Merton's whose work articulated a bridge between the older and newer programs in sociology of science. See T. Gieryn, "Boundary-work and the demarcation of science from non-science," *American Sociological Review* 48 (1983): 781–95.

no longer viewed as an autonomous force that was set on an appropriate track by, among other things, religious developments in Britain and Holland. Switchman explanations can now be used intensively to explain internal developments in normal scientific fields. As mentioned earlier, sociologists of scientific knowledge often apply the Duhem–Quine "underdetermination thesis" to argue that observational evidence cannot by itself restrict the field of relevant theoretical explanations to a single possibility. Therefore, something besides the nature of the evidence or pure logical possibilities is responsible for "switching" consensual interpretations onto the particular theoretical tracks they take. For Barnes, a causal explanation can be constructed whenever the historical evidence shows that particular social interests or other factors predisposed an acceptance, rejection, or disregard of one or another possible theoretical interpretation.

Second, the differentiation and stability of a boundary between science and nonscience is itself a continual social construction that can be explained by such factors as social consensus, the distinctive socialization of scientists, and scientists' ability to persuade key elites and members of the public to accept the authority of science as a basis for unquestioned belief. Consequently, the task for the sociology of scientific knowledge is no longer to examine the social influences that operate across the boundary from "society" to "science" but to examine how the boundary is itself a product of the social organization of scientific activity.

As in the case of the British challenge to Mannheim, the strong programmatic treatment of Merton's sociology expanded the topics and explanatory methods of the sociology of knowledge without radically altering the program.[70] However, the practical effect of this reorientation was radical insofar as it motivated a challenge to the rhetoric supporting the autonomy of science. This challenge implicated the institutional grounds claimed by the sociology of science itself.

The autonomy of science. Merton, like Mannheim, was careful to distinguish the social and historical conditions that gave rise to scientific innovations

[70] R. J. Anderson, J. A. Hughes, and W. W. Sharrock argue that the strong program's mode of explanation is little different from "old-fashioned functionalism" and that it is open to many of the same criticisms. Anderson and his colleagues point out that the demonstration of causality is typically made by showing homologies between particular scientific theories and other beliefs extant in the social milieu in which the theories originated. In a functionalist explanation, abstract homologies (e.g., between basic dimensions of Puritan belief and the ethos of science) are used to demonstrate that the milieu supported or motivated the promulgation and acceptance of the theory. The strong program rewrites the congruence arguments in functionalism into stronger causal idioms, but the task of demonstrating and defending connections between particular abstract formulations of "belief" and "knowledge" faces many of the same intractable problems. See R. J. Anderson, J. A. Hughes, and W. W. Sharrock, "Some initial difficulties with the sociology of knowledge: a preliminary examination of 'the strong programme'," *Manchester Polytechnic Occasional Papers,* no. 1, 1987.

from the process of innovation within the specialized disciplines. But contrary to what is sometimes argued, Merton and his followers did not ignore the "esoteric content" of scientific activity, nor did they define the natural sciences as asocial enterprises.[71] Instead, they characterized modern science as a distinctive institution, whose normative "ethos" and reward system were conducive to the relatively unencumbered pursuit of esoteric knowledge. The question was not "What social conditions give rise to justified true belief?" but "What institutional conditions are necessary to produce and certify knowledge claims that sometimes conflict with religious and political authority?" Merton recognized that science has often been pressed into the service of political, economic, and religious interests, but he claimed that the conflicts and ethical dilemmas that arise under such conditions testify to the normative expectation that science should be an unencumbered pursuit of knowledge for its own sake. Merton did not overtly make an ontological or epistemological claim about what distinguishes science from other institutions; instead, he presented a functional argument regarding how "standardized social sentiments about science" give rise to and support the historically distinctive ethos of science.[72] Nevertheless, his account of the optimal institutional conditions for nurturing "pure" scientific development implied a view of scientific rationality that Barnes and Bloor attacked.

Merton's articles on the ethos of science were written in the late 1930s and early 1940s.[73] In one sense, these articles were an extension of his thesis on the ethical values conducive to the development of the seventeenth-century English science. As noted earlier, Merton developed his argument along Weberian lines: Whereas the Puritan ethic gave rise to science, in the modern era, scientific activity becomes a relatively autonomous historical development and an "end" in itself. But in contrast with Weber's image of the "steel-hard cage" of industrial society, Merton's autonomous science is less tinged with ominous implications:

> Three centuries ago, when the institution of science could claim little independent warrant for social support, natural philosophers were likewise

[71] See Merton, "Paradigm for the sociology of knowledge," p. 37, where he briefly mentions studies of "the ways in which the cultural and social context enters into the conceptual phrasing of scientific problems." He adds further (p. 39) that "vestiges of any tendency to regard the development of science and technology as *wholly* selfcontained and advancing irrespective of the social structure are being dissipated by the actual course of historical events." For an example of "Mertonian" research that, in its fashion, deals with the "content" of science, see Bernard Barber and Renée Fox, "The case of the floppy-eared rabbits: an instance of serendipity gained and serendipity lost," *American Journal of Sociology* 64 (1958): 128–36.

[72] Robert K. Merton, "The normative structure of science," chap. 13 of Merton, *The Sociology of Science,* quotation from p. 268. Originally published under the title "Science and technology in a democratic order," *Journal of Legal and Political Science* 1 (1942): 115–26.

[73] Ibid. Also see Robert K. Merton, "Science and the social order," *Philosophy of Science* 5 (1938): 321–37.

> led to justify science as a means to the culturally validated ends of economic utility and the glorification of God. The pursuit of science was then no self-evident value. With the unending flow of achievement, however, the instrumental was transformed into the terminal, the means into the end. Thus fortified, the scientist came to regard himself as independent of society and to consider science as a self-validating enterprise which was in society not of it.[74]

Merton was also concerned about the more immediate situation in Nazi Germany.[75] Merton reacted to the exodus of Jewish scientists from Germany and the Nazi domination of scientific and scholarly activity by contending that a democratic social structure encouraged the pursuit of "pure" (i.e., basic) science. His account provided a more abstract version of Mannheim's proposals regarding the social conditions conducive to value-free intellectual discourse (an "unanchored, relatively classless stratum" in the universities). By utilizing the Parsonian conceptual framework to define a distinctive constellation of four "institutional imperatives" for modern science, Merton avoided some of the concrete difficulties associated with Mannheim's proposal. As in many general typologies, the categories overlap and reinforce one another, and together they portray a coherent picture of science:

Universalism. This is "the canon that truth-claims, whatever their source, are to be subjected to *pre-established impersonal criteria:* consonant with observation and with previously confirmed knowledge."[76] This norm does not guarantee objectivity; rather, it fosters a preobjective commitment to meritocratic institutional procedures for sharing and evaluating research results.

Communism. Merton later renamed this as *communalism,* perhaps to avoid the obvious political connotations. Either term meant that the "substantive findings of science are a product of social collaboration and are assigned to the community. They constitute a common heritage in which the equity of the individual producer is severely limited."[77] This norm is implicated by the

[74] Merton, "The normative structure of science," p. 268.

[75] Bernard Barber mentions another relevant circumstance, which was a reaction to the "massive critique" of the ideology of "pure science" launched in the 1930s by Boris Hessen and a group of "scientific humanists" who were inspired by Hessen. Hessen was a member of the Russian delegation at the Second International Congress on the History of Science, held in London in 1931. The paper he read there, "The social roots of Newton's 'Principia,'" argued that even "pure" science had social origins and consequences and that the agenda of scientific research should be more closely related to a broad conception of social progress. The Parsonian–Mertonian emphasis on the differentiation of social institutions and the normative framework of a relatively autonomous scientific institution were ways to take into account the "social aspects" of science without "going too far" in the direction of a socialist program. See B. Barber, *Social Studies of Science* (New Brunswick, NJ: Transaction Publishers, 1990), pp. 3 ff.

[76] Merton, *The Sociology of Science,* p. 270.

[77] Ibid., p. 271.

institutional convention of "eponymy" in which scientists' property rights are limited to the prestige and esteem from having phenomena, theories, proofs, measuring units, and the like named after them (e.g., Baade's star, the Heisenberg principle, Gödel's proof, volts, curies, roentgens, Tourette's syndrome, etc.). Such named "products" are disseminated openly and used freely, and the reward system of science encourages the rapid publication of results rather than secrecy and hoarding.

Disinterestedness. Merton emphasizes that this norm is enforced through "a distinctive pattern of institutional control of a wide range of motives which characterizes the behavior of scientists."[78] He distinguishes the institutionally sanctioned behavior of scientists from any inherent individual virtues. Scientists conform to strict standards of conduct not because they are superior individuals but because it is in their interest to avoid fraud, cultism, informal cliques, and trivial and spurious claims. The enforcement mechanism in this instance is "the public and testable character of science . . . [which] it may be supposed, has contributed to the integrity of men of science."[79]

Organized skepticism. This is "both a methodological and an institutional mandate . . . [for the] temporary suspension of judgment and the detached scrutiny of beliefs in terms of empirical and logical criteria" sometimes leading to conflicts between scientific and other institutional systems of belief. "The scientific investigator does not preserve the cleavage between the sacred and the profane, between that which requires uncritical respect and that which can be objectively analyzed."[80]

To an extent, this constellation of norms is modeled after a Parsonian reading of Weber's ideal–typical account of bureaucracy, with its emphasis on universalism, specialized competency, the impersonality and communal property of the office, and the institutionalization of meritocratic standards for adjudicating competition.[81] The main difference is Merton's emphasis on substantive institutions and practices for testing experimental findings, rewarding achievements with nonmonetary credit, and communicating and

[78] Ibid., p. 276.
[79] Ibid.
[80] Ibid., p. 277.
[81] Max Weber, *Economy and Society,* vol. 2 (Berkeley: University of California Press, 1978); Talcott Parsons, *The Structure of Social Action,* vol. 2 (New York: Free Press, 1937), pp. 506 ff. Merton developed the Weberian ideal type in his model of bureaucracy: "Bureaucratic structure and personality," chap. 8 of *Social Theory and Social Structure,* enlarged ed., pp. 249–61). See James March and Herbert Simon, *Organizations* (New York: Wiley, 1958), chap. 3, for a discussion and critique of the Merton–Weber model.

verifying specialized researches. Like Weber's account, Merton's is readily criticized as an idealized version that overlooks the personalistic and factional machinations in "actual" organizations. Merton was careful to identify the norms as ideal standards rather than descriptions of actual behavior, but he nevertheless was criticized on this point in an important paper by Barry Barnes and R. G. A. Dolby:

> These norms have from time to time been professed by scientists. The sociologist must distinguish professed norms from the patterns of positively sanctioned behaviour; these professed norms are in themselves incapable of providing real guidance for action. Merton can point to examples of his norms in what scientists say, but he does not produce any evidence of behaviour modified by these norms.[82]

This and similar criticisms covered at least three related aspects of Merton's functionalist approach:

1. The norms were stated so abstractly that it was unclear how they were relevant to specific instances of scientists' conduct. Merton derived the norms from biographies and memoirs of scientists, and it seemed likely that such writings rhetorically exaggerated scientists' commitments to rational and otherwise honorable conduct.
2. Merton's definition of the norms incorporated a coherent picture of scientific methodology based on early-twentieth-century philosophy of science. He supposed that under the appropriate institutional circumstances, the process of generating and verifying discoveries would lead to a progressive accumulation of theories and technological applications. The Kuhnian picture of revolutionary discontinuity, although endorsed by the Mertonians,[83] complicated their version of a unitary scientific method guided by independent standards of verification and transcendent norms of rationality. Consequently, questions about how communities of scientists distinguished among incommensurable theories and also how normal science remained stable in the face of the possibility of alternative paradigmatic commitments were placed on the agenda for the sociology of science. Social factors were no longer limited to sources of facilitation, interference, or resistance to inherently rational scientific innovations.
3. Merton and his colleagues liked to claim that the sociology of

[82] Barnes and Dolby, "The scientific ethos," pp. 12–13.

[83] Barber, *Social Studies of Science*, p. 246, highlights his and Merton's endorsement of Kuhn by mentioning the following anecdote: "In the early 1960s, when Tom Kuhn, an old friend of ours, was trying to get the University of Chicago Press to publish his *Structure of Scientific Revolutions* . . . he asked us to write supporting letters."

science was "self-exemplifying."[84] In their view, specialized professional journals, peer review processes, policies of academic freedom, and promotion on the basis of merit were necessary features of modern scientific institutions. Such institutional arrangements supposedly functioned to ensure the efficient circulation and testing of results with minimal interference from nonscientific interests. Since professional sociology and, more specifically, the sociology of science also included specialized journals, peer review, professional associations, and so forth, the Mertonians suggested that there was reason to suppose that the institutional requisites for scientific progress in sociology were in place. In one of many reiterations of this theme, Barber recites a forward-looking history:

> Gradually, very gradually, the sociology of science has achieved something of the scientific status it has longed for. Cole and Zuckerman have recently shown, using citation data, the great increase in cognitive consensus in the field from the early period 1950–54 to the period 1970–73. And besides a decent amount of cognitive consensus, the sociology of science has achieved all the essential characteristics of an institutionalized scientific field: regularized university courses of instruction, special journals, special funding agencies, special professional associations, and specialized scholarly conferences.[85]

This way of elaborating the institutional grounds for a science of sociology became problematic in light of the post-Kuhnian emphasis on practical and conceptual consensus. In Kuhn's account, consensus was bound to disciplinary-specific complexes of theory, experimental procedure, and instrumentation; it did not "naturally" emerge from the free exercise of reason guided by general rules of method. Despite its academic trappings and conspicuous regard for rules of method, professional sociology was, in Kuhn's terms, a "preparadigm" discipline, as there was no immediate prospect of an internal

[84] For instance, in his "Author's preface" to *The Sociology of Science,* Merton mentions that "the sociology of science exhibits a strongly self-exemplifying character: its own behavior as a discipline exemplifies current ideas and findings about the emergence of scientific specialties" (p. ix).

[85] Barber, *Social Studies of Science*, p. 247. Barber cites Jonathan Cole and Harriet Zuckerman, "The emergence of a scientific specialty: the self-exemplifying case of the sociology of science," pp. 139–74, in Lewis A. Coser, ed., *The Idea of Social Structure: Papers in Honor of Robert K. Merton* (New York: Harcourt Brace Jovanovich, 1975). This optimistic style of "reflexive" assessment is also expressed by Thackray, who suggests that the "distance traveled" in the sociology of science is indicated by the following measure: "Prevailing standards of devotion to, and performance in, the quantitative study of science may be seen from two recent books on stratification and competition in the physics community. One offers 44 tables in its 174 pages of text, the other 42 tables in 261 pages. Examples could be multiplied" (A. Thackray, "Measurement in the historiography of science," p. 21).

consensus among practitioners on fundamental matters of theory, fact, and appropriate practice. Despite what Barber claims about the "cognitive consensus" in the sociology of science, after the functionalist paradigm lost its stranglehold on American sociology, it became increasingly difficult to claim that its offspring in the sociology of science exemplified a stable scientific field. As Stephen Turner observes:

> The self-exemplification of the rise of Mertonian sociology of science was announced at a time when the project was collapsing; after that time it was evident that the project had been sustained not so much by "merit," as the theory supposed, as by connections, power, and patronage, and by its willingness to forego hard questions about the content of science.[86]

The arguments against Merton's framework did not entirely demolish their target. This was not because of any flaw in Barnes's, Bloor's, or Mulkay's reasoning but because Merton and his defenders had devised what Garfinkel calls a "specifically vague" theoretic account that was capable of evading or absorbing an impressive range of criticisms. The attack on Merton was most effective when focused on his early articles on "norms," but Merton later argued that he never claimed that the norms acted as unequivocal standards guiding all scientific conduct. Instead, he contended, priority races and related competitions among scientists create dilemmas about the enactment of normatively appropriate conduct.[87] So for instance, the norm of communalism does not "tell" scientists just when to disseminate results of their experiments, and scientists may legitimately hold back "incomplete" or possibly spurious results until they are judged ready for critical evaluation in the scientific community.

Merton was not deterred by the possibility that "counternorms" could be formulated in dialectical opposition to each of the four norms. His modified framework easily permitted conflicting normative commitments to coexist in a functional system. Although Merton's account, especially of disinterestedness, strongly promoted science and scientists and implied an archaic view of the gentlemanly pursuit of "pure" knowledge, to say that Merton elaborated an ideology or a rhetoric rather than a description of science does not fully displace his functionalist argument.[88] One can easily revise Merton's account by "bracketing" his argument on behalf of scientific autonomy, so that the argument becomes an immanent "account" supporting the autonomy of science, rather than a transcendental description of the

[86] Stephen Turner, "Social constructionism and social theory," *Sociological Theory* 9 (1991): 22–33, quotation from pp. 27–28.
[87] Merton, "The ambivalence of scientists," pp. 383–412, and "Behavior patterns of scientists," pp. 325–43 of *The Sociology of Science*.
[88] See Michael Mulkay, "Norms and ideology in science," *Social Science Information* 5 (1976): 637–56.

place of science in society. Once this is done, the norms become rhetorical themes serving the creation and maintenance of a movable border between science and the rest of society. To say that this rhetoric reflects particularistic interests within the scientific community is readily absorbed into Merton's conception of functionalism, since his framework permits "function" to be defined in reference to particular groups or classes within a larger social unit.[89]

Although translating the passive modality of "function" into the more active voice of rhetorical "strategy" entails more than a minor adjustment to the Mertonian conceptual scheme, it does not entirely disrupt it. Some of Merton's former students have developed just such a constructionist–functionalist hybrid by retaining aspects of Merton's arguments in a more explicit emphasis on discourse and agency.[90] The advantage of such left-Mertonianism is that it preserves the functionalist form of argument without a positivistic commitment to verification, crucial tests, and the like. Evaluative standards, rewritten as themes for rhetorical strategies, now serve a more temporary and fallible autonomy and honorific status of science and particular groups of scientists. Moreover, the criticisms by Barnes and Dolby to the effect that the norms can be contested and do not describe actual behavior can now be incorporated in a more flexible and differentiated functional framework.

It might fairly be argued that Merton anticipated virtually all of the criticisms and empirical alternatives to his program that emerged from Britain and elsewhere since the 1970s. (To say this, of course, is to identify "Merton" with the dutiful work of his followers and interpreters, whose incessant efforts maintained Mertonian sociology's central place in the academic field.) As stated earlier, Merton's paradigm for the sociology of knowledge included investigations of how the social context of science influences "conceptual phrasing" of scientific knowledge, and in another essay, he called for ethnographic studies of scientific practices.[91] In a defense of Merton against criticisms to the effect that he and his followers ignored the "content" of science, Norman Storer stated that the lack of emphasis on the esoteric contents of scientific research was a temporary lag in the research program but not an essential hole in the Mertonian conception of science.[92]

[89] Merton, "Manifest and latent functions."

[90] See Thomas Gieryn, "Boundary-work and the demarcation of science from nonscience"; also his "Distancing science from religion in seventeenth-century England," *Isis* 79 (1988): 582–93; and Susan Cozzens and Thomas Gieryn, eds., *Theories of Science in Society* (Bloomington: Indiana University Press, 1990).

[91] Merton, "Paradigm for the sociology of knowledge," pp. 37 ff.; Merton, "Forward" in Bernard Barber, *Science and the Social Order* (Glencoe, IL: Free Press, 1952).

[92] Norman W. Storer, "Introduction," pp. xi–xxxi, in Merton, *The Sociology of Science*.

Storer also updated the normative framework in light of the Kuhnian "revolution" by saying that it applied to stable periods of normal scientific progress rather than to revolutionary irruptions of paradigms.

To cite these defenses and revisions of Merton's program does not imply that the strong program's criticisms were inappropriate or ineffective. Even though Merton's all-encompassing theoretical proposals could be read retrospectively to anticipate the new developments, the Mertonians simply were not doing very much to develop studies along some of the lines advocated by proponents of the strong program. As Merton's various pronouncements on the self-exemplifying character of the sociology of science indicated, he and his followers operated with apparent confidence that they were on the right track. In that context, some of the more effective arguments against Merton's program simply ignored the entire corpus of Mertonian studies or summarily dismissed them by saying that they focused only on scientists and institutions rather than the content of science. Practically speaking, it did not matter very much that these arguments slighted Merton's achievements and that they expressed a great deal of confusion about what could be meant by the "content" of science, since what resulted was a set of independent initiatives and fresh approaches that enlivened at least one area of sociology.

The consolidation of a strong program in the sociology of knowledge

Bloor, Barnes, and other British sociologists of knowledge drew on a variety of sources to supplement and broaden the earlier programs in the sociology of knowledge. Their strong program in the sociology of knowledge retained Mannheim's basic two-step form of demonstration while modifying it to cover science and mathematics. With suitable modifications of Mannheim's terms, adherents to the strong program sought to show the following:[93]

1. Although scientists and mathematicians may act in accordance with the immanent logic of theory, their actions are not unequivocally determined by the "nature of things" or "pure logical possibilities."[94] On the contrary, the emergence and crystallization of scientific paradigms is influenced at many decisive points by intra- and extratheoretical "social" factors of the most diverse sort.

[93] See the passage from Mannheim, *Ideology and Utopia*, p. 267, quoted earlier.

[94] This paraphrase of Mannheim does not take into account that the expressions "pure logical possibilities" and "the nature of things" no longer contrast with "social" explanations; instead, they should be comprehended as part of an explanatory rhetoric used in natural as well as social science accounts.

2. The influence of social factors on the concrete content of scientific and mathematical knowledge is of more than peripheral importance. Social interests are associated with both "extrascientific" inducements and affiliations and "intrascientific" membership in one or another faction in a scientific field. These various interests give rise to the persuasive tactics, opportunistic strategies, and culturally transmitted dispositions that influence the content and development of scientific knowledge.

Sociologists of scientific knowledge who adhere to the strong program often accomplish Step 1 with the aid of arguments from the philosophy of science regarding the underdetermination of theories by facts and the theory-ladenness of observation, and they use more general skeptical arguments about the relationship between signs and meanings.[95] Following Kuhn, they tend to view historical controversies as particularly illuminating phenomena.[96] Their descriptions of controversies demonstrate that consensus is essentially fragile, that controversies end without being definitively settled by the facts alone, and that stable scientific fields often include disgruntled members who ascribe the consensus in their fields to "mere" conformity. Historical and ethnographic documentation of such matters provides the necessary leverage for contesting the unequivocal determinacy of the "nature of things" or "pure logical possibilities" and demonstrating the contingent nature of consensus in particular disciplines.

Step 2 is elaborated by using diverse sources from sociology, anthropology, and the philosophy of language. Bloor, for instance, often uses Durkheim's basic method for linking the symbolic content of religious ritual and magical belief to the structural divisions within the tribe.[97] He and Barnes update Durkheim's secondhand anthropology by using Mary Douglas's cognitive anthropology and particularly her "grid-group" scheme for linking the

[95] Theory-laden perception is discussed by N. R. Hanson, *Patterns of Discovery* (Cambridge University Press, 1958). The "underdetermination" thesis is credited to Pierre Duhem, *The Aim and Structure of Physical Theory* (Princeton, NJ: Princeton University Press, 1954); and W. V. O. Quine, "Two dogmas of empiricism," in his *From a Logical Point of View*, 2nd ed. (Cambridge, MA: Harvard University Press, 1964). An emphasis on the rhetorical use of scientific demonstration is presented in Feyerabend, *Against Method.* An often-cited skeptical argument about the determinacy of prediction is given in Nelson Goodman, *Fact, Fiction, and Forecast* (Indianapolis: Bobbs-Merrill, 1973). For a concise account of the use of the underdetermination and theory-ladenness theses in the sociology of scientific knowledge, see Karin Knorr-Cetina and Michael Mulkay, "Introduction: emerging principles in social studies of science," pp. 1–18, in Karin Knorr-Cetina and Michael Mulkay, eds., *Science Observed: Perspectives on the Social Study of Science* (London: Sage, 1983).

[96] The most sustained and explicit use of controversies for this purpose is in the "empirical relativist program" associated with Harry Collins and a number of colleagues (who are or were once) at the University of Bath.

[97] David Bloor, "Durkheim and Mauss revisited: classification and the sociology of knowledge," *Studies in the History and Philosophy of Science* 13 (1982): 267–97.

properties of a group to the cognitive style of its members' beliefs and arguments. Barnes, Bloor, and Collins also use Mary Hesse's "network" approach to the organization and entrenchment of culturally specific classificatory schemes.[98] This approach enables a demonstration of nonarbitrary (i.e., relational) variations among the configurations of similar semantic domains in different knowledge communities.

Bloor and Barnes were forthright in claiming that their program was capable of developing *causal explanations* of the contents of particular scientific "beliefs." As I elaborate in Chapter 5, this claim has been criticized for confusing an analysis of meaning with a causal explanation.[99] It is clear, however, that proponents of the strong program redefined what Mannheim and Merton regarded as causal explanations when they broadened Mannheim's method of demonstration. The effort to strengthen the sociology of knowledge by applying it to the most exact forms of reasoning and practice had the effect of collapsing Mannheim's threefold distinction among scientific, social scientific, and ordinary systems of knowledge.

For those who accepted the strong program in the sociology of knowledge, the only remaining contrast was a pragmatic or "social" one between the relationally privileged beliefs in the sciences and various other popular and esoteric belief systems. Even this distinction became fragile when the terms used for describing ideological conflicts were applied to controversies in scientific fields. And as soon as the epistemic privilege of science was defined as an entirely social matter, the door was opened for critical analyses of the narrow social composition of the scientific estate and of particular interests ascribed to members of that estate. Nevertheless, Bloor and Barnes and, in a different way, Collins were clear about placing the sociology of scientific knowledge squarely within the domain of empirical science. As I describe in the next chapter, their claims drew fire from the "new" sociology of scientific knowledge, and a great deal of confusion remained concerning the epistemological, critical, and reflexive implications of social explanations of the contents of science.

[98] See Mary Hesse, *The Structure of Scientific Inference* (London: Macmillan, 1974). For an exposition of this, see Barry Barnes, "On the conventional character of knowledge and cognition," in Knorr-Cetina and Mulkay, eds., *Science Observed*, pp. 19–51.

[99] Writing decades before Bloor and Barnes, Felix Kaufmann (*Methodology of the Social Sciences* [New York: Humanities Press, 1944], p. 16) defines "the genetic fallacy" as a matter of "confounding the analysis of meanings with causal explanation of facts." This is related to what Gilbert Ryle calls "category mistakes." See Ryle, *The Concept of Mind* (Chicago: University of Chicago Press, 1949), pp. 16 ff. It is doubtful that proponents of the strong program would grant strong authority to this "fallacy" and the distinction on which it rests.

CHAPTER 3

The rise of the new sociology of scientific knowledge

If anything holds together the various programs in the "new" sociology of scientific knowledge that have emerged since the early 1970s, it is their commitment to a "radical" view of scientific knowledge. As the many current debates in the field indicate, however, there is little agreement about what this means: Is the radicalism primarily epistemological in scope, or should it also be overtly political and aligned with older traditions of ideology critique? Does the target of radical critique include conventional theories and methods in the social sciences as well as the natural sciences? It is also not at all clear, despite its avowed radicalism, that the strong program articulates a decisive break with prior traditions in the sociology of knowledge. In this chapter, I argue that despite their supposedly radical commitments, the new sociologies of science use some familiar social science idioms and explanatory strategies, and they run into some familiar pitfalls associated with the role of ordinary language in philosophical and sociological investigations. In subsequent chapters I invoke Wittgenstein's later philosophy of language and Garfinkel's ethnomethodology to advocate a more complete break with the conventional views of language and social scientific practice that continue to be advanced in the sociology of knowledge.

As R. J. Anderson, J. A. Hughes, and W. W. Sharrock argue in an ethnomethodological critique of the strong program, some of the familiar puzzles and complaints about the older sociology of knowledge apply no less forcefully to the "new" programs.[1] Given the greater ambitions of the strong program, the debates about it tend to be especially strident, and it is easy to overlook the continuity between those debates and the controversies that have surrounded the sociology of knowledge from its outset. As

[1] R. J. Anderson, J. A. Hughes, and W. W. Sharrock, "Some initial problems with the strong programme in the sociology of knowledge," *Manchester Polytechnic Occasional Papers,* no. 1, 1987. Among the older criticisms they discuss is A. Child's "The problem of imputation resolved," *Ethics* 55 (1944): 96–109. To this, one could add Alexander von Schelting's review of Mannheim's *Ideology and Utopia,* 2nd ed., in *American Sociological Review* 1 (1936): 664–74.

always, the debates follow familiar philosophical lines, with the critics of the sociology of knowledge advancing rationalist and/or realist lines of argument, and the defenders taking cultural relativist and social constructivist positions.[2]

Many philosophers like to treat the strong program as a kind of "straw" position, as though the entire corpus of studies reflected a clear-cut relativistic epistemology. Sociologists of knowledge are no less prone to paint equally undifferentiated portraits of "philosophers." These rhetorical simplifications often result in a highly polarized conception of the explanatory claims in the sociology of knowledge, so that the "new" sociologists of knowledge can seem to be claiming that the particular "contents" of scientific theories are arbitrary and unfounded and that they merely "reflect" partisan ideological interests.[3] Even though Bloor's and Barnes's conception of determination is nonreductionist, implying no explicit threat to the integrity of the systems of knowledge explained, in the heat of debate it is

[2] Among the many philosophers' critiques of the sociology of knowledge are Larry Laudan, *Progress and Its Problems: Towards a Theory of Scientific Growth* (Berkeley and Los Angeles: University of California Press, 1977), chap. 7; Allan Franklin, *The Neglect of Experiment* (Cambridge University Press, 1986); and Franklin, *Experiment Right or Wrong* (Cambridge University Press, 1990). Mario Bunge, "A critical examination of the new sociology of science, Part 2," *Philosophy of the Social Sciences* 22 (1992): 46–76; and Robert Nola, "The strong programme for the sociology of science, reflexivity and relativism," *Inquiry* 33 (1990): 273–96. A concise defense of the strong program is that by Barry Barnes and David Bloor, "Relativism, rationalism and the sociology of knowledge," pp. 21–47, in M. Hollis and S. Lukes, eds., *Rationality and Relativism* (Oxford: Blackwell Publisher, 1982). Also see Bloor's "Afterward: attacks on the strong programme," pp. 163–85 of the 2nd ed. of his *Knowledge and Social Imagery* (Chicago: University of Chicago Press, 1991). Not all philosophers line up squarely against the strong program and its spin-offs in the sociology of science. Joseph Rouse, Steve Fuller, Edward Manier, Thomas Nickles, and Ian Hacking, among others, are more favorably inclined toward the new sociology of science.

[3] Paul Roth and Robert Barrett, in "Deconstructing quarks," *Social Studies of Science* 20 (1990): 579–632, address the question of "arbitrariness," but in a way that compounds existing confusions concerning the subject. In their critical discussion of Andrew Pickering's *Constructing Quarks* (Chicago: University of Chicago Press, 1985), Roth and Barrett try to distinguish between arbitrary "social" conventions and the conventional understandings shared by physicists, which in their view are "as reasonable a facsimile of 'the truth' as could well be sought" (p. 597). For an example of "arbitrary" conventions, they give the example of a rule in the traffic code to drive on one or the other side of the street. Although it may seem obvious that this rule varies as one crosses the short distance from Calais to Dover, in another sense it is not so arbitrary. For drivers acting within a local system of traffic in which a rule to drive on one side or the other is embedded, the rule is far from arbitrary, and a violation can have immediate material consequences that can be no less disastrous than for a chemistry student to ignore the "laws" of chemistry while conducting an experiment. Without having established a technical warrant for doing so, Roth and Barrett speak as though they were physicists, repeatedly using the pronoun *we* when discussing what physicists know and accept. By doing so they rhetorically set up the dramatic difference between the arbitrary rule in a particular traffic code (viewed comparatively) and the nonarbitrary commitments of physicists (viewed noncomparatively). If viewed as an expression of "our" commitments (in which "we" are members of a coherent community), a rule is far from arbitrary.

easy for protagonists from both sides to argue as though the sociology of knowledge offered an attack on the validity of scientific knowledge.

A more politicized sense of radicalism appears when the strong program's critiques of Merton's overzealous pronouncements on the integrity of the "men of science" are enlisted to support the currently fashionable suspicions about "establishment" science, technical reason, objective discourse, and the like. Although there may be good reasons for such suspicions, it is not at all clear that the strong program lends any more support for them than did Mannheim's relationism or Merton's functionalism. Far from advocating antiscientific tendencies, Barnes and Bloor express strong commitments to sociological realism and scientism. For instance, Bloor makes the following summary statement:

> Throughout the argument I have taken for granted and endorsed what I think is the standpoint of most contemporary science. In the main science is causal, theoretical, value-neutral, often reductionist, to an extent empiricist, and ultimately materialistic like common sense. This means that it is opposed to teleology, anthropomorphism and what is transcendent. The overall strategy has been to link the social sciences as closely as possible with the methods of other empirical sciences. In a very orthodox way I have said: only proceed as the other sciences proceed and all will be well.[4]

A further damper on tendencies to treat the strong program as a unilaterally critical philosophy of science is the repeated insistence by its enthusiasts that its claims are grounded not in arguments but in an accumulation of empirical studies.[5]

Despite efforts by constructivist sociologists to distinguish their approach to the sociology of knowledge from philosophical relativism or idealism, they continue to be caught up in the realist–constructivist debate. The familiar themes from the debate continue to arise in exchanges among sociologists of knowledge and philosophers and in debates among proponents of different factions within the sociology of knowledge.[6] These debates

[4] Bloor, *Knowledge and Social Imagery,* p. 156.

[5] The best-known argument of this sort is Steven Shapin's "History of science and its sociological reconstructions," *History of Science* 20 (1982): 157. Shapin argues that the cumulative weight of the empirical studies in the sociology of scientific knowledge should deter further debate about the possibility of successfully accomplishing such studies. Also see H. M. Collins, "An empirical relativist programme in the sociology of scientific knowledge," pp. 85–114, in K. Knorr-Cetina and M. Mulkay, eds., *Science Observed: Perspectives on the Social Study of Science* (London: Sage, 1983). See especially p. 86, where Collins asserts that research programs "are best generated out of practice and example, and best proclaimed and systematized with at least some degree of hindsight." Collins mentions that Bloor's proposals were nevertheless very influential, even though they ran ahead of much of the empirical research in the field.

[6] Examples of such exchanges can be found in Hollis and Lukes, eds., *Rationality and Relativism.* For debates within the sociology of science, see A. Pickering, ed., *Science as Practice and Culture* (Chicago: University of Chicago Press, 1992).

often draw more attention from academic audiences than do any of the particular studies cited in the arguments.

The strong program's policies

The various "schools" and "programs" in the sociology of science that have emerged since the early 1970s were never entirely united and have grown increasingly factious. Nevertheless, as an expository device, it is still useful to cite the guiding principles for the "strong program" in the sociology of knowledge proposed by David Bloor. For the moment, I will put aside the question of whether these principles do in fact guide the various historical and ethnographic studies affiliated with the program.[7]

> 1. It would be causal, that is, concerned with the conditions which bring about belief or states of knowledge. Naturally there will be other types of causes apart from social ones which will cooperate in bringing about belief.
> 2. It would be impartial with respect to truth and falsity, rationality or irrationality, success or failure. Both sides of these dichotomies will require explanation.
> 3. It would be symmetrical in its style of explanation. The same types of cause would explain, say, true and false beliefs.
> 4. It would be reflexive. In principle its patterns of explanation would have to be applicable to sociology itself. Like the requirement of symmetry this is a response to the need to seek general explanations. It is an obvious requirement of principle because otherwise sociology would be a standing refutation of its own theories.[8]

These policies have been adopted in many studies in the sociology and social history of science, and they have also provided targets for numerous criticisms.[9] Bloor's causalist proposals are not entirely accepted in the

[7] Laudan argues that on some points the relationship between the principles and the research is very doubtful. See Larry Laudan, "The pseudo-science of science?" *Philosophy of the Social Sciences* 11 (1981): 173–98. To an extent Bloor agrees that the *principles* are not intended to provide the basis for the "strength" of the program. See David Bloor, "The strengths of the strong programme in the sociology of knowledge," *Philosophy of the Social Sciences* 11 (1981): 206.

[8] Bloor, *Knowledge and Social Imagery*, pp. 4–5.

[9] These critiques include Laudan, "The pseudo-science of science?"; Stephen Turner, "Interpretive charity, Durkheim, and the 'strong programme' in the sociology of science," *Philosophy of the Social Sciences* 11 (1981): 231–44; Steve Woolgar, "Interests and explanation in the social study of science," *Social Studies of Science* 11 (1981): 365–94; Anderson, Hughes, and Sharrock, "Some initial problems with the strong programme in the sociology of knowledge"; and Jeff Coulter, *Mind in Action* (Oxford: Polity Press, 1989), chap. 2.

sociology of scientific knowledge,[10] but his recommendations regarding impartiality and symmetry (Principles 2 and 3) continue to be advocated in all the major lines of constructivist and discourse-analytic inquiry. Although Bloor's reflexivity requirement makes sense as a warning against transcendentalist excess, the application of that requirement to existing studies in the sociology of science can be puzzling if not paradoxical. The following review of the four policies mentions some of the difficulties with their intelligibility and application. In later chapters, I elaborate on these and related difficulties associated with the "new" sociologies of scientific knowledge.

Causality

As I stated in Chapter 2, Bloor's and Barnes's conception of causality does not radically break with the more traditional explanatory approaches in the sociology of knowledge. Instead, it subsumes various classical modes of sociological explanation under a broadened conception of causality. These classic forms of explanation include Durkheim's argument that categorical differentiations in the "sacred" realm reflect divisions among "men" in the tribe,[11] Weber's "switchman" explanation, and Mannheim's two-step method for demonstrating the social determination of knowledge.

The strong program is original in its application of these modes of explanation to modern scientific and mathematical theories and practices, and some studies in the program supplement them with more recently developed methods of semantic, semiotic, and ethnographic analysis. Particular studies, like Donald MacKenzie's explanation of the social commitments fueling the controversy between Pearson and Yule over methods of statistical association,[12] are more intensive in their focus on the particular social interests associated with individual scientists and epistemic communities than were the older functionalist accounts on the relation between knowledge and social structure. But as I contended earlier, this is consistent with an action-centered modification of Mertonian functionalism. Some proponents of the new sociology of science adopt phenomenological, ethnomethodological, and Wittgensteinian conceptual themes. However, as I will argue in Chapter 5, they tend to assimilate Wittgenstein's and Garfinkel's writings into modes of sociological explanation that show little cognizance

[10] Programmatic statements and debates on these issues are presented in the collection edited by Knorr-Cetina and Mulkay, *Science Observed.*

[11] See David Bloor, "Durkheim and Mauss revisited: classification and the sociology of knowledge," *Studies in the History and Philosophy of Science* 13 (1982): 267–97.

[12] Donald MacKenzie, *Statistics in Britain 1865–1930* (Edinburgh: Edinburgh University Press, 1981).

of the more radical anticausalist and antiepistemological implications of those writings.

Because of the breadth of Barnes's and Bloor's program of causal sociological explanation, it is not entirely clear what is implied by saying that scientific knowledge is "determined" by social context. Moreover, it remains unclear how a sociologist can give a nonpartisan account of a group's collective understandings and how those understandings contribute to a historical process. To begin with, the term *knowledge* is often used so broadly that it is difficult to specify just what is to be explained by the sociology of knowledge. Knowledge can include all sorts of behavioral manifestations, testimonies, and textual products of a group's activities, and it is far from easy to select a definite constellation of these to represent a group's epistemic commitments. Even when representative expressions or documents can be identified, further problems beset any effort to distinguish the relevant antecedents, correlates, and consequences of their contents. Similar methodological problems face virtually all areas of empirical sociological research, but they are exacerbated in this case by the peculiar connotations of the ordinary concept of knowledge. To claim to "know" something is to assert that it is nonnegotiable, or at least less negotiable than matters of "belief" or "opinion." A research program that defines a particular group's "collective knowledge" by treating it no differently than "public opinion" or "shared belief" must discount or downgrade the asymmetric validity claims asserted by members of the group.[13] Consequently, causal explanations of knowledge are likely to be resisted by the subjects of study (if they are given any say about the matter), since these subjects may conclude that their own validity claims have not been taken seriously enough. Accordingly, the technical difficulties associated with sociology-of-knowledge explanations can be compounded by intractable conflicts with the epistemic communities studied. As I argue throughout this and the following chapters, none of the current programs in the sociology of scientific knowledge has escaped the familiar confusions and conflicts engendered by older variants of sociology of knowledge explanation.

Symmetry and impartiality

I have lumped together symmetry and impartiality, since both policies propose that all theories, proofs, or facts should be treated as "beliefs" to be explained socially. Both of these methodological policies are related to Mannheim's nonevaluative general total conception of ideology in which "no judgments are pronounced as to the correctness of the ideas to be

[13] See Coulter, *Mind in Action*, pp. 36 ff.

treated." The point of the policy is to establish sociological or conventionalist explanations by displacing a priori assumptions about the immanent development of "rational" or "true" beliefs. (Recall the modification of Mannheim's two-step program of explanation outlined at the end of the last chapter: [1] the demonstration that a scientific field does not actually develop in accordance with immanent laws or the "nature of things" and [2] the elaboration of social interests that are associated with both "extrascientific" inducements and affiliations and "intrascientific" membership in one or another faction in a scientific field.)

As I understand the symmetry postulate, it does not require sociologists to make identical explanations of the historical sequelae to, for example, Roentgen's "discovery" of X-rays and Blondlot's "illusion" of N-rays. Instead, it prohibits teleological explanations that treat contingent historical outcomes as grounds for explaining how a discovery was made or an illusion was exposed. Determinations that X-rays were "discovered," whereas N-rays were a spurious product of "pathological science," might best be treated as *historical judgments* generated from a more comprehensive field of negotiations and disputes.[14] To say that such judgments explain the historical events to which they apply is like using the verdict of a criminal trial to explain the process through which it was reached ("The defendant was found guilty because she was guilty!").

Although the symmetry and impartiality postulates may effectively circumvent what Bloor and Barnes call *teleological* explanations, their application to particular cases can raise some thorny problems. These problems have to do with the question of how a sociologist can describe a controversial episode without trading on the endogenous vocabularies, reasons, and justifications that scientists employ to advance or debunk particular claims in their specialized fields. This, of course, is a familiar problem that has haunted the sociology of knowledge from its outset. Jürgen Habermas gives the following account of the problem:

> Agreement and disagreement, insofar as they are judged in the light of reciprocally raised validity claims and not merely caused by external factors,

[14] For a highly interesting though asymmetric account of rejected science, see Ivar Langmuir, "Pathological science," *General Electric R&D Center Report,* no. 68-C-035, Schenectady, NY. In a brief account of the N-ray affair, Bloor (*Knowledge and Social Imagery,* pp. 29–30) speaks of the rays' "spurious" character, and like Langmuir he cites problems in Blondlot's experimental procedures. So to a large extent, Bloor's account incorporates a vocabulary for describing Blondlot's experiments that presumes that Blondlot and his assistants had erroneously "believed" they had found a natural kind of radiation. For an approach to the negotiation of discovery claims that is perhaps more compatible with the strong program's precepts, see Augustine Brannigan, *The Social Basis of Scientific Discoveries* (Cambridge University Press, 1981). Also see Malcolm Ashmore, "The theatre of the blind: starring a Promethean prankster, a phoney phenomenon, a prism, a pocket and a piece of wood," *Social Studies of Science* 23 (1993): 63–106.

> are based on reasons that participants supposedly or actually have at their disposal. These (most often implicit) reasons form the axis around which processes of reaching understanding revolve. But if, in order to understand an expression, the interpreter must *bring to mind the reasons* with which a speaker would if necessary and under suitable conditions defend its validity, he is *himself* drawn into processes of assessing validity claims.[15]

Briefly summarized, the problem concerns the difficulty of maintaining independence between the methodological strategies used in the sociology of knowledge and the immanent modes of criticism that arise in the fields studied. This difficulty had several related aspects and consequences:

1. When applying the Duhem–Quine and related epistemological theses to particular cases, a confusion of the following sort can be engendered: The demonstrable historical fact that a particular theory was accepted without "unequivocal" support by the experimental evidence may be confused with a particular criticism to the effect that the experimenters accepted the results prematurely, failed to take account of relevant alternatives, or unfairly dismissed rival claims without rigorously testing them. When used in a specific explanatory account, the general philosophical thesis regarding the inevitable underdetermination of theories by finite experimental evidence can suggest that something improper was going on in cases in which it is shown that a particular controversy was brought to a close not from rigorous testing but from various "social" pressures and "vested interests" arising from an adjudicating community.[16]

2. Using the symmetry and impartiality theses to level the playing field before taking account of various contending theories and experimental practices may seem to promote a vanquished or marginal theory at the expense of the victorious or established program.[17] For instance, Harry

[15] Jürgen Habermas, *The Theory of Communicative Action*, vol. 1: *Reason and the Rationalization of Society*, trans. Thomas McCarthy (Boston: Beacon Press, 1984), p. 115.

[16] Wes Sharrock and Bob Anderson ("Epistemology: professional scepticism," pp. 51–76, in G. Button, ed., *Ethnomethodology and the Human Sciences* [Cambridge University Press, 1991]) identify a common tendency in sociological arguments, which is to compare the communal judgments being analyzed (whether scientific or commonsensical) with the sorts of strict epistemic standards that skeptical philosophers demand of their interlocutors. When situated judgments are shown to fall short of such standards, this seems to call for an explanation. But for Sharrock and Anderson, no such explanation is called for when none is demanded in the relevant circumstances.

[17] For example, see Evelleen Richards, "The politics of therapeutic evaluation: the vitamin C and cancer controversy," *Social Studies of Science* 18 (1988): 654. Richards adopts the policy of symmetry, but her treatment of a controversy between what she calls "establishment" biomedical research institutions and proponents of vitamin C therapy for cancer implicitly bolsters the case for the latter in the face of the dismissal of their claims by the former. Richards and two colleagues later argued against the possibility of taking a neutral stance toward the controversies studied and opted for a value-committed position against "dominant ideologies." See Pam Scott, Evelleen Richards, and Brian Martin, "Captives of controversy: the myth of the neutral social researcher in contemporary scientific controversies," *Science, Technology, and Human Values* 15 (1990): 474–94.

Collins and Trevor Pinch attempt to treat symmetrically the controversies surrounding demonstrations of the "paranormal" phenomenon of psychokinesis. Their study describes how skeptical scientists launched a vigorous attack in which they enlisted the aid of a professional magician to expose the trickery of "spoon benders" like Uri Geller. Collins and Pinch argue that the skeptics were far from disinterested, since they presumed that psychokinesis was fraudulent when they set up their experimental "tests" and interpreted the results. According to the underdetermination thesis, because no finite amount of testing can provide absolute proof of a theory, the particular force of Collins and Pinch's argument derives from their documenting the unusually partisan and ethically questionable conduct during the controversy.[18] They tell of strange and dishonest practices used by protagonists on both sides of the controversy. Collins and Pinch were not trying to vindicate claims about paranormal phenomena. However, since psychokinesis was already highly suspect, their symmetrical treatment of the controversy had the rhetorical effect of downgrading the relative status of the "establishment" scientists' claims without further damaging the already dubious claims of the parapsychologists.[19]

3. To the extent that members of scientific communities fail to give "full" consideration to actual or possible claims that are radically at odds with their own procedures and theoretical commitments, they may seem to be acting in an arbitrary way. But as Wittgenstein points out, to fall short of such transcendental standards does not necessarily indicate arbitrariness:

> All testing, all confirmation and disconfirmation of a hypothesis takes place already within a system. And this system is not a more or less arbitrary and doubtful point of departure for all our arguments: no, it belongs to the essence of what we call an argument. The system is not so much the point of departure, as the element in which arguments have their life.[20]

An "impartial" examination of such a system can create the impression that members have set up an arbitrary boundary that restricts entry to alternative ways of thinking and acting. The medium in which "arguments have their life" then takes on a political cast, as though it were erected through explicit decisions and deliberate machinations.

4. General vocabularies for describing science, its products, and the mistakes and misuses associated with it (e.g., terms like *discover, invent, evidence, interpretation, artifact, hoax*) have familiar partisan and asymmetric uses in the natural and social sciences. Although sociologists of science

[18] H. M. Collins and T. J. Pinch, *Frames of Meaning: The Social Construction of Extraordinary Science* (London: Routledge & Kegan Paul, 1982).

[19] Collins (personal communication) mentioned that parapsychologists tended to read his and Pinch's book (ibid.) as support for their cause, although this was certainly not the case for the various critics of parapsychology discussed in the book.

[20] Ludwig Wittgenstein, *On Certainty*, ed. G. E. M. Anscombe and G. H. von Wright (Oxford: Blackwell Publisher, 1969), sec. 105.

occasionally try to use nonevaluative vocabularies when describing or explaining particular cases, it is not at all clear that they can do so.[21]

5. Descriptions of particular experiments, simulations, and theoretical models necessarily use members' accounts of procedures, results, and judgmental criteria. These accounts can be extremely difficult to understand and even more difficult to assimilate into historians' and social scientists' narratives. It remains to be seen how experimental practices can be glossed in social studies of science without relying on the partisan terms and locally informed evaluations of what was tested, what may have been ignored, and what was adequately resolved. Sociologists assume an immense burden when they claim to convey technically adequate descriptions while remaining aloof from the partisan commitments infused with the techniques described.

Reflexivity

Unlike the ethnomethodological version of reflexivity discussed in Chapter 1, Bloor's version is more of a criterion for establishing the "scientific" status of his own program. His reflexivity requirement is somewhat similar to Mannheim's and Merton's efforts to apply the sociology of knowledge to their own programs. As explained in the previous chapter, Mannheim tried to secure pragmatic authority for the sociology of knowledge by arguing that its unique historical and institutional situation enabled a relatively value-free assessment of diverse modes of knowledge. Merton made a somewhat bolder claim to the effect that the professionalized subdiscipline of sociology of science "exemplified" the properties of a maturing scientific specialty. But unlike Bloor, Mannheim and Merton both subscribed to the view that rational modes of communication emerge when the appropriate institutional conditions are established. Consequently, their "reflexive" analyses of their own disciplinary programs acted to support scientistic claims about the immanent development of those programs. Although such reflexive arguments were self-serving and regressive, they were internally consistent.[22] Bloor's provision for reflexivity raised more difficult problems, since he no

[21] Even Callon and Latour, for all their semiotic sophistication, seem entranced by the prospect of establishing a "symmetrical vocabulary." See M. Callon and B. Latour, "Don't throw the baby out with the Bath school! A reply to Collins and Yearley," pp. 343–68 in A. Pickering, ed., *Science as Practice and Culture* (Chicago: University of Chicago Press, 1992.

[22] I mean this in a very weak sense. As Alexander von Schelting argues, Mannheim's proposals to evaluatively compare particular ideologies from a "superparticular" standpoint presuppose an extraordinary validity criterion that Mannheim is unable to satisfy. The best he can do is to invoke the position of the socially free-floating intelligentsia, and as von Schelting points out, this supposes that *"the fact that a conception comes out of the brain of a socially unbound intellectual is the guarantee of its validity."* See von Schelting, "Review of *Ideologie and Utopie*," *American Sociological Review* 1 (1936): 664–74, quotation from p. 673, emphasis in original. But if Mannheim's account is "weak" in the sense that it offers no guarantee of validity and provides only a pragmatic communicational ground for a discourse in which diverse ideas will not be ruled out of order on the basis of a priori ideological

longer could pretend to ground the sociology of knowledge by claiming that appropriate conditions were in place for the immanent development of scientific rationality. Moreover, it is hard to imagine how proponents of the strong program could reflexively examine their own contributions while remaining impartial about their truth or falsity, rationality or irrationality, and success or failure. And if they were to attain such reflexive transcendence, it is doubtful that such a heroic achievement would exemplify "science" as it is usually practiced.[23]

Bloor avoids a reflexive conundrum by de-emphasizing matters of principle. In his rejoinder to Larry Laudan's criticism that the strong program's principles do not define a recognizable mode of scientific practice, Bloor turns to psychology and gives the following explanation of how sociologists of knowledge emulate the sciences:

> [Laudan] has failed to see that I am an inductivist. He consistently tries to understand my position through a haze of deductivist assumptions. . . . It is as if action only seems intelligible to my critic if it can be made out to follow from stated principles. I have no such bias. The student of the piano may not be able to *say* what features are unique to the playing of his teacher, but he can certainly attempt to emulate them. In the same way we acquire habits of thought through exposure to current examples of scientific practice and transfer them to other areas. Indeed some thinkers such as Kuhn and Hesse believe that this is exactly how science itself grows. Thought moves inductively from case to case. My suggestion is simply that we transfer the instincts we have acquired in the laboratory to the study of knowledge itself.[24]

This perhaps serves to get Bloor out of a tight spot in his argument with Laudan, but it raises some further problems. How have "we" acquired the "instincts" of the laboratory? Although Bloor certainly understands mathematics, his sociology of mathematics is not "mathematical,"[25] and the

commitments, and if Mannheim can demonstrate that these conditions hold for the sociology of knowledge, then his reflexive proposals can at least be consistent. Similarly, Merton's provisions for a "self-exemplifying" sociology of science may be superficial, and they certainly do not guarantee validity, but they are consistent with the way he analyzes other scientific communities.

[23] Gilbert and Mulkay, for example, vividly document how participants in a biochemistry controversy were considerably less than impartial when assessing the truth and falsity of the contending theories. See G. Nigel Gilbert and Michael Mulkay, *Opening Pandora's Box: A Sociological Analysis of Scientists' Discourse* (Cambridge University Press, 1984). As von Schelting points out ("Review," p. 674), "a high degree of 'vital interestedness' in a problem, and even in a definite kind of solution, may also provide a comparatively high probability of cognitive success in some cases. 'Social attachment' and 'vital interestedness,' like jealousy, can render clear-sighted as well as blind."

[24] Bloor, "The strengths of the strong programme in the sociology of knowledge," p. 206. This is a reply to arguments made in Laudan, "The pseudoscience of science?" pp. 180–81. See chap. 2 of Bloor's *Knowledge and Social Imagery* for an elaboration of his social psychology of "sense experience, materialism, and truth."

[25] In Chapter 5 I discuss an example of sociological research that is "mathematical" in a strong sense: Eric Livingston, *The Ethnomethodological Foundations of Mathematics* (London: Routledge & Kegan Paul, 1986). It is indicative that in his review of that volume Bloor complains that Livingston pays insufficient attention to social forms of explanation. See D. Bloor, "The living foundations of mathematics," *Social Studies of Science* 17 (1987): 337–58.

sociologists of science who have conducted laboratory ethnographies have not tried to emulate the practices they observe in the way that a novice tries to pick up technical skills in a laboratory. Sociologists and historians of science generally go to considerable trouble to avoid falling into the "unreflective" habits of the laboratory members they observe.[26] They work with archives and other texts, and they write discursive arguments, so it is unclear how they could inductively transfer the "instincts" of the laboratory to their literary practices.

Bloor seems to assume that these instincts are habits of individual thought that can be abstracted from the unique ensembles of instrumentation, embodied technique, and vernacular discourse in the laboratory. Perhaps, following Bruno Latour and Steve Woolgar's findings on the prevalence of "literary inscription" in laboratory science, the relevant instincts are embedded in the practices of writing "scientifically."[27] But it would hardly favor a critical and analytic approach to scientists' literary practices to say that sociologists of science habitually emulate the practices they analyze. For these and other reasons, "reflexivity" has become something of a battleground in recent years among various British sociologists of science. As I explain later, this requirement for the strong program has turned into a distinct research program in the sociology of knowledge.

The strong program's progeny, siblings, and close relatives

The family of studies associated with the strong program is a loose and extended one, and its lines of ancestry are far from "pure." Sibling rivalries

[26] The clearest example of this is Latour and Woolgar's policy of assuming the attitude of a "stranger" in their ethnography of a laboratory at Salk Institute: *Laboratory Life: The Social Construction of Scientific Facts* (London: Sage, 1979; 2nd ed., Princeton, NJ: Princeton University Press, 1986). In their historical study of the dispute between Boyle and Hobbes on the relevance of experiments in natural philosophy, Steven Shapin and Simon Schaffer adopt Latour and Woolgar's "stranger" strategy as a precaution against adopting a "member's" perspective on the "self-evident" character of the events they describe. See their *Leviathan and the Air Pump: Hobbes, Boyle, and the Experimental Life* (Princeton, NJ: Princeton University Press, 1985), p. 6. Harry Collins (*Changing Order: Replication and Induction in Scientific Practice* [London: Sage, 1985], chap. 3) gives a rich account of his and a colleague's efforts to build a laser, and his book with Trevor Pinch (*Frames of Meaning: The Social Construction of Extraordinary Science* [London: Routledge & Kegan Paul, 1982] describes parapsychology experiments in which the authors took part. Nevertheless, there is no clear sense in which Collins's relativistic sociology incorporates the habits of the laboratory other than as a critically examined source of subject matter.

[27] Latour and Woolgar *(Laboratory Life)* claim that an interest in "inscription" does identify their practices with the scientists they study, but by the time they wrote the postscript to their 1986 edition, they had moved away from their earlier scientistic version of reflexivity. Karin Knorr-Cetina and Klaus Amann ("Image dissection in natural scientific inquiry," *Science, Technology & Human Values* 15 [1990]: 260) dispute the centrality of literary inscription in laboratories and argue that visual imaging is more important.

have grown intense as some of the maturing subprograms have reached adolescence, while at the same time the former hostility with the Mertonian clan has been defused by a partial intermarriage of themes and research initiatives (and, I think, by the mutual recognition that the two clans were never so distant in the first place). The program has also mingled with explicitly politicized and highly critical treatments of objective science. Although Barnes, Bloor, and Michael Mulkay did not give an obvious political cast to their proposals, in recent years some of their arguments have been adopted in politicized critiques of the scientific and medical establishment.[28]

Although the social study of science is still a relatively small field, I cannot pretend to cover the entire literature. As is the case for other scholarly fields, the production of writing far outstrips anyone's ability to read a substantial portion of it. Fortunately, there is sufficient duplication, at least at a programmatic level, to enable a reader to gain a fairly confident grasp of the literature and its divisions without having to read all of it. An unfortunate effect of this is that particular studies tend to serve as guideposts and "citation magnets" that index a heterogeneous range of researches. Having said this, I will now perpetrate the usual acts of violence by laying out a scholastic typology.

The continuing strong program

The strong program has traveled well beyond Edinburgh, and its initiatives have influenced studies from a variety of perspectives. The studies most closely aligned with the program are case studies of particular historical developments. One of the best known of the more recent studies is Andrew Pickering's *Constructing Quarks: A Sociological History of Particle Physics*.[29] As its title suggests, the study reviews the series of theoretical and experimental developments since the 1960s that culminated in the establishment of what Pickering calls the "quark/gauge theory worldview." This worldview is populated by new theoretical entities, including quarks, which are said to be fundamental constituents of protons and neutrons. Gauge theory uses the concept of "charm" to explain the coherent nexus of new entities and forces, and it provides an incentive for particle physicists to pursue funding for increasingly massive and powerful instruments to "penetrate" more deeply the inner structure of matter.

[28] See Donna Haraway, *Primate Visions: Gender, Race, and Nature in the World of Modern Science* (New York: Routledge & Kegan Paul, 1989); Haraway, *Simians, Cyborgs, and Women* (New York: Routledge & Kegan Paul, 1991; Evelyn Fox Keller, *Reflections on Gender in Science* (New Haven, CT: Yale University Press, 1985).

[29] Pickering, *Constructing Quarks.*

In line with the strong program's two-step method of demonstration, Pickering contests what he calls "the scientist's version" of the immanent development of a series of experiments supporting the new theories about the composition of matter. He cites the familiar philosophical arguments regarding the underdetermination of theories by experimental facts, and he also states that the "facts" themselves are "deeply problematic."[30] This, he says, is because the factual status of experimental data depends on fallible judgments about whether or not the relevant equipment was functioning properly, effective controls were made, and the relevant signals were correctly discriminated from noisy backgrounds. Moreover, the "factual" sense and meaning of the experimental data are found through the use of models, analogies, and simulations that align the data with theoretical preconceptions. Pickering contends that the relation between theory and experimental data is one of "tuning" or "symbiosis" rather than the independent verification of theory by means of facts. His historical account demonstrates the "potential for legitimate dissent" on questions of experimental procedure and theoretical interpretation of data. He describes the debates among different research groups and uses their discrepant accounts as a basis for demonstrating the multiplicity of possible interpretations of the relevant experimental events and their theoretical implications. To explain how scientists managed to make experimental interpretations and theory choices, he introduces a concept of "opportunism in context," a way of describing how scientists pursue the particular experimental–interpretive pathways that enable them to exercise their professional skills and follow up the most "interesting" of the available theoretical developments.

Pickering's study is distinguished by its close attention to experimental practices and instrumentation. He discusses the available designs for bubble-chamber apparatus, methods for interpreting traces of subatomic particles, and computer simulation procedures used in experiments on "weak neutral currents." Pickering's training as a physicist was indispensable for this procedure, since it enabled him to make claimably "legitimate" counterfactual assessments of what the experiments he reviews could have demonstrated. This competency permits him to avoid engaging in the kind of armchair relativism in which arguments about possible theoretical alternatives are used for considering the rationality of judgments made on a particular occasion. So in a sense, Pickering's account is also a "scientist's version," albeit one that expresses a set of theoretical and methodological commitments different from those made by members of the research groups he

[30] Ibid., p. 6.

examined. Pickering's is not the only version of this sort, and his view of both the history of physics and the "physics of physics" has been challenged by Peter Galison and Allan Franklin.[31]

Pickering's pragmatic focus is consistent with the trend in social studies of science toward descriptions of experimental instrumentation, technique, and analysis.[32] The more abstract, theory-based conception of knowledge familiar from earlier sociohistorical studies is gradually turning into a more particularistic conception of the material sites, artifacts, and techniques of "knowledge production."[33] The focus is more intensive and "internal" (in the nonrationalist sense), as the aim is to identify the pragmatic strategies and informal judgments made at the worksite when researchers sort through "messy" arrays of data and decide whether their equipment is working properly. In this respect, the strong program has converged with other programs, including the "empirical relativist program," laboratory studies, and ethnomethodological studies of scientific work.

The empirical relativist program

The empirical relativist program, a constructivist program, is closely affiliated with the strong program. It is sometimes known as the Bath school because H. M. Collins and his current and former students at the University of Bath are the major contributors.[34] Their studies tend to focus on contemporary scientific controversies, and they attempt to give symmetrical descriptions of the incommensurable positions, theory-laden experimental practices, and nonrational (or extrarational) methods for reaching closure on the disputed matters. Many of the Bath school's studies use empirical cases as a basis for criticizing traditional philosophical conceptions of the role of replication (or crucial tests) in experimental science. In their view, the concept of replication is problematic because replication is attempted less

[31] Peter Galison, *How Experiments End* (Chicago: University of Chicago Press, 1987); Allan Franklin, "Do mutants have to be slain, or do they die of natural causes? The case of atomic parity-violation experiments," chap. 8 of his *Experiment Right or Wrong*.

[32] See for instance, Shapin and Schaffer, *Leviathan and the Air Pump;* David Gooding, "How do scientists reach agreement about novel observations?" *Studies in History and Philosophy of Science* 17 (1986): 205–30.

[33] See Andrew Pickering's introduction to his edited collection, *Science as Practice and Culture* (Chicago: University of Chicago Press, 1992). Also see the special issue on artifact and experiment, *Isis* 79 (1988): 369–476.

[34] The Bath perspective is represented by Collins, *Changing Order;* Trevor Pinch, *Confronting Nature: The Sociology of Solar Neutrino Detection* (Dordrecht: Reidel, 1986); D. L. Travis, "Replicating replication? Aspects of the social construction of learning in planarian worms," in H. M. Collins, ed., *Knowledge and Controversy: Studies of Modern Natural Science,* special issue of *Social Studies of Science* 11 (1981): 11–32.

often than is usually imagined, and when scientists do try to replicate others' results, they often modify the original equipment and procedures to suit their own programmatic interests. Moreover, unless a practitioner is already familiar with the techniques and equipment described, a written report of observational procedures can rarely be used as self-sufficient instructions for reproducing the observation. Methods accounts are written to respect proper canons of scientific reportage and not to describe what scientists actually do, and scientists who repeatedly obtain "good" results in an experiment are often unable to explain how they get them. Rather than trying to replicate techniques from written instructions, researchers often prefer to recruit personnel from other labs in which the techniques are established, and when other scientists fail to replicate the findings, the original experimenters typically complain that their procedures were not correctly followed.[35] Consequently, according to Collins, assessments of whether or not a scientist has replicated an experiment cannot be extricated from judgments about the plausibility of the experimental outcomes. Untested assumptions about the competency and credibility of the experimenter, the adequacy of the experimental design, and the strength and meaning of the experimental evidence all combine to reinforce the acceptance or rejection of particular experimental demonstrations.

Although the "empirical relativists" place less emphasis on causality than do Barnes and Bloor, they use similar argumentative strategies.[36] They compose descriptions of the relevant experiments and document them with quotations from practitioners on both sides of a controversy in order to demonstrate (1) that experimental data do not, by themselves, determine when an experiment will count for or against a given theory and (2) that negotiations among a "core set" of researchers investigating a controversial phenomenon determine when the matter will be counted as "closed." A "core set" is the relatively small number of researchers (or research labs) who take an active part in generating and resolving a scientific controversy. The empirical relativist approach is "empirical" in the sense that it uses published and unpublished testimonies by members of the core set to document relational configurations of theoretical commitment and experimental practice.

For example, in his account of the controversies surrounding Joseph

[35] This list pulls together aspects of what Collins calls "the experimenters' regress," along with related features from the Bath school's and other studies on the local organization of experimental practice.

[36] The argumentative strategy that Pickering uses in *Constructing Quarks* could very easily be listed under the empirical–relativist program rather than the strong program. The main difference is that Pickering relies more exclusively on a historical approach to recent and contemporary physics rather than observations at the worksite and interviews with practitioners.

Weber's experiments on gravity radiation, Collins is able to demonstrate dramatic discrepancies between Weber's assessment of the experiments and his critics' versions of them.[37]

Weber designed a relatively simple and yet delicate gravity detector consisting of a massive aluminum bar suspended inside a vacuum chamber. He linked this "antenna" to an arrangement of electrical instruments for amplifying and measuring vibrations resonating through the aluminum mass. Weber tried to insulate the antenna as much as possible from all known sources of vibration from electrical, magnetic, thermal, acoustic, and seismic sources. He could not eliminate thermal "noise," but he assumed that with the appropriate controls in place, such noise would register as relatively random background fluctuations. Having taken into account the characteristic manifestations of such noise, Weber claimed to detect a number of especially high peaks on the detector's chart recorder, and he suggested that these gave evidence of gravity waves. This claim was treated skeptically by other members of the "core set," partly because, according to Collins (p. 83), the gravitational energies that would be necessary to generate vibrations of such magnitude were way out of line with theoretical calculations of "the amount of energy that was being generated in the cosmos." In response to such criticisms, Weber modified his apparatus by placing two detectors one thousand miles apart so that he could examine data from the two detectors for simultaneous peaks. Thereafter, he refined his observations by claiming to discover periodicity in the succession of peaks, and he claimed that the data indicated a coherent extragalactic source of radiation. This announcement touched off efforts by a number of other scientists to "replicate" the experimental results. The replications did not support Weber, however, and within a few years his results were "nearly universally disbelieved."[38]

Collins interviewed many of the scientists involved in the dispute over Weber's experiments and was able to elicit several points of fundamental disagreement. These disagreements included the very question of whether Weber's experiments had indeed been replicated. Collins (p. 85) quotes one scientist as saying that "everybody else is just doing carbon copies" of Weber's apparatus, whereas Weber complains (p. 86) that "it is an international disgrace that the experiment hasn't been repeated by anyone with [the original] sensitivity." As Collins reconstructs the issue, the question of what counts as a replication turns on a complex assessment of the phenomenon

[37] This controversy is reviewed in Collins, *Changing Order*, chap. 4, and my subsequent citations refer to this source. Collins gives a more elaborate account in his articles, "The seven sexes: a study in the sociology of a phenomenon, or the replication of experiments in physics," *Sociology* 9 (1975): 205–24, and "Son of the seven sexes: the social destruction of a physical phenomenon," *Social Studies of Science* 11 (1981): 33–62.

[38] Collins, *Changing Order*, p. 81.

being detected. The components of the apparatus, the sensitivity of the measuring instruments, and the procedures for controlling backgrounds all incorporate preconceptions about the phenomenon and its relation to the materials being used to detect it. These preconceptions enter into decisions about which materials can be substituted for one another without essentially changing what the apparatus measures, how sensitive a detector must be in order to resolve the relevant evidence, and what degree of precaution is required to take account of one or another extraneous source. According to Collins (p. 87), key discrepancies between Weber's and his critics' decisions on all of these matters were not resolved on the basis of "scientific" reasons alone, since the relevant arguments included a morass of judgments on questions of personal honesty, technical competence, institutional association, style of presentation, and nationality. Nevertheless, the controversy did not last for very long. Researchers at the several labs where gravity-wave experiments were attempted eventually concluded (although for different reasons) that Weber was wrong. Although he kept trying to vindicate his earlier experiments, Weber no longer was able to get funding, and for all practical purposes his claims about gravity waves became a dead issue.

Collins (p. 90) acknowledges that "it almost goes without saying that the almost uniformly negative results of other laboratories were an important point," but he argues that none of the alleged tests of Weber's experiments unequivocally falsified them. He observes (p. 91) that not only did Weber find fault with his critics' evidence but the critics also found fault with one another's procedures and findings. Therefore, "given the tractability, as it were, of all the negative evidence, it did not *have* to add up so decisively." Moreover, Collins contends, Weber weakened his own position when he adopted a method for calibrating the apparatus recommended by his critics. By doing so he caved in to his critics' assumptions about what gravity waves must be like. This weakened his findings and truncated his initial sense of what those findings could document. "Making Weber calibrate his apparatus with electrostatic pulses was one way in which his critics ensured that gravitational radiation remained a force that could be understood within the ambit of physics as we know it. They ensured physics' continuity – the maintenance of the links between past and future."[39]

There are many striking parallels between the gravity-wave controversy in the early 1970s and the much publicized "cold fusion" affair in 1989–90.[40] Like Weber, Stanley Pons and Martin Fleischmann reported results that were way out of line with accepted theory. Nevertheless, a number of theorists immediately set about speculating how existing theory might account for

[39] Ibid., pp. 105–06.

[40] Bart Simon, "Voices of cold (con)fusion: pluralism, belief and the rhetoric of replication in the cold fusion controversy" (M.A. thesis, University of Edinburgh, 1991).

their findings. Pons and Fleischmann also used a relatively simple apparatus that still was complicated enough to sustain considerable debate about whether any particular replication attempt was, indeed, reproducing the original conditions under which the positive "effect" was observed. The debate was also very contentious, and an entire range of personal, institutional, and technical assessments were pronounced by the various parties involved. (In this instance, a tremendous amount of publicity added fuel to the fire.) And as some would argue, despite the rapid accumulation of negative results (after a few initial reports of replications), there was no single or decisive disproof.[41] All of the alleged tests were, or could be, criticized on methodological grounds, and in many cases, the experimenters were transparently partisan in their assessments of the theoretical possibility of cold fusion, the status of electrochemistry compared with nuclear physics, and (less openly) the status of the University of Utah.[42]

Collins convincingly claims that the "closure" of the gravity-wave controversy did not rest on a single "crucial test," demonstrating that Popperian falsificationism only gives an idealized account of what the core set of scientists resolved through their extended and relatively unrestricted dispute. However, it is not entirely clear whether Collins's account demonstrates that the content of physics was socially determined in this case. Recall Mannheim's criterion for sociology of knowledge demonstrations:

> The historical and social genesis of an idea would only be irrelevant to its ultimate validity if the temporal and social conditions of its emergence had no effect on its content and form. If this were the case, any two periods in the history of human knowledge would only be distinguished from one another by the fact that in the earlier period certain things were still unknown and certain errors still existed which, through later knowledge, were completely corrected.[43]

[41] An indication of the contentiousness of particular "crucial tests" is the lawsuit threatened by an attorney representing Pons and Fleischmann. According to a newspaper article (David Stipp, "Cold-fusion scientists' lawyer tells skeptic to retract report or face suit," *Wall Street Journal,* June 6, 1990, p. B4), the threatened lawsuit concerned a paper in the March 29 issue of *Nature* by physicist Michael J. Salamon that reported "a finding that many scientists regard as a final crushing blow to the claims that cold fusion exists." Salamon described his own attempt to replicate the Pons–Fleischmann experiment, and he claimed that his measurements of neutron emissions gave no evidence of cold fusion. In a letter to Salamon, Pons and Fleischmann's lawyer "asserted that Mr. Salamon had published a 'factually inaccurate' report on cold fusion that had caused 'undue ridicule and negativism' about the phenomenon." This letter not only refueled the controversy over cold fusion, but it also became a subject of controversy in its own right. The newspaper article quotes a number of physicists who denounced resorting to a legal resolution as a deplorable contravention of "the spirit of free academic inquiry."

[42] Much of my account of the "cold fusion" affair is based on discussions with Guido Sandri, College of Engineering, Boston University.

[43] Mannheim, *Ideology and Utopia,* p. 271.

In the aftermath of the gravity-wave controversy, Weber's experiments failed to have a lasting impact on the continuity of physics; indeed, as Collins contended, Weber's discrepant claims have thus far been treated rather cruelly by history. Weber's critics used "the continuity of physics" as a resource for overriding his claims. Although the episode might have turned out differently, by all accounts the relevant contents of physics survived the potential transformations implicated by Weber's experimental results. Moreover, the case could be argued that Weber's claims were erroneous even if the error was not decisively demonstrated by any single test or argument. Even though Collins can imagine that physics could have changed, the most he can argue is that existing physical theory was not changed by Weber's experiments.

In another sense, however, Collins's study problematizes any assessment of what would or would not be relevant to the "ultimate validity" of existing physics. His study demonstrates that the validity of Weber's experimental results was never subjected to an ultimate test, nor given the contingencies in the testing situation, could it have been given such a test. The question of the "ultimate validity of ideas" is simply off the table, and so Mannheim's criterion for establishing the social determination of ideas cannot be meaningful. Consequently, the opponents of the sociology of knowledge have an impossible burden if, to paraphrase Mannheim, they must show that "the temporal and social conditions" supporting the continuity of physics "had no effect on its content and form." The possibility is ruled out a priori, and not by the accumulating evidence from the sociology of science.

Laboratory studies

In the late 1970s, several sociologists and anthropologists undertook ethnographic investigations of laboratory practices.[44] In contrast with previous

[44] For a review of the first cohort of ethnographies of scientific labs, see Karin Knorr-Cetina, "The ethnographic study of scientific work: towards a constructivist interpretation of science," in Knorr-Cetina and Mulkay, eds., *Science Observed,* pp. 115–40. Knorr-Cetina discusses published and unpublished manuscripts from six of the earliest laboratory ethnographies, including her own study, *The Manufacture of Knowledge: An Essay on the Constructivist and Contextual Nature of Science* (Oxford: Pergamon Press, 1981); Latour and Woolgar, *Laboratory Life;* Michael Lynch, "Art and artifact in laboratory science: a study of shop work and shop talk in a research laboratory," (Ph.D. diss., University of California at Irvine, 1979), later published under the same title (London: Routledge & Kegan Paul, 1985); Sharon Traweek, "Culture and the organization of the particle physics communities in Japan and the United States," paper presented at the conference on Communications in Scientific Research, Simon Fraser University, 1981 (Traweek's study was later published under the title *Beam Times and Life Times: The World of Particle Physics* [Cambridge, MA: Harvard University Press, 1988]); John Law and Rob Williams, "Putting facts together: a study of scientific persuasion," *Social Studies of Science* 12 (1982): 535–58; Michael Zenzen and Sal Restivo, "The mysterious morphology of immiscible liquids: a study of scientific practice," *Social Science Information* 21 (1982): 447–73; Doug McKegney, "The

investigations of laboratory activities, these ethnographies were based on sustained observations of daily routines in specific laboratory settings.[45] The scope of investigation included the day-to-day shoptalk and methodic practices at particular laboratories, as well as the written and other forms of communication between various research groups and outside agencies. The earliest studies were, for the most part, conducted independently of one another by sociologists and anthropologists from at least five different countries who observed activities in subfields of biology, biochemistry, neurobiology, wildlife ecology, chemistry, and high-energy physics. As might be expected, the studies did not all tell the same story, but the most prominent of them supported social constructivist views of scientific activity. These studies claimed that "direct observation of the actual site of scientific work (frequently the scientific laboratory)" decisively demonstrated the social determination of even the most technical "contents" of science.[46] Claims about "direct" observations of "actual" practices have been muted in more recent discussions, but the naive enthusiasm with which the earliest studies announced their findings was important because of the way it attracted attention to them.

research process in animal ecology," paper presented at the conference on The Social Process of Scientific Investigation, McGill University, Montreal, 1979. Collins and Pinch's *Frames of Meaning* could be added to this collection, since it involved participant observational research. Because it was the first book in this collection to be published, many reviewers treat Latour and Woolgar's ethnography as the first "lab study." Knorr-Cetina's review avoids speculating about which of the studies got started the earliest, and she emphasizes that several of them began independently of one another. Unfortunately, in the "Postscript to second edition (1986)" of their book (Princeton, NJ: Princeton University Press, 1986), Latour and Woolgar retrospectively take credit for the widespread misimpression of their originality (p. 275): "When the first edition of *Laboratory Life* appeared in 1979, it was surprising to realise that this was the first attempt at a detailed study of the daily activities of scientists and their natural habitat. The scientists in the laboratory were probably more surprised than anyone that this was the only study of its kind." This claim should indeed be a source of surprise. As Latour elsewhere acknowledges ("Will the last person to leave the social studies of science please turn on the tape-recorder?" *Social Studies of Science* 16 [1986]: 541–48), his and Woolgar's study was by no means the only attempt in the late 1970s to study detailed laboratory practices.

[45] Ludwik Fleck's autobiographical account of his involvement in the development of the Wasserman test was perhaps the earliest account of this kind. See his *The Genesis and Development of a Scientific Fact* (Chicago: University of Chicago, 1979; originally published in 1935). Also see James Senior, "The vernacular of the laboratory," *Philosophy of Science* 25 (1958): 163–68; Bernard Barber and Renée Fox, "The case of the floppy-eared rabbits: an instance of serendipity gained and serendipity lost," *American Journal of Sociology* 64 (1958): 128–36; W. D. Garvey and Belver C. Griffith, "Scientific communication: its role in the conduct of research and creation of knowledge," *American Psychologist* 26 (1971): 349–62; and Jerry Gaston, *Originality and Competition in Science* (Chicago: University of Chicago Press, 1973).

[46] Knorr-Cetina, "The ethnographic study of scientific work," p. 117 (emphasis in original). Although the variant of constructivism advanced by Latour, Woolgar, and Knorr-Cetina incorporates aspects of ethnomethodology, as I argue in Chapter 5, studies by Garfinkel, Livingston, and myself do not follow the constructivist line.

Consistent with their intensive case-study approach, the laboratory ethnographies did not advance causal explanations; instead, they articulated a much more action-centered descriptivist approach to the "construction of facts." Their descriptions emphasized the contrast between the situated and improvisational performance of actual practices in "messy" practical and interactional circumstances versus rationally reconstructed experimental reasoning in textbooks and research reports. Laboratory ethnographies were by no means the first studies of science to note such differences, but their descriptive accounts added vivid detail to previous discussions.[47]

To an extent, the ethnographies advanced familiar themes from the sociology of knowledge, and as always, the first item on the agenda was the Mannheimian injunction to demonstrate "that the process of knowing does not actually develop historically in accordance with immanent laws, that it does not follow only from the 'nature of things' or from 'pure logical possibilities'." Only now the demonstration included a much more intensively focused history of laboratory projects and communicational exchanges. The very vocabulary used in the studies to describe scientific work as "construction" or "fabrication," and scientific reality as "artifact," suggested a discrepancy between the inherent "nature of things" and the material resources and products of laboratory work. Karin Knorr-Cetina expresses this point of view as follows:

> The constructivist interpretation is opposed to the conception of scientific investigation as descriptive, a conception which locates the problem of facticity in the relation between the products of science and an external nature. In contrast, the constructivist interpretation considers the products of science as first and foremost the result of a process of (reflexive) fabrication. Accordingly, the study of scientific knowledge is primarily seen to involve an investigation of how scientific objects are produced in the laboratory rather than a study of how facts are preserved in scientific statements about nature.[48]

Much was made of the artificiality of the specimen materials (e.g., specially bred or genetically engineered laboratory animals and microorganisms, and ingredients purchased from commercial suppliers) and of the many levels of interpretive, interactive, and instrumental mediation between scientists' accounts and the "natural" objects and facts described in those accounts.

[47] Contrasts between what scientists do and how they report on experiments are discussed in Barber and Fox, "The case of the floppy-eared rabbits"; Gerald Holton, *The Scientific Imagination: Case Studies* (Cambridge University Press, 1978); and Peter Medawar, "Is the scientific paper fraudulent? Yes; it misrepresents scientific thought," *Saturday Review,* August 1, 1964, pp. 42–43.

[48] Knorr-Cetina, "The ethnographic study of scientific work," p. 118–19.

Latour and Woolgar, for instance, argued that the laboratory researchers they observed during their ethnography at the Salk Institute did not investigate things in themselves; rather, the lab scientists examined "literary inscriptions" produced by technicians working with recording instruments: "Between scientists and chaos there is nothing but a wall of archives, labels, protocol books, figures, and papers." They summarized their argument as:

> Thus, in emphasising the process whereby substances are *constructed,* we have tried to avoid descriptions of the bioassays which take as unproblematic relationships between signs and things signified. Despite the fact that our scientists held the belief that the inscriptions could be representations or indicators of some entity with an independent existence "out there," we have argued that such entities were constituted solely through the use of these inscriptions.[49]

Latour and Woolgar also made the rather astonishing claim that "once the end product, an inscription, is available, all the intermediary steps which made its production possible are forgotten. The diagram or sheet of figures becomes the focus of discussion between participants, and the material processes which give rise to it are either forgotten or taken for granted as being merely technical matters."[50] They added that these steps and mediations are not irreversibly forgotten, since disputes between researchers and research groups can motivate efforts selectively to "deconstruct" the literary traces and documents and to relate them back to their practical origins.

Two significant features of this claim were (1) that scientific work is largely a *literary* and *interpretive* activity and (2) that scientific facts are constructed, circulated, and evaluated in the form of written *statements.* Latour and Woolgar went so far as to state (bold print in the original), **"A fact is nothing but a statement with no modality - M - and no trace of authorship."** (Since the bold statement I have just quoted includes "modality" only as an object of reference, and the only trace of the statement's authorship is that it appears on page 82 of *Laboratory Life,* I suppose that it is as good an example as any of a "fact.")

As Latour and Woolgar defined them, modalities are qualifying phrases or other markers of temporal or local reference (e.g., "these data *may* indicate that . . ."; "I *believe* this experiment shows that . . ."). They are clear-cut examples of "indexical" expressions. A "fact" is constructed when a community of researchers comes to use and accept a statement without qualification whereas an "artifact" is a statement including modalities. Latour and Woolgar presented an ideal–typical schema describing the transformation of state-

[49] Latour and Woolgar, *Laboratory Life,* pp. 245, 128.

[50] Ibid., p. 63. The complexities that are glossed over in this assertion can be appreciated by reading chaps. 4 and 7 of M. Lynch, *Art and Artifact in Laboratory Science.*

ments into facts, and in the case they reconstructed, they observed that laboratory practices, in the context of various machinations in a wider sociotechnical field, resulted in a demodalized statement: "TRF is Pyro-Glu-His-Pro-NH_2."[51]

By the late 1970s when Latour made his visits to the Salk Institute, this statement was taken for granted as a fact, and it provided the scaffolding for an entire research program. He and Woolgar reconstructed a series of publications, citation networks, disputes among rival labs, and negotiations among co-members of a research team. They used various schematic accounts of these heterogeneous communications to represent the dense existential conditions under which modalities were stripped from previously qualified and contentious statements about the hormonal factor, TRF. Accordingly, Latour and Woolgar situated Mannheim's distinction between the existential conditioning of knowledge and the "ultimate validity" of natural facts in an account of how "a statement became transformed into a fact and hence freed from the circumstances of its production."[52] In their view, such a statement acted simultaneously as a fact, cut loose from its generative social origins, and a *social construction* whose origins had been "forgotten" once the fact was used as a "black box." Accordingly, they interpreted the stability and apparent independence of a fact as itself a "constructive" accomplishment.[53]

Latour and Woolgar drew out some very pleasant implications of their study for social science. By placing "statements" and "literary inscriptions" in the foreground of their story, they were able to identify a familiar kind of terrain for making good literary sense of arcane laboratory activities. Although they mentioned the massive equipment and "hidden skills [that] underpin literary inscription,"[54] their textual analysis "reflexively" equated

[51] Latour and Woolgar, *Laboratory Life*, p. 147. TRF is a formula for "Thyrotropin Releasing Factor." Latour and Woolgar reconstruct the successive changes in the way that this hormone was described between 1962 and 1969.

[52] Garfinkel, Lynch, and Livingston ("The work of a discovering science") address a similar relationship between what they call the "night's work" and the "Independent Galilean Pulsar," although they make clear that neither of these can be described as an "account."

[53] At times, Latour and Woolgar's argument exhibits the worst qualities of a functionalist account, in which the analyst, having decided what is essential to the organization of a culture, manages to discount any counterevidence or explicit denial of its significance and relevance. For instance, when confronted with scientists' rejection of the anthropological "observer's" claim that the participants' factual "beliefs" are due to textual persuasion, Latour and Woolgar assert (p. 76): "The function of literary inscription is the successful persuasion of readers, but the readers are only fully convinced when all sources of persuasion seem to have disappeared. In other words, the various operations of writing and reading which sustain an argument are seen by participants to be largely irrelevant to 'facts,' which emerge solely by virtue of these same operations." In an appreciative review essay, Ian Hacking ("The participant irrealist at large in the laboratory," *British Journal for the Philosophy of Science* 39 [1988]: 277–94), offers a better sense of what Latour and Woolgar might have been claiming when they argued that TRH was "created" as a fact in 1969. At the same time, Hacking enables readers to grasp that the story can just as readily support a kind of realism concerning TRH.

natural science research with the sorts of literary and interpretive activities that social scientists are accustomed to doing. Moreover, they were able to claim that the supposed privileges of the natural sciences derived not from the indubitable significance of their discoveries but from the expensive equipment and institutional maneuvers that transform naturalistic statements into practically unassailable texts.

Latour and Woolgar (pp. 256–7) also give an amusing rejoinder to the derisive argument that their account of the lab is soft and not very credible compared with the lab participants' understanding of their work:

> In order to redress this imbalance [of credibility], we would require about a hundred observers of this one setting, each with the same power over their subjects as you [lab scientists] have over your animals. In other words, we should have TV monitoring in each office; we should be able to bug the phones and the desks; we should have complete freedom to take EEGs; and we would reserve the right to chop off participants' heads when internal examination was necessary. With this kind of freedom, we could produce hard data.[55]

By emphasizing how the "facts" of biochemistry were "nothing but" statements to be written, read, cited, and circulated in a research community, Latour and Woolgar contributed to contemporary efforts in the humanities and social sciences to broaden the application of textual and hermeneutic approaches. What could be better than to demonstrate that the "hard sciences" were themselves interpretive and literary enterprises, in which the "author," the "theory," the "nature," and the "public" all are effects of the text?[56]

In a strange way, however, Latour and Woolgar presented a kind of left-handed variant of the picture of language and practical action presented in a logical–positivist philosophy of science. By linking Latour and Woolgar to the logical positivists, I am not accusing them of being closet positivists. Instead, I am arguing that their explicitly antipositivistic approach provides a mirror image of many of the elements of the logical–positivist "picture" of language and practical action. Like the logical positivists, Latour and Woolgar (1) tried not to use preconceptions about "external reality" when they reconstructed the genesis of scientific facts, (2) treated scientific activities as operations on statements, (3) defined scientific facts as state-

[54] Latour and Woolgar, *Laboratory Life*, p. 245.
[55] Ibid., pp. 256–57.
[56] Ibid., p. 150, n. 8. The idea of literary inscription is especially attractive, as it uses an ethnography of praxis to license the sort of literary work familiar to humanists and social scientists. Donna Haraway, for instance, frequently cites Latour and Woolgar when setting up feminist literary (re)interpretations of primatology stories. See her *Cyborgs, Simians, and Women* (London: Routledge & Kegan Paul, 1991).

ments generated through operations on other statements, and (4) equated statement forms with epistemic relations.[57]

The major difference (and it was a major one), was that Latour and Woolgar insisted that the operations they described could not be encompassed by any system of formal logic. They also supported their argument with an empirical demonstration of how the genesis and acceptance of a fact depended on an extensive and contingent network of negotiations and machinations. But like the logical positivists, Latour and Woolgar fell into some of the grammatical "snares" that Wittgenstein identified in his later writings. The bleaching bones of the many casualties caught by such snares can be found along the well-worn trail of attempts to construct analytic languages purified of worldly preconceptions. Latour and Woolgar undertook a more modest variant of this when they attempted to describe scientists' activities without buying into the scientists' preconceptions about the relevant objects of study:

> We have attempted to avoid using terms which would change the nature of the issues under discussion. Thus, in emphasising the process whereby substances are *constructed,* we have tried to avoid descriptions of the bioassays which take as unproblematic relationships between signs and things signified. Despite the fact that our scientists held the belief that the inscriptions could be representations or indicators of some entity with an independent existence "out there" we have argued that such entities were constituted solely through the use of these inscriptions. . . . Interestingly, attempts to avoid the use of terminology which implies the preexistence of objects subsequently revealed by scientists has led us into certain stylistic difficulties. This, we suggest, is precisely because of the prevalence of a certain form of discourse in descriptions of scientific process. We have therefore found it extremely difficult to formulate descriptions of scientific activity which do *not* yield to the misleading impression that science is about *discovery* (rather than creativity and construction). It is not just that a change of emphasis is required; rather, the formulations which characterise historical descriptions of scientific practice require exorcism before the nature of this practice can best be understood. (p. 128)

To accomplish such exorcism, Latour and Woolgar present their ethnography from the point of view of a fictional "observer" who sees what is going on in the lab without being taken in by the scientists' beliefs in an unseen biochemical order of things. The observer describes just what he finds intelligible in the lab: the traces, texts, conversational exchanges, ritualistic activities, and strange equipment. An illustration of this sort of description is

[57] For an account of the "instincts" associated with positivism and an elaboration of the distinctive variant of logical positivism associated with the Vienna circle and its following, see Ian Hacking, *Representing and Intervening* (Cambridge University Press, 1983), pp. 41 ff.

given in a later discussion by Woolgar, when he offers a "native's" description, followed by an "ethnographer's" redescription, of a simple item: equipment

> A pipette is a glass tube with the aid of which a definite volume of liquid can be transferred. With the lower end in the liquid, one sucks the liquid up the tube until it reaches a particular level. Then, by closing the top end with finger or thumb to maintain the vacuum, the tube can be lifted and the measured volume of liquid within it held. Release of the vacuum enables the liquid to be deposited in another beaker, etc.

> Here and there around the laboratory we find glass receptacles, open at both ends, by means of which the scientists believe they can capture what they call a "volume" of the class of substance known as a "liquid." Liquids are said to take up the shape of the vessel containing them and are thought to be only slightly compressible. The glass objects, called "pipettes," are thought to retain the captured "volume" and to make possible its movement from one part of the laboratory to another.[58]

Woolgar goes on to say that the second account, despite its laborious efforts to avoid subscribing to the lab members' "beliefs" about the properties of liquids, glass, and the like, continues to include unexamined terms (like "glass" and "vessel") that could themselves be subjected to redescription. He leaves open the question of whether an ethnographer can fully succeed in bracketing all native preconceptions, especially when such "native preconceptions" include canonical assumptions about scientific method shared by professional anthropologists and sociologists. Despite these "reflexive" considerations, Woolgar does not question the impulse to "step back" from the setting studied, and both he and Latour continue to seek "to explain the science . . . without resorting to any of the terms of the tribe."[59]

The problem is that most of the terms of the tribe are our terms as well, since (1) they are integrally part of the discursive grammar through which scientists' activities are conducted and made intelligible *in situ* and (2) they are embedded in social science vocabularies for making coherent descriptions and explanations of tribal activities of diverse kinds.[60] Latour and Woolgar's search for an impartial ethnographic "metalanguage" recalls the Vienna Circle's search for a neutral observation language to describe the "raw" sensory data from which inferences about unseen entities, temporal

[58] Steve Woolgar, *Science: The Very Idea* (Chichester: Ellis Horwood; and London: Tavistock, 1988), p. 85.

[59] Latour, *The Pasteurization of France*, pp. 8–9. Also see Latour, *Science in Action* (Cambridge, MA: Harvard University Press, 1987), p. 13, where "the first rule of method" includes the proviso that "we will carry with us no preconceptions of what constitutes knowledge."

[60] See W. W. Sharrock and R. J. Anderson, "Magic, witchcraft, and the materialist mentality," *Human Studies* 8 (1985): 357–75.

relationships, and conceptual identities can be derived. As Wittgenstein pointed out, although descriptions couched in such language may be strictly correct, they can exhibit some curious properties.

> There is a way of looking at electrical machines and installations (dynamos, radio stations, etc., etc.) which sees these objects as arrangements of copper, iron, rubber, etc. in space, without any preliminary understanding. And this way of looking at them might lead to some interesting results. It is quite analogous to looking at a mathematical proposition as an ornament. – It is of course an absolutely strict and correct conception; and the characteristic and difficult thing about it is that it looks at the object without any preconceived idea (as it were from a Martian point of view), or perhaps more correctly: it upsets the normal preconceived idea (runs athwart it).[61]

Latour and Woolgar's man-from-Mars "observer" expresses no qualms about running athwart the preconceived ideas of the scientists studied. However, like many "radical" undertakings in sociology of science and ethnomethodology, *Laboratory Life* owes an unacknowledged debt to early-twentieth-century philosophy of science. My saying this may strike many readers as a very unfair characterization, since Latour and Woolgar devote so much of their argument to combating various realist and rationalist philosophies, histories, and sociologies of science.[62] What I am contending, however, is not that they are insufficiently skeptical about science or that their ethnography expresses unwarranted commitments to social realism but that their *inverted image* of scientific realism retains the grammatical framework of the inverted system. Rather than intensively attacking the entire picture of science, language, and representation that they oppose, they present a photographic negative or mirror image that retains much of its basic outline. A clear sense of this inversion can be gained by examining a series of Latour and Woolgar's summary statements:

[61] Ludwig Wittgenstein, *Zettel,* trans. G. E. M. Anscombe (Berkeley and Los Angeles: University of California Press, 1970), sec. 711.

[62] Indeed, in their "Postscript to second edition (1986)," p. 279, Latour and Woolgar maintain their antirealist vigilance by dismissing a criticism I had made of their man-from-Mars "observer." See M. Lynch, "Technical work and critical inquiry: investigations in a scientific laboratory," *Social Studies of Science* 12 (1982): 499–534. They argue that I express "a commitment to an (actual) objective character of the technical practices and the real worldly objects of study." Although I would not deny treating laboratory practices as "real" – that is, as witnessable, describable, and not imagined, in the ordinary sense of the word *real* – I do not buy their argument that this commits me to the "realist" side of a polarized debate between "realists" and "constructivists." In any event, this is beside the point, since my criticism was that Latour and Woolgar's ethnography could not live up to its *own* commitment to revealing the "social construction" of actual laboratory work. I was not saying that I could specify a more "objective" version; in fact, in his review of my book, Latour chides me for being so forthcoming as to admit to the limits of my approach. The fact that Latour and Woolgar's "Postscript" backs away from some of the more naive claims originally made by their ethnography indicates to me that they got the point of the sort of criticism I gave.

> The difference between object and subject or the difference between facts and artefacts should not be the starting point of the study of scientific activity; rather, it is through practical operations that a statement can be transformed into an object or a fact into an artefact. . . . By observing artefact construction, we showed that reality is the *consequence* of the settlement of a dispute rather than its *cause*.
>
> . . . [I]f reality is the consequence rather than the cause of this construction, this means that a scientist's activity is directed, not toward "reality," but toward these operations on statements.[63]

Although Latour and Woolgar certainly do not subscribe to a metaphysical view concerning the primacy of "ideas" in determining what can be known or perceived, their account suggests a kind of "textism," a principled emphasis on the centrality of "statements" and "traces" in a dualistic account of the relations of signs to referents. As they describe them, the collaborative projects, textual and material resources, and struggles for symbolic "credit" in scientific fields all are directed to transforming "statements" into taken-for-granted facts. It is as though laboratory work were primarily directed to fashioning and refashioning "statements" and that what any statement is doing were a secondary product of direct operations on the statement's form.[64]

This antirealist picture of language implies a radical separation between the form of a statement and its pragmatic use. If only for the sake of

[63] Latour and Woolgar, *Laboratory Life*, pp. 236–37. A polemical and pedagogical use of "inversion" also pervades Latour's highly popular text *Science in Action* (Cambridge, MA: Harvard University Press, 1987). In an explication of the argument in that text, Jon Guice ("A tiny breathing space: methodological localism, social studies of science and Bruno Latour," unpublished paper, Department of Sociology, University of California at San Diego, 1991) identifies "inversion" as one of Latour's primary textual devices. Latour illustrates his arguments with a Janus-faced figure whose more mature face enunciates the voice of "ready-made science" and whose youthful face speaks for "science in action." The bearded figure recites such principled assertions as "science is a *cause* of collective action," "nature is a *cause* of closure," "science is a *cause* of projects," and his clean-shaved counterpart turns these expressions around by saying that science and nature are the *consequence* of collective action, closure, and projects. This presentational device is overtly symmetrical, as both voices are granted an equivalent role, but only the voice of "science in action" enunciates what Latour proposes to be newsworthy. The voice of "ready-made science" presents established philosophical and sociological positions, which Latour treats as the academic common sense that his kind of social study of science confronts and problematizes. Although Latour's book has proved to be a marvelously successful teaching device, it conserves a Cartesian framework that Latour might otherwise be inclined to discard.

[64] Latour and Woolgar (*Laboratory Life*, p. 80) acknowledge that they may be oversimplifying matters by focusing on formal components of statements: "There are, of course, those who argue that this kind of determinate relationship between context and a particular interpretation of a statement does not exist. For our purposes, however, it is sufficient to note that changes in the type of statement provide the *possibility* of changes in the fact-like status of statements." It is not entirely clear that this weaker argument entirely avoids the problems associated with treating facts as statements and defining complex activities as though they were oriented to the production and stabilization of "statements" per se.

argument, Latour and Woolgar state their case by setting up a contrast between the "reality out there" about which scientists have no direct experience and the directly observable communications, writings, inscriptions, instruments, and maneuvers in a social field through which scientists establish "facts." They speak of the relationship between statements and reality as though it was a matter of linear causality: Either reality is the *cause* of constructive operations on statements, or reality is the *consequence* of such construction. This language of causality and unidirectionality is, perhaps, not to be taken literally. In a later discussion of "inversion," Woolgar tries to clarify the issue with the following statement of policy:

> Our first policy is to be critical of the relationships which construe a unidirectional connection between two elements of the representational couple. We need to take issue both with the idea that elements of the couple are distinct and with the notion that the object is prior (or antecedent) to the representation. Inversion asks that we consider representation as *preceding* the represented object.[65]

As I point out in Chapter 5, this criticism of the relationship between "two elements of the representational couple" does not go far enough. To say that representation *precedes* the represented object retains the dualistic picture of words and meaning that Wittgenstein and Garfinkel attacked. Although Latour and Woolgar's argument is no longer a straightforwardly causal one, it still suggests an initial separation of "representation" and "represented object" that is mediated by some sort of practical determination. This also holds for the view of "indexicality" promoted by Latour and Woolgar. While proposing to extend Garfinkel's concept to cover natural science discourse, they assume that indexicality is a ubiquitous representational "problem":

> The implication is that scientific expressions are no better able to yield a determinacy of meaning than any employed in "nonscientific" or common sense contexts. Garfinkel's (1967) discussion can also be read as supporting this conclusion. In a related manner, a number of continental semioticians have recently begun to extend the tools of literary analysis to the study of rhetoric in a wide number of areas. . . . For semioticians, science is a form of fiction or discourse like any other, one effect of which is the "truth effect," which (like all other literary effects) arises from textual characteristics, such as the tense of verbs, the structure of enunciation, modalities, and so on. Despite the enormous difference between Anglo-Saxon studies of the ways in which indexicality is repaired and continental semiotics, they hold in common the position that scientific discourse has no privileged status. Science is characterized neither by an ability to escape indexicality, nor by the absence of rhetorical or persuasive devices.[66]

[65] Woolgar, *Science: The Very Idea*, p. 36.
[66] Latour and Woolgar, *Laboratory Life*, p. 184, n. 2.

In Chapter 1, I gave a very different account of "indexicality," emphasizing that to say "all expressions are indexical" carries no definite implication about the certainty or uncertainty of linguistic communication. Indexicality does not necessarily imply that the meaning or intelligibility of particular utterances is "problematic." Instead, it implies that words or isolated statements do not "contain" unequivocal meanings and that understanding and determinate reference are achieved through situated uses of indexical expressions. To say that "science is a form of fiction" may effectively counter philosophical arguments to the effect that scientific propositions somehow contain a "special" form of logical grammar. But once we agree that an examination of propositional forms does not enable us to distinguish scientific from fictional statements, it no longer tells us anything about science or fiction to say that science is a form of fiction. Moreover, in the discussion of "truth effects" toward the end of the passage, Latour and Woolgar speak of "textual characteristics" as though they were a source of determinate "effects." As I understand it, Garfinkel and Sacks's discussion of indexical expressions makes an entirely different point, that the stability of sense, relevance, and meaning does not arise from the forms of propositions but from the circumstances of their use.[67]

Part of the difficulty of assessing constructivist studies is that their terminology often suggests a more clear-cut metaphysical stance than their arguments support. So for instance, although the language in many of the passages quoted earlier suggests that statements somehow cause truth effects or that constructive activity "creates" reality, the detailed examples and arguments in the studies portray something more akin to phenomenological "constitution." By couching descriptions of practical actions in familiar explanatory idioms, constructivists capitalize on misreadings to the effect that the "contents" of science are caused by social activities or that semiotic "actants" are equivalent to the more conventional species of sociological "actors." Such misreadings are virtually guaranteed when the studies are presented to audiences of social scientists, and this undoubtedly has contributed to the excitement and controversy generated by the more widely read studies. The writing is "clear" to readers who have limited acquaintance with the relevant philosophical and literary–theoretic sources, and the quasi-causal language seems startling and counterintuitive. Consequently, the better-known constructivist studies have fueled a great deal of interest and controversy, whereas the central claim that scientific facts are constructions remains ambiguous.

[67] Garfinkel and Sacks, "On formal structures of practical actions." To link Garfinkel and Sacks's version of indexicality to an "Anglo-Saxon" tradition is rather odd, since they criticize analytic philosophy and give original development to philosophical initiatives coming from phenomenology and existentialism, as well as Wittgenstein's later writings.

As I have suggested, although constructivist studies retain the basic form of sociology-of-knowledge explanations, they give no definitive answer to Mannheim's question: Are social processes of innovation "to be regarded merely as conditioning the origin or factual development of ideas (i.e., are they of merely genetic relevance), or do they penetrate into the 'perspective' of concrete particular assertions?" The assertion that constructive activity (or representation) precedes the establishment of an objective fact is not necessarily inconsistent with saying that such activity was "merely of genetic relevance." The definition of a fact as a statement stripped of all traces of sociohistorical perspective does not overtly conflict with what Mannheim says. The only difference is that constructivists define the very stability of a fact and the very absence of particularistic modalities in the factual statement as themselves results of construction. But unless we maintain the contrasting possibility of an "unconstructed reality," the claim that particular scientific facts owe their genesis to constructive activities holds little significance for the traditional programs in sociology of knowledge. As long as a particular fact remains in use in a discipline as a stable and unquestioned foundation for further inference and action, an account of the social history of its construction has no immediate relevance to the acceptance and use of the fact as an objective reality. Under such circumstances, there is no demonstrable difference between saying that a stable and accepted fact is a "stabilized construction" or a "correct statement about reality." As Collins, Latour, and Woolgar each acknowledge, the relativist or constructivist emphases in their studies are matters of methodological policy. Their studies do not empirically demonstrate that "scientific facts are constructed," since this is assumed from the outset. It would be more accurate to say that they demonstrate that a constructivist vocabulary can be used for writing detailed descriptions of scientific activities. Nevertheless, the descriptive language often suggests stronger forms of empirical demonstration, and readers often treat these studies as though they show that natural science is not as pure a form of activity as we once thought and that the validity of widely accepted scientific theories and empirical laws should now be called into question.

The crisis in relativist and constructivist studies

Despite, or indeed because of, the success of ethnographic and empirical relativist studies of science, many of the originators of these approaches have moved on to other modes of study.[68] "Laboratory studies" quickly tran-

[68] Laboratory studies continue to be published. Some of the more recently published studies are Sharon Traweek, *Beamtimes and Lifetimes: The World of High Energy Physics* (Cambridge, MA: Harvard University Press, 1988); Klaus Amann and Karin Knorr-Cetina, "The fixation of (visual) evidence," pp. 85–121, in M. Lynch and S. Woolgar, eds., *Representation in Scientific Practice* (Cambridge, MA: MIT Press, 1990); Kathleen Jordan and Michael Lynch, "The sociology of a genetic engineering technique: ritual and rationality in the performance

scended the laboratory and gave rise to a number of spin-off programs in science and technology studies. Latour, Woolgar, Mulkay, Collins, Pinch, and some of the other proponents of ethnographic and "empirical relativist" studies found new interests and consolidated their earlier studies into more abstract arguments. Mulkay and Woolgar became increasingly preoccupied with "discourse" and "reflexivity," and Latour devised a general approach to "science in action" based mainly on historical studies of scientific and technological innovation.[69] A number of sociologists of scientific knowledge borrowed themes from the sociology of science to "colonize" the sociology and social history of technology, health economics, and social problems research.[70] Although a few sociologists continue to develop original ethnographic studies, more often the themes and research strategies from constructivist studies are used for interpreting archival materials and constructing historical modes of demonstration.[71]

This development may seem strange in light of my arguments that laboratory studies gave dubious and ambiguous explanations of the "contents" of science. It makes sense, however, if we take account of the combination of success and inconclusiveness attributed to these studies by many social scientists:

1. The earliest constructivist studies quickly claimed victory over the intractable "contents" of science, even if they left some doubt about what exactly might be meant by such "contents." Social scientists were now free to consolidate the lessons from ethnographic studies to launch more comprehensive treatments of larger networks or fields of scientific endeavor.[72]
2. At the same time, persisting practical and interpretive difficulties discouraged efforts to devote inordinate attention to singular labo-

of the plasmid prep," pp. 77–114, in A. Clarke and J. Fujimura, eds., *The Right Tools for the Job: At Work in 20th Century Life Sciences* (Princeton, NJ: Princeton University Press, 1992); Joan Fujimura, "Constructing 'do-able' problems in cancer research: articulating alignment," *Social Studies of Science* 17 (1987): 257–93.; and Alberto Cambrosio and Peter Keating, "'Going monoclonal': art, science and magic in the day-to-day use of hybridoma technology," *Social Problems* 35 (1988): 244–60.

[69] Bruno Latour, *Science in Action* (Cambridge, MA: Harvard University Press, 1987).

[70] John Law, ed., *Power, Action, and Belief: A New Sociology of Knowledge?* (London: Routledge & Kegan Paul, 1986); Wiebe Bijker, Thomas Hughes, and Trevor Pinch, eds., *The Social Construction of Technological Systems* (Cambridge, MA: MIT Press, 1987); Steve Woolgar and Dorothy Pawluch, "Ontological gerrymandering: the anatomy of social problems explanations," *Social Problems* 32 (1985): 214–27; Malcolm Ashmore, Michael Mulkay, and Trevor Pinch, *Health and Efficiency: A Sociology of Health Economics* (Milton Keynes: Open University Press, 1989).

[71] See, for example, Shapin and Schaffer's *Leviathan and the Air Pump;* and Susan Leigh Star, *Regions of the Mind: British Brain Research, 1870–1906* (Stanford, CA: Stanford University Press, 1989).

[72] Latour and Knorr-Cetina took the lead in this centrifugal movement: Latour, "Give me a laboratory and I will raise the world," pp. 141–70, in Knorr-Cetina and Mulkay, eds., *Science*

ratory practices. At least two often-recited arguments gave sociologists good reasons not to conduct such studies:

a. The earlier laboratory studies used empiricist claims that did not withstand "reflexive" scrutiny. The rhetorical force of the claim that these studies were based on the direct observation of actual practices did not hold up very well under the sort of skeptical scrutiny advocated by those same studies.
b. Laboratory studies ignored the "broader" phenomena of interest to sociologists. This argument was often used by sociologists committed to "macro" levels of analysis, but sociologists of science who were sensitive to the criticism tried in various ways to "take account" of communities and institutions outside the laboratory, and this tended to detract from the attention paid to singular practices within laboratory settings.

These arguments were used by proponents as well as critics of laboratory ethnographies, and so they did not necessarily express a distaste for the entire genre of studies.[73] Aside from the criticisms, ethnographic studies remain exceedingly difficult to undertake, as they are not supported by the professional archives, established literatures, and disciplinary communities that facilitate familiar modes of historical and sociological scholarship. Just a few of the pragmatic difficulties faced by would-be laboratory ethnographers are the following:

1. Access to cutting-edge research is difficult. Not only can it be hard to gain permission to "hang out" in labs, but the research also is exceedingly technical, requiring extensive tutorials in a variety of skills and subjects that sociologists usually prefer to avoid. This problem is not unique to ethnographies of science, though it is especially pertinent.
2. "Social" phenomena are inextricably bound to "thick" technical talk and action. To demonstrate these phenomena requires tutoring one's audience in the competence systems in which the actions are embedded. Even then, readers are likely to treat "thick" description as "opaque" description or "tedious reportage."[74]

Observed; and Knorr-Cetina, "The ethnographic study of scientific work," esp. pp. 132–33, on "the transepistemic connection of research."

[73] See M. Lynch, "Technical work and critical inquiry: investigations in a scientific laboratory," *Social Studies of Science* 12 (1982): 499–534; Latour and Woolgar, "Postscript to second edition (1986)," pp. 277–78; Bruno Latour, "Postmodern? No. Simply amodern! Steps towards an anthropology of science," *Studies in the History and Philosophy of Science* 21 (1990): 145–71.

[74] A nice specimen of how some historians reacted to dense ethnographic accounts of scientific practices is provided by Christopher Lawrence's review of M. Lynch's *Art and Artifact in Laboratory Work,* in *Isis* 79 (1988): 473.

3. Career demands in academic social science – including undergraduate curricula, relations with departmental colleagues, and responsibility for keeping up with the literature in sociology – do not provide established scholars with much incentive for undertaking studies of disciplines more demanding than sociology. Intensive ethnographic research is more suited to dissertation research, and since the early 1980s there have been very few graduate students in the field.

Rather than undertake the difficult, time-consuming, and epistemologically suspect tasks of ethnography, many sociologists of science have preferred to take refuge in offices and libraries. There they can act as if they are observing "science in action" while engaging in more respectable academic pursuits: sifting through historical archives and secondary sources, composing scholarly syntheses of the diverse literatures in the sociology of science and related areas, and performing close textual analyses.[75] The move into history of science, technology studies, and studies of "discourse" has attracted larger constituencies of readers, and the literary turn in the humanities and social sciences has helped valorize familiar academic preferences and habits. The current prestige accorded to textual approaches provides little incentive to undertake the "merely empirical" and epistemologically naive work of "observing actual practices."

The common refrain to go "beyond the laboratory walls" supposes that there is little left to do inside the lab, and in many cases it encourages the kind of comprehensive sociology of science that was fostered by the Mertonian program. Only now, this might better be called *left-Mertonianism,* because the more forceful arguments are concerned with the way that funding, state imperatives, and other "broadly social" agendas influence the local sites of scientific practice, and an action-centered approach provides a locus for articulating such trans-epistemic influences.[76]

The reflexive turn

With the rising popularity of literary and interpretive approaches in the human sciences, a number of sociologists of science broke away from the attempt to demonstrate a reflexive grounding for the sociology of scientific knowledge and began to examine reflexivity as a phenomenon in its own right. Instead of trying to emulate a version of science or devise pragmatic criteria for validity, Woolgar, Malcolm Ashmore, Mulkay, and others began to investigate "our own ability to construct objectivities through representa-

[75] If what I am saying here is a criticism, readers should note that it is not one from which I exempt my own work. Witness this volume.

[76] For a clear example of this approach, see Chandra Mukerji, *A Fragile Power: Scientists and the State* (Princeton, NJ: Princeton University Press, 1989).

tion. These representational activities include the ability to adduce evidence, make interpretations, decide relevance, attribute motives, categorize, explain, understand and so on."[77] Although these themes identify general epistemic activities, "reflexive" sociologists did not propose them as grounds for validity, nor did they try to emulate how natural scientists adduce evidence, make interpretations, and so forth. Quite the opposite. In order to make these activities *analyzable,* Woolgar and his colleagues suggested that sociologists step back from an "unreflexive" effort to describe and explain scientific practices. In their view, the object of study should be the discursive and interpretive practices through which "objective" accounts are produced, whether in science or sociology of science.[78]

This is perhaps a more consistent application of Bloor's impartiality postulate than Bloor had in mind, since it proposes that sociologists should be no less impartial about their own validity claims than they are about the claims made by practitioners in the scientific fields studied. In effect, the reflexivity postulate becomes an extremely strong injunction to act in accordance with the Mertonian norm of "disinterestedness," but rather than securing scientific authority it tends to estrange the reflexive sociologist from any program of objective description and explanation.

As a way of breaking out of a regressive attempt to emulate positive science, Woolgar, Mulkay, Ashmore, Pinch, and others occasionally resorted to (mildly) Derridian disruptions of conventional social science writing practices. They did so by embedding critical interlocutors in their texts, by composing plays in which sociologists of science were the characters, and by writing texts in which the authorized scientific "voice" was disrupted and parodied.[79] Although some readers might conclude that such exercises provide an effective *reductio ad absurdum* of the sociology of scientific knowl-

[77] Steve Woolgar, *Science: The Very Idea* (Chichester: Ellis Horwood; London: Tavistock, 1988), p. 93.

[78] Howard Horwitz discusses similar "self-reflective" tendencies among proponents of "the 'new' historicism" ("'I can't remember': skepticism, synthetic histories, critical action," *South Atlantic Quarterly* 87 (1988): 787–820. Horwitz (p. 799) argues that "critical self-reflection can enjoy no greater cognitive authority than any other scrutiny . . . the fact that cognition of the self can occur is precisely the mark that the subject is 'divided': it can never know itself as a whole but only through and as an image of itself, only in a form different from itself. The subject is thus constituted and known only discursively and in history." In other words, "critical self-consciousness" does not mark a different kind of empirical critical inquiry; it simply shifts the subject. To this it should be added that Woolgar and Ashmore typically do not examine "their own" arguments; rather, they critically "reflect" on the arguments made by their colleagues and close rivals in the sociology of science. Aside from the occasionally odd verbal maneuver, they engage in a respectable (and sometimes valuable) form of academic criticism.

[79] See the various contributions to Steve Woolgar, ed., *Knowledge and Reflexivity: New Frontiers in the Sociology of Knowledge* (London: Sage, 1988). Also see Malcolm Ashmore, *A Question of Reflexivity: Wrighting the Sociology of Scientific Knowledge* (Chicago: University of Chicago Press, 1989); Michael Mulkay, *The Word and The World: Explorations in the Form of Sociological Analysis* (London: Allen & Unwin, 1985); and Mulkay, "Looking backward," *Science, Technology, and Human Values* 14 (1989): 441–59.

edge, the serious aim was to expose the discursive practices that supposedly remain hidden when sociologists use them as explanatory resources. Mulkay gives the following rationale for these discursive practices:

> The phrase "new literary forms" is better than, say, "new analytical language," because what was needed . . . was not a new vocabulary for writing about social life, but new ways of organizing our language which would avoid the implicit commitment to an orthodox epistemology that was built into the established textual forms of social science. In an attempt to address the self-referential nature of SSK's central claims and to display the ways in which analysts' claims are moulded by their use of specific textual forms, I began to employ multi-voice texts in which both analytical claims and textual form could become topics of critical discussion in a natural manner. Texts of this kind made it possible, I found, to replace the unitary, anonymous, socially removed authorial voice of conventional sociology with an interpretative interplay within the text as a result of which the voices involved became socially located and their constructive use of language became available for comment both within the text and beyond.[80]

Consequently, historiographic and ethnographic exercises became occasions for examining how "we" conduct "our" inquiries about "them." Even though this version of "reflexivity" may be accused of having taken leave of its senses, it did not entirely take leave of Mannheim's nonevaluative general total conception of knowledge. Instead, it attempted to take Mannheim's "total conception" to its ultimate limit.

Postconstructivist trends

Since the publication of Latour's widely cited text *Science in Action,*[81] discussions of the sociology of scientific knowledge have acquired a distinctively French accent. In part, this follows the belated influence on sociology of "deconstructionist" and "discourse-analytic" approaches. In addition, many American sociologists with limited acquaintance with Continental philosophy and literary theory have treated Latour's and his colleagues' "actor-network" theory[82] as a novel and innovative supplement in sociology to symbolic interactionist and, to a lesser extent, functionalist approaches.

[80] Michael Mulkay, "Preface: the author as a sociological pilgrim," pp. xiii–xix in Mulkay, *Sociology of Science: A Sociological Pilgrimage* (Bloomington: Indiana University Press, 1991), quotation from p. xvii.

[81] Latour, *Science in Action.*

[82] Latour is careful to credit Michel Callon, his colleague at l'Ecole nationale supérieure des mines, Paris, with coequal status as a founder of the approach associated with their school. Callon has published several influential articles in English, some of which are coauthored by Latour, but because Latour's books are far more familiar to the English-speaking world, his name has become an emblem for the collaboratively developed approach and, indeed, for the entire field of "new" sociologies of scientific knowledge. For an overview of technical aspects of this approach, see Michel Callon, John Law, and Arie Rip, eds., *Mapping the Dynamics of Science and Technology* (London: Macmillan, 1986).

Typically, American sociologists adopt de-radicalized versions of Latour's approach that miss its critical implications for sociology, and their references to "Latour" often ignore the strong linkage to themes and vocabularies from semiotics, hermeneutics, and existential philosophy that are alien to the dominant traditions of North American empirical social science.[83]

Latour builds on his and Woolgar's ethnographic study as well as a large body of historical studies of scientific and technological innovation, and he constructs a theory of action that attempts to link laboratory practices to a wider field of technical–political negotiations. Perhaps the clearest example of Latour's approach is his historical textual study of Pasteur.[84] This study announces a radical break from previous traditions in the sociology of knowledge in at least two ways: First, Latour explicitly disavows the disciplinary commitments of sociology and presents a study that is neither a social–historical description of Pasteur and Pasteurism nor a philosophical argument regarding the foundations of science. Instead, Latour says that he wants to conduct a philosophical investigation and an empirical field study "under the same roof."[85]

Moreover, Latour says he wants to investigate a field in which "social context" and the "contents" of science are not yet differentiated. In addition, he disdains any attempt to explain scientific innovations by using either "cognitive content" or "social context" as coherent explanatory factors. He comments further that relevant distinctions between social and technical factors, context and content, science and nonscience, and so forth are produced in the fields of negotiation that effectively create "scientific" or "technical" innovations. Accordingly, he moves Mertonian or Edinburgh school-type explanations into the field he seeks to investigate. As Latour and Michel Callon acknowledge, their policies in this regard are partially congruent with ethnomethodological commitments and with various "reflexivist" or discourse-analytic treatments.[86] At the same time, they attempt to compre-

[83] For example, see Kay Oehler, William Snizek, and Nicholas Mullins, "Words and sentences over time: how facts are built and sustained in a specialty area," *Science, Technology, and Human Values* 14 (1989): 258–74.

[84] Latour, *The Pasteurization of France*. Another often-cited exemplar is Michel Callon, "Some elements of a sociology of translation: domestication of the scallops and of the fishermen of St. Brieuc Bay," in J. Law, ed., *Power, Action, Belief: A New Sociology of Knowledge?* (London: Routledge & Kegan Paul 1986), pp. 196–229. For an application of this approach to technology studies, see John Law, "On the methods of long-distance control: vessels, navigation, and the Portuguese route to India," pp. 234–63, in Law, ed., *Power, Action, and Belief*. Programmatic statements of Latour's and Callon's position include M. Callon and B. Latour, "Unscrewing the big Leviathan: how actors macrostructure reality and how sociologists help them to do so," pp. 277–303, in K. Knorr and A. Cicourel, eds., *Advances in Social Theory and Methodology: Toward an Integration of Micro and Macro Sociologies* (London: Routledge & Kegan Paul, 1981); B. Latour, "Give me a laboratory and I will raise the world," pp. 141–70, in K. Knorr-Cetina and M. Mulkay, eds., *Science Observed*.

[85] Latour, *The Pasteurization of France*, p. 252, n. 8.

[86] Callon ("Some elements of a sociology of translation," p. 225, n. 3) mentions that ethno-

hend a broader field of actions and agencies than the laboratory practices and discursive phenomena described in laboratory studies.

Having disavowed sociological theory and method, Latour turns to semiotics, and particularly the formal approach to narrative developed by A. J. Greimas.[87] Although Latour's approach is far from a technical exercise in Greimasian semiotics, it follows the general outlines of a semiotic approach and uses selected terms from the technical vocabulary. He sets up his study by selecting a set of nineteenth-century texts and commentaries in which references to "Pasteur" and "Pasteurism" were made. In his account, "Pasteur" is a textual signifier, and Latour tries to trace the way in which this signifier was inserted into a developing story that wove together a coherent and yet heterogeneous network of entities and agencies. These entities and agencies are associated with domains such as daily life on the farm, sexual practices and personal hygiene, the architecture and therapeutic regime of the clinic, sanitary conditions in the city, and the microscopic entities and causal relations demonstrated in Pasteur's laboratory. Through a complex series of interventions and machinations, Pasteur's laboratory and the microbial "agencies" demonstrated through the disciplinary regime of the laboratory become an obligatory point of passage for translating privileged accounts of the disparate effects of microbial agents into solutions for problems concerning diseased farm animals, prostitution and its associated ills, epidemics, and urban sanitation.

In a sometimes deliberate maneuver, Latour upgrades his textual analysis of the semiotic "actant" "Pasteur" into a substantive narrative about Pasteur's historical actions. Although Latour can be accused of playing fast and loose with semiotic vocabularies, his conflation of textual analysis and historiography is consistent with contemporary semiotic theory. Despite their programmatic warnings to the contrary, Latour's and Callon's accounts are readily understood in terms of more conventional social–historical concepts of power, social influence, and Machiavellian strategy.[88] To an extent, such misreadings work to their advantage, since they enable Latour and Callon to present historical narratives about spatially and historically extended distributions and deployments of events and activities (thus transcending the well-known limitations of "micro" studies of laboratory activities) while at the same time disavowing sociological realism in favor of formal semiotics.

In the second radical break from previous traditions, both Latour and

methodologists have also taken into account "the simultaneous construction of scientific facts and social context." Also see Latour, *The Pasteurization of France*, p. 253, n. 15.

[87] A. J. Greimas and A. Courtes, *Semiotics and Language: Analytical Dictionary* (Bloomington: Indiana University Press, 1983).

[88] Latour repeatedly disavows such sociological readings of his conceptual vocabulary, and yet his historical narratives are difficult *not* to read as, for example, realistic accounts of how a person named Pasteur managed to build alliances and proselytize a particular research program.

Callon also disavow any a priori distinction between human and nonhuman agents of scientific and technological development. This maneuver has drawn a great deal of attention and criticism, and much of the controversy can be ascribed to their ambitious effort to develop a semiotic theory into a full-blown ontology.[89]

Latour and Callon speak of "actors" as both nonhuman and human entities and forces. As Latour puts it, "I use 'actor,' 'agent,' or 'actant,' without making any assumptions about who they may be and what properties they are endowed with . . . they can be anything – individual ('Peter') or collective ('the crowd'), figurative (anthropomorphic or zoomorphic) or nonfigurative ('fate')."[90] In his study of efforts by a group of scientists to devise methods for propagating scallops in St. Brieuc Bay, Callon includes scallops, gulls, wind and ocean currents, fishermen, and scientists among the various actants involved in the story.[91] Latour includes Pasteur, farmers, clinicians, cows, and microbes among the heterogeneous collection of "actors" in his account. His indiscriminate use of "actor" also applies to the set of predicates typically assigned to "actors" in sociological theories of action: "If I use the words 'force,' 'power,' 'strategy,' or 'interests,' their use has to be equally distributed between Pasteur and those human or nonhuman actors who give him his strength."[92] Latour thus retains familiar terms of sociological description and analysis while backing away from any exclusively "sociological" connotations of those terms. In contrast with formal semiotics, in which the grammatical concept of "actant" is less easily confused with the familiar sociological concept of "actor," Latour's semiotic history deliberately plays on the ambiguities and apparent absurdities created by translating technical semiotic vocabularies into sociohistorical descriptors. His narratives consequently invite the very sociological misreadings and appropriations that he programmatically disavows.

Latour's account is an original and clever rewriting of the Pasteur story. But like his and Woolgar's focus on the transformation of "statements" in the course of laboratory research, it is indebted to a highly formalistic understanding of linguistic reference. Although microbes, scallops, ocean currents, and scientists all can fulfill the grammatical role of actant in a semiotic system, one would have to be completely "bewitched" by grammar to suppose that this endows each of these actants with comparable ontological

[89] Simon Schaffer likens Latour's ontology to the nineteenth-century conception of "hylozoism." See S. Schaffer, "The eighteenth Brumaire of Bruno Latour," *Studies in History and Philosophy of Science* 22 (1991): 174–92.

[90] Latour, *The Pasteurization of France,* p. 252, n. 11.

[91] Callon, "Some elements of a sociology of translation." Also see Law, "On the methods of long-distance control," in which ship designs, trade winds, ocean currents, and sailors are described as relevant "actors" in the story.

[92] Latour, *The Pasteurization of France,* p. 252, n. 10.

status. Although Latour and Callon clearly do not intend to suggest that all actors are equivalent, they presume that they can move through linguistic fields in a kind of presuppositionless operation. As Latour (p. 12) comments, "I use history as a brain scientist uses a rat, cutting through it in order to follow the mechanisms that may allow me to understand at once the *content* of a science and its *context*." He attempts neither to reduce this history to a sociological explanation nor to adopt the terms of the Pasteurian tribe. Even if Latour's semiotics does provide a neutral point of departure for respecifying the historical texts he analyzes, it seems clear that his social science readers cannot resist believing that he is telling a somewhat eccentric sociological story.[93]

Latour's and Callon's "actor-network" approach is currently the most radical and interesting of the postconstructivist sociologies of scientific knowledge. As I pointed out earlier, a number of other proponents of constructivism have undertaken textual or discourse-analytic approaches. Influenced by ethnomethodological and other approaches to language use and practical actions, Woolgar, Mulkay, Steven Yearley, John Law, Karin Knorr-Cetina, and others have conducted a diverse range of studies on the production and use of texts, visual representations, conversation, and the interrelations among them.[94] In addition, constructivist approaches have linked up with more explicitly politicized critiques of science, particularly those associated with feminist sociology and epistemology. It remains to be seen whether the sociology of scientific knowledge clearly supports feminist critiques of "objective science," since it can be argued that such critiques retain a determinate picture of scientific and technical ideology that is problematized in constructivist studies.[95] But at least when considered abstractly, the presumption that natural science has been shown to be "social," even in its most detailed contents, opens the door for more specific arguments and demonstrations regarding the "gendered" nature of scientific facts.

[93] In their critique of Latour and Callon, Collins and Yearley understand references to nonhuman "actors" to be endorsing the ontological discriminations made by the scientists studied, and they demand that their French colleagues tell a more consistently sociological story. Collins and Yearley appear to be confusing Latour's and Callon's semiotic vocabulary with ontological predications, but this confusion is understandable given the conflation I discussed earlier. See H. M. Collins and S. Yearley, "Epistemological chicken," pp. 301–26, in A. Pickering, ed., *Science as Practice and Culture* (Chicago: University of Chicago Press, 1992).

[94] See M. Mulkay, J. Potter, and S. Yearley, "Why an analysis of scientific discourse is needed," pp. 171–204, in Knorr-Cetina and Mulkay, eds., *Science Observed*. Also see the papers in the edited collections by Lynch and Woolgar, eds., *Representation in Scientific Practice* (Cambridge, MA: MIT Press, 1990); and Pickering, *Science as Practice and Culture*.

[95] Some of the contentious issues on this issue are raised in a polemical exchange published in *Social Studies of Science* 19 (1989): Evelleen Richards and John Schuster, "The feminine method as myth and accounting resource: a challenge to gender studies and social studies of science," pp. 697–720; Evelyn Fox Keller, "Just what is so difficult about the concept of gender as a social category?" pp. 721–24; and Richards and Schuster, "So what's not a social category? Or you can't have it both ways," pp. 725–30.

The various "new" sociologies of science make up a fragmented field, and the strong program's agenda to give sociological explanations of the content of science is riven with debate on each of its key terms. These debates concern (1) the possibility of discriminating "social" factors from "cognitive" or "technical" factors, (2) the ability to give causal explanations of how social context affects scientific developments, (3) the identification of the relevant "contents" of science, (4) the discrimination of "science" from nonscience, and (5) whether or not items (1) through (4) should be treated as members' achievements in a discursive field rather than sociologists' analytic tasks. The new sociology of science has turned inward and is beset by multiple rivalries and controversies while at the same time its achievements continue to be heralded as grounds for a social epistemology.[96]

Especially in North American sociology, a conventional history[97] is now recited in many of the participants' reviews of one another's studies. Such members' histories typically credit the new sociologists of science with having penetrated the recesses of the laboratory and having demonstrated that in its every detail scientific knowledge is created, not discovered. At the same time, so the story goes, these studies are limited by their attention to what can be observed within the walls of the laboratory, and they fail to take into account the "larger" contexts in which scientific construction takes place. What is needed, we are told, is an effort to explain scientists' local practices by reference to structural sources of institutional power, ideology, and funding; only then will the new sociology of science provide the kind of comprehensive and normative foundation that can enable us to rebuild a more egalitarian science.[98]

This narrative has captivated many of the participants in the field, and it can inspire an almost religious mission to take science off its pedestal and deliver it to the people. Curiously, it has fostered a kind of left-Mertonian revival, in which the kind of universal functionalism that Merton once criticized[99] is replaced by a conception of systematic action in which local actions are functionally related to particularistic values, interest groups, and rhetorical tropes. This enables sociologists to appeal to a more comprehen-

[96] Steve Fuller (*Social Epistemology* [Bloomington: Indiana University Press, 1988]) tries to solve the problems of the normative philosophy of science by recasting the programmatic aims of the new sociology of science.

[97] By "conventional history" I mean an account of a common history that is generated by interested participants in that history. See David Bogen and Michael Lynch, "Taking account of the hostile native: plausible deniability and the production of conventional history in the Iran-contra hearings," *Social Problems* 36 (1989): 197–224.

[98] Such an argument is succinctly presented in William Lynch and Ellsworth Fuhrman, "Recovering and expanding the normative: Marx and the new sociology of scientific knowledge," *Science, Technology, and Human Values* 16 (1991): 233–48.

[99] Merton, "Manifest and latent functions," chap. 3 of *Social Theory and Social Structure*, enlarged ed., esp. pp. 84 ff.

sive set of commitments (presumably shared with readers) that enables normative criticisms to be launched.

As I read it, however, the story credits laboratory studies with too much success while at the same time it invites us to forget why those studies were undertaken in the first place. The "new" sociologies of science were, and in some cases still are, motivated by reactions against normative philosophy of science and structural–functionalist sociology. They enabled a more differentiated conception of the "actual" practice of science that does not follow from a unitary model of scientific method. The aim was not to attack "science" per se, since in principle, the studies attempted to suspend a priori understandings of what science is while examining particular cases of observation, experiment, and theoretical controversy. Not surprisingly, this turned out to be difficult to do, and none of the existing laboratory studies delivered the sort of "thick description" that would resolve the realist–constructivist debate; indeed, it misconceives the nature of that debate to figure that empirical studies could have resolved it.[100] So rather than supplying empirical ammunition for policy studies and politicized critiques, the effort to observe, describe, and explain "actual" scientific practice *in situ* takes us back to the drawing board, where we are left to reconsider what it means to produce observations, descriptions, and explanations of something "actual." This, as I understand it, is ethnomethodology's agenda.

Ethnomethodological studies of work

I mentioned ethnomethodological studies of work in Chapter 1, and in later chapters I elaborate further on them, so I will discuss these studies only briefly here. Although sociologists of scientific knowledge used programmatic initiatives and themes from ethnomethodology and the two areas share a number of cognate interests and issues, ethnomethodological studies of scientific work are an independent development.

In the early 1970s, Garfinkel proposed a program of ethnomethodological studies of work, with the aim of investigating what he called the "missing what" in analytic studies of occupations and professions. Briefly, he argued that sociologists who study the various arts and sciences of practical action

[100] As Wes Sharrock and Graham Button point out ("The social actor: social action in real time," pp. 137–75, in G. Button, ed., *Ethnomethodology and the Human Sciences* [Cambridge University Press, 1991]), the idea of "thick description" derives from Gilbert Ryle ("The thinking of thoughts," in *Collected Papers*, vol. 2 [London: Hutchinson, 1971]), although it is most often attributed to Clifford Geertz. A "thick" description not only is more detailed than a "thin" one and it not only concerns what can be directly witnessed on some occasion, but it also incorporates a "member's" localized recognition of the actions described, for example, like moves in a game, gestures rather than motions, and actions within a developing colloquy.

typically investigate "social" aspects of, for instance, music, without addressing how musicians manage to play music together. Similarly, when they investigate activities in the legal professions, sociologists tend to describe various "social" influences on the growth and development of legal institutions while taking for granted that lawyers write briefs, present cases, interrogate witnesses, and engage in legal reasoning. In contrast, Garfinkel stated, ethnomethodology investigates the work-specific competencies through which musicians make music together, or lawyers conduct legal arguments, in and as collaboratively produced and coordinated actions.[101]

Garfinkel's program provided an incentive for investigating the "contents" of scientific and mathematical practices, although the initiative in this case did not follow from a desire to extend or strengthen Mannheim's sociology of knowledge. From the outset, Garfinkel and his colleagues expressed no interest in explaining scientific facts in reference to the "social context" of their production, nor did they try to construct comprehensive models of the various fields of activities and institutional conditions that make up the context of any particular innovation. Their objective instead was to examine how scientific discoveries and mathematical proofs are produced and "extracted" from the disciplinary-specific *Lebenswelt* of the laboratory project or the mathematics lesson.

Again, their aim was not to explain "discovery" as a matter of "social construction" but to try to gain a better understanding of scientific work than can be derived from reading biographies of scientists or from reconstructing experiments and proofs. Largely by coincidence, ethnomethodological studies of laboratory practices became associated with the studies of "laboratory life" by Latour and Woolgar, Knorr-Cetina, and other constructivists, and in the past decade this association has resulted in a partial convergence on topics and problem areas. This convergence enables a fairly precise specification of a set of cognate issues and points of debate. I have already raised many of these issues, and I pursue them further in later chapters, but in brief outline, they include the following:

1. The "problem" of reflexivity. This includes the various ways in which the language and practices through which sociologists investigate, describe, and explain other practices become intertwined

[101] A collection of ethnomethodological studies of work was published in 1986 (H. Garfinkel, ed., *Ethnomethodological Studies of Work* [London: Routledge & Kegan Paul]), although many of the papers in the collection had been written several years earlier. Garfinkel's program linked up with the Wittgenstein-inspired ethnomethodological investigations conducted by Wes Sharrock and some of his colleagues and students at the University of Manchester (this approach and its convergence with Garfinkel's program is represented in Button, ed., *Ethnomethodology and the Human Sciences*). Another line of development, regarding the practical and situated production and use of technology, is represented in Lucy Suchman, *Plans and Situated Actions* (Cambridge University Press, 1987).

with the language through which "members" conduct their affairs, produce arguments, and "analyze" one another's conduct.

2. The "conflation" of epistemological questions and methodological issues. This concerns the breakdown of Mannheim's programmatic separation between ideology-critique and sociology-of-knowledge explanations. This breakdown occurs whenever sociological descriptions use ordinary language expressions that imply criticism, skepticism, or acceptance of the "methods" or "beliefs" described.
3. The search for a "neutral" or "nonevaluative" observation language. This search continues the legacy of Mannheim's nonevaluative general total conception of ideology, and it is embodied in the strong program's policies of symmetry and impartiality. The ethnomethodological policy of "indifference," described in Chapter 5, offers an alternative to sociology-of-knowledge proposals to "step back" from the fields of study investigated.
4. The ultimately "undecidable" nature of distinctions between social and technical "factors," science and nonscience, or fact and construction. As pointed out earlier, sociologists of science have recently taken the position that these distinctions are "negotiated" within the fields of action that produce "scientific" innovations. It is not always clear what is meant by "negotiation" or "boundary work," and the implications for explanatory programs in the sociology of science remain to be worked out. In both these respects, ethnomethodological studies of practical action and language use can deepen existing understandings.

In my view, all of these problems point to a need to introduce into the new sociology of scientific knowledge more sophisticated conceptions of language use and practical action. Although indebted on several key points to phenomenological, Wittgensteinian, and ethnomethodological initiatives, sociologists of scientific knowledge retain a familiar admixture of commitments to objectivistic inquiry. These commitments include the idea that sociologists of science must step back from the commitments and linguistic usages in the fields they study, that descriptive "metalanguage" should be independent of the discourse described, that indexicality and reflexivity generally inhibit representation and communication rather than facilitate them, and that the stable conceptual apparatus of sociology (or formal semiotics) is adequate to the task of explaining the contents of other fields. What is seldom realized by sociologists of science is that these commitments are, or at least can be, criticized from the standpoint of ethnomethodology, Wittgenstein's later philosophy, and phenomenology.

I have proposed such a criticism here, but because there is no single ethnomethodological, Wittgensteinian, or phenomenological "position" on matters of interest to the sociology of scientific knowledge, my criticism will disclose what I think such a position should be. Rather than, for instance, promoting ethnomethodology at the expense of sociology of knowledge, my intention is to argue for a postanalytic ethnomethodology. In a sense, my critique anticipates a convergence between ethnomethodology and the sociology of science to make up a kind of "epistemic sociology."[102] This development does not encompass all, or even the major part, of research in the two fields, since many ethnomethodologists and sociologists of science remain committed to programs in scientific sociology that do not contribute to this agenda. But if I have diagnosed the situation correctly, many of the current debates and confusions in constructivist sociology of science are symptoms of a deep and ambivalent struggle with established commitments to a science of society. Given the virtues associated with science, and the absence of respected alternatives within the social sciences, it is difficult simply to put aside the pretenses and contradictions inherent in a "scientific" sociology of scientific knowledge. Nevertheless, it is becoming increasingly possible to imagine a unique mode of investigation that addresses familiar themes from epistemology (or the history of ideas more generally): a mode of investigation that is neither armchair philosophy nor methodologically driven sociology, one that examines empirical evidence as a spur to the imagination rather than as proof of hypotheses and that treats practices of observation, representation, measurement, and argumentation as social phenomena to be investigated rather than abstract methodological guarantees. I am not, of course, the first to imagine this possibility. As stated in Chapter 1, this was the vision that led Garfinkel to develop ethnomethodology, and to a large extent it has animated the "new" sociology of scientific knowledge. However, despite programmatic claims and sales pitches to the contrary, it is a vision that has yet to have much impact on the disciplinary agenda in sociology. Sociology's fate continues to ride on its status as a science, and given the many threats to the discipline's standing as a legitimate academic profession, attempts to disown sociological scientism may appear to be disloyal or even treacherous. Nevertheless, unless we are to continue to invest hope in the yet-to-be-redeemed "progress" of scientific sociology, a complete break with scientism may be warranted.

[102] See Coulter, *Mind in Action*, chap. 1, for a general discussion of "epistemic sociology."

CHAPTER 4

Phenomenology and protoethnomethodology

Edmund Husserl's ambition was to account for the achievements of the mathematical natural sciences without attributing those achievements to a naturalistic foundation, and his effort to do so created a legacy that ethnomethodology and the new sociology of science have taken up and transformed into empirical research programs. Few ethnomethodologists and sociologists of science today mention Husserl, perhaps because his effort to develop a "science" of the life-world based on a transcendental foundation was long ago repudiated in both Continental and Anglo-American philosophy. The disregard of Husserl is doubly unfortunate, however. First, the assumption that Husserl is irrelevant to contemporary research in ethnomethodology is belied by Garfinkel's continued injunctions to his students to "misread" Husserl from the standpoint of their projects at hand.[1] Although Alfred Schutz is usually considered to be the phenomenologist most relevant to contemporary sociological research, an argument can be made that he delivers a rather weak version of the Husserlian critique of the natural sciences. Second, the problems that motivated Husserl's effort to build a transcendental foundation for his analysis of the life-world continue to haunt empirical sociology. Indeed, it can be argued that a tendency toward transcendental analysis remains implicit whenever social scientists (including ethnomethodologists) employ one or another variant of a distinction between "common sense" and "analytic" understandings of social practices.

In this chapter, I begin by treating Husserl's genealogy of natural science as a precursor to Michel Foucault's and Garfinkel's postphenomenological investigations. This is a very partial and idiosyncratic reading of Husserl and the phenomenological tradition, and it is constructed retrospectively in light of more recent developments in ethnomethodology and the sociology of science. I will then examine how Schutz integrated Husserl's critique of natural science with Felix Kaufmann's less radical views on scientific methodology.

[1] In his lectures at UCLA in the early 1980s and more recently in a presentation at Boston University, Garfinkel advised "misreadings" of Husserl, Heidegger, and Merleau-Ponty ("The curious seriousness of professional sociology," delivered to the Colloquium for the Philosophy of Social Sciences, Boston University, December 1989). Garfinkel has repeatedly praised the depth and cogency of Husserl's writings, and he identifies ethnomethodology's research agenda with Husserl's genealogy of the *Lebenswelt*.

Schutz's writings on the rationality of the natural and human sciences and his exemplary explications of the everyday life-world had a prominent influence on Garfinkel's and Aaron Cicourel's "protoethnomethodological" treatments of commonsense knowledge. Schutz treated science primarily as a cognitive or theoretical activity, separate from the "world of daily life" within and beyond the walls of the laboratory. Protoethnomethodology retains Schutz's cognitivism, along with his presumption of an analytic vantage point, separate from the "natural attitude" of ordinary practical activity. In my view, protoethnomethodology is not simply the historical precursor of ethnomethodology, it is generated whenever ethnomethodologists invoke a mythology of "science" as the antithesis of "commonsense knowledge." Moreover, protoethnomethodology stands as the professionally acceptable form of a "discipline" that inhibits the development of a postanalytic ethnomethodology.

The phenomenological genealogy of natural science

In *The Crisis of European Sciences and Transcendental Phenomenology,* Husserl accounts for Galileo's "invention" of the *mathesis universalis* by disclosing the praxiological foundations of mathematical natural science.[2] He holds Galileo responsible for a "surreptitious substitution of the mathematically substructed world of idealities for the only real world, the one that is actually given through perception, that is ever experienced and experienceable – our everyday life-world."[3]

As Husserl reconstructs this genealogy, Galileo inherited the ancient legacy of geometry along with the platonic view of geometric forms as ideal and permanent essences lying behind the world of appearances. Husserl traces the ideal forms – the perfectly straight line, dimensionless point, angular figure, and regular curve – back to the protogeometric practices of surveying and measuring. The surveyor's measuring instruments embody relatively "pure" lines, scales, curves, and angles, which act as templates for shaping and polishing surfaces or reckoning material alignments and lengths. The purified shapes represented in Euclidean geometry not only enable descriptive or mapping functions; they also provide generative models for extrapolating, predicting, and planning yet-to-be realized architectures. The elaboration of a constructed object, built environment, or project of action

[2] Edmund Husserl, *The Crisis of European Sciences and Transcendental Phenomenology,* trans. David Carr (Evanston, IL: Northwestern University Press, 1970).

[3] Ibid., pp. 48–49. Husserl's account of "Galileo's invention" is a philosophical allegory rather than a historical description. Whether or not Galileo actually was responsible for the "achievement" that Husserl attributed to him is largely irrelevant to the phenomenology of scientific practice he proposes.

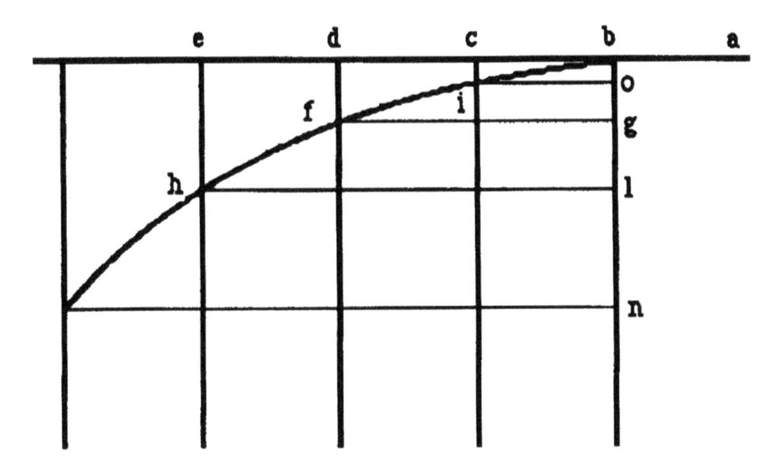

Figure 4.1. Galileo's diagram for the motion of projectiles. H. Crew and A. de Salvio, trans., *Dialogues Concerning Two New Sciences* (1665) (New York: Dover, 1954).

thus takes shape on an approximately axiomatic foundation and in approximate accord with the purified forms and established theorems of geometry.

For Husserl, Galileo's mathematization of nature takes its point of departure from the work of constructing built environments in accord with Euclidean forms. For example, the curvilinear pathway for a projectile represented in Figure 4.1 was, for Galileo, a graphic expression of a mathematical law in nature. Although Husserl does not question the facticity of the relationships expressed by the laws of classical physics, he does challenge their naturalized genealogy. In Husserl's analysis, a smooth and relatively "perfect" curve such as the one in Figure 4.1 expresses a "forgotten" genealogy. He viewed such curves as the end products of an iterative "polishing" through which experimental actions, equipment, measures, and mathematical analyses are brought together. Only after the phenomenal elements in the experimental field are stabilized through a disciplined and repetitive praxis does the mathematical law become apparent as what was always the case for projectiles and analogous material phenomena.

Husserl argues that the praxiological coupling of material relations with mathematical forms eventually was hypostatized into a "scientific" nature indifferent to human historicity and purpose. Henceforth, it was the goal of a Galilean science to use mathematics to discover the inherent structure of the universe, a structure that was always and already fit to be measured. For Husserl, there can be no essential demarcation between the artisan's craft and the mathematical relations "discovered" in the phenomenal field constructed through that craft. Accordingly, the geometric rendering in Figure 4.1 is a construction built up over time in a life-world of practices.

According to Husserl, the apparent correspondence between mathematical forms and natural properties is less a ground for rational certainty than a mystery at the heart of Galilean science. This is because any such correspondence, established as a singular achievement in the course of a scientific project, secures its self-evidence only as long as it rests on the unexplicated foundation of the intuitively given surrounding world. As complex worldly interventions, acts of measurement and calculation in the physical sciences presuppose the stability and meaningfulness of the life-world *(Lebenswelt)* in which they are secured. The calculative techniques of any science depend on an intuitive grasp of the distinctive subject matter of that science.

> One operates with letters and with signs for connections and relations (=, ×, +, etc.), according to *rules of the game* for arranging them together in a way essentially not different, in fact, from a game of cards or chess. Here the *original* thinking that genuinely gives meaning to this technical process and truth to the correct results (even the "formal truth" peculiar to the formal *mathesis universalis*) is excluded.[4]

Husserl did not believe that he could secure the foundations of natural science until he had explained the "original thinking" that gave rise to the presumed isomorphism between calculative "games" and the subject matter of a science. He sought to retrieve the "lost" praxiological foundation of the natural sciences by explicating the phenomenal field of the life-world. His phenomenological science of the life-world was based on the preliminary results of such an explication. Although I will not go into the details of Husserl's conceptual apparatus, for our purposes it is sufficient to point out that he never entirely abandoned the idea that scientific truths could be traced back to a unitary experiential foundation. Rather than treating technical rules of the game in the specialized sciences as the basis for a universalized grasp of the intuitively given structures of the life world, Husserl subordinated these "games" to the acts of a transcendental consciousness.[5]

A different, although thematically congruent, view of the history of mathematical physical science can be gained by replacing Husserl's philoso-

[4] Ibid., p. 46.

[5] Husserl's transcendental ego is reciprocally related to the Galilean *mathesis,* even though he is critical of Galileo's naturalistic genealogy. Boyle's experimental program, as Steven Shapin describes it ("Robert Boyle and mathematics: reality, representation, and experimental practice," *Science in Context* 2 [1988]: 23–58) , may be reciprocal to a different "phenomenology" of practical actions. Instead of respecifying a grand theoretic maneuver through which mathematical ideals become the essence of nature, the task would be to respecify the "matters of fact" that Boyle generates through his technical and literary procedures. The "games" in this case would be Wittgensteinian rather than Husserlian, since they would be related through family resemblance rather than emanating from a foundational "pole" of transcendental consciousness.

phy of perceptual consciousness with an account of the representational crafts and conventions that organize the intersubjective fields in which perception takes place. Samuel Edgerton, for instance, makes a convincing case for how the scientific revolution was set up by an earlier "rediscovery" of linear perspective by Brunelleschi and Alberti in fifteenth-century Florence.[6] Edgerton documents how the skills of the "artisan-engineer" Brunelleschi combined with the optical theory of Alberti to articulate a novel set of representational conventions. The artisan-engineers of Florence "were called on for two requisite talents: skill in mathematics and the ability to draw." In a development roughly akin to Husserl's account of the "primal geometrer's" craft of "polishing" the artifact to align its surface with the limit forms of geometry,[7] the artisan-engineers were attuned to "pure" geometric forms in the course of their constructive praxis. Brunelleschi invented a mirror device for perspectival painting, and Alberti later articulated the principles through which the lines of sight disclosed by the design and operation of the device could be abstracted from the practical situation and used in various mathematical operations.

Edgerton observes that art historians no longer treat pre-Renaissance art as a "naive" form of representation. In part, they have been sensitized by nonperspectival modernist art to see how the medievalist renderings may be faithful to phenomenological experience. In comparing two paintings of the city of Florence, the first from the medieval and the second from the Renaissance period, Edgerton writes:

> The painter of the earlier picture did not conceive of his subject in terms of spatial homogeneity. Rather, he believed that he could render what he saw before his eyes convincingly by representing what it felt like to walk about, experiencing structures, almost tactilely, from many different sides, rather than from a single, overall vantage. In the Map with a Chain [the Renaissance/linear perspective painting] the fixed viewpoint is elevated and distant, completely out of plastic or sensory reach of the depicted city. In the [medieval] fresco, on the other hand, jutting building corners, balconies, and rooftops are thrust out and huddled toward the viewer from both sides of the picture.[8]

The medieval painting in its own way is a faithful rendering, as it recollects a familiar field of practical relevancies rather than giving a snapshot (or

[6] Samuel Y. Edgerton, *The Renaissance Rediscovery of Linear Perspective* (New York: Harper & Row, 1975). My explication of Edgerton's study in this section ignores his and other art historians' attribution of linear perspective to the operations of a "mind's eye." My reading is far from incompatible with Edgerton's historical description, as is evidenced by our collaborative investigation of contemporary "image processing" in astronomy. See M. Lynch and S. Y. Edgerton, "Aesthetics and digital image processing representational craft in contemporary astronomy," pp. 184–220, in G. Fyfe and J. Law, eds., *Picturing Power: Visual Depiction and Social Relations* (London: Routledge & Kegan Paul, 1988).

[7] Husserl, *The Crisis,* p. 376.

[8] Edgerton, *The Renaissance Rediscovery of Linear Perspective,* p. 9.

camera obscura) view of a field from a momentarily assumed perspective. The viewer's motility is implicated in the medieval device of "split representation," which Edgerton (p. 14) describes as "the propensity to represent three-dimensional objects as if split apart and pressed flat, so that the picture shows more sides and parts of the object than could possibly be seen from a single viewpoint." The linear perspectival painting is neither more nor less "objective" than the medieval rendering; rather, it organizes a different field of "objective" and "subjective" relations, with the fixed "point of view" and "line of sight" acting as a center from which the details of a momentary scene are projected.

Alberti's optical treatise articulates a plane geometry for painters in which each "point" in the field is simultaneously "a sign," a sign being "anything which exists on a surface so that it is visible to the eye."[9] Alberti's *signum,* a textual "figure" or "mark," is "something tangible, like a dot on a piece of paper."[10] The features of the image appear to be displaced outward onto the canvas. Sign-referent relations take the place of the point-by-point correspondence between visual image and object. Moreover, the painter's construction of a plane of signs is seen to take place in a concrete geometrical field. Painting thus becomes a kind of embodied mathematics, using hybrid objects (e.g., dots and marks) that concretely approximate geometric limit forms. The plane that a painter composes is organized as an empirical graph in which a grid of lines link up "like threads in a cloth."[11]

The devices of linear perspective coordinate the literary spaces and embodied practices. They also assimilate a historically specific field of optical instruments, representational technologies, a theory of optics, methods of cartography, and as Edgerton (p. 37) indicates, the practical arts of measurement employed in the marketplace.[12] The fixed point, convergence of rays, hyperrealism, and point-by-point correspondences between object and image make up a veritable epistemology: an account of the mechanisms of vision, an account of their truth, and a set of precautions and correctives for establishing their limits and rectifying their errors.[13] Galilean science takes this pragmatic–semiotic system one step further, attributing the plane of signs and the grid on which it is inscribed to nature's authorship.

Husserl may have overgeneralized the extent to which Galileo's successors presumed the *mathesis universalis.* According to Steven Shapin, Robert Boyle maintained a strict distinction between the nonmathematical empirical manifold and the experimental devices he used for collecting, framing,

[9] Ibid., p. 80.

[10] Ibid.

[11] Ibid.

[12] See Bruno Latour, "Visualisation and cognition," *Knowledge and Society* 6 (1986): 1–40, for a summary of the large literature on the methods of printing and circulating texts that aided and abetted this practical construction of a field of objectified relations.

[13] Bacon's protoexperimental program includes numerous correctives of this sort. See Francis Bacon, "The new organon," pp. 39–248, in J. Spedding, R. L. Ellis, and D. D. Heath, eds., *The Philosophical Works of Francis Bacon* (London, 1858 [1623]), vol. 4.

and representing pressures, volumes, and specific gravities.[14] According to Shapin, mathematics was not essential to the ontotheological view that Boyle promulgated; instead, for Boyle the incalculable diversity of the phenomenal world in which the experimenter operates testifies to God's distance from the human condition, and the mathematicians' idealizations are human constructions rather than evidences of God's plan.

Svetlana Alpers makes a related argument about seventeenth-century Dutch art and science, saying that the Dutch assumed a descriptivist attitude toward the visible world, which contrasted with the "mathematical" orientation of the post-Renaissance Italians.[15] Nevertheless, a mathematics was built into devices like the camera obscura that, according to Alpers, Dutch artists may have used for projecting images of the landscapes onto a surface that they then traced with meticulous care. If that is so, the geometries of linear perspective were embedded in the instrumentation, even if they were not identified with the essential "reality" of the natural manifold.

The psychologist J. J. Gibson articulates a similar view of the relation between geometrical devices and naturalistic investigations, arguing that an "orthodox theory of perception," dating back at least to Johannes Kepler's optics, presumed a set of relations between an external world and a perceived image that framed modern epistemological discussions and psychological researches:

> The germ of the theory as stated by him was that everything visible radiates, more particularly that every point on a body can emit rays in all directions. An opaque reflecting surface . . . becomes a collection of radiating point sources. If an eye is present, a small cone of diverging rays enters the pupil from each point source and is caused by the lens to converge to another point on the retina. The diverging and converging rays make what is called a *focused pencil* of rays. The dense set of focus points on the retina constitutes the retinal image. There is a one-to-one projective correspondence between radiating points and focus points.[16]

A key feature of this theory of vision is its integration of a particular representational schema, a kind of mathematical analysis, and particular technological designs:

[14] Shapin, "Robert Boyle and mathematics."

[15] Svetlana Alpers, *The Art of Describing: Dutch Art in the 17th Century* (Chicago: University of Chicago Press, 1983).

[16] James J. Gibson, *The Ecological Approach to Visual Perception* (Hillsdale, NJ: Erlbaum, 1986), pp. 58–59. Kepler's was only one of a long line of theories of optics that Edgerton (*The Renaissance Rediscovery of Linear Perspective,* chap. 5) traces back to the early Greeks. The image of the cone of diverging rays goes back to the stoics, and Euclid developed the association between "rays" of light and geometrical lines. Although there were often debates about the direction of the rays (i.e., whether they were projected from the eye outward or the eye received them passively from light reflected from the surfaces of objects), the nature of what they transmitted, and the means by which visual contents were communicated from the eye to the soul, many of the basic elements of the "orthodox theory" were in place well before Kepler.

> This theory of point-to-point correspondence between an object and its image lends itself to mathematical analysis. It can be abstracted to the concepts of projective geometry and can be applied with great success to the design of cameras and projectors, that is, to the making of pictures with light, photography. . . . But this success makes it tempting to believe that the image on the retina falls on a kind of screen and is itself something intended to be looked at, that is, a picture.[17]

Gibson calls this the "little man in the brain" theory of perception and contends that it remains "one of the most seductive fallacies in the history of psychology" (p. 60). Gibson advances his own psychological theory of perception, and he does not offer a full-blown phenomenology of vision, but his arguments enable a transformation of Husserl's critique of naturalistic epistemology into an investigation of the experimental interventions and material props through which a particular view of nature is established and supported. Like Husserl, Gibson points to the praxiological origins of a set of "natural" relations, but he places more emphasis on the perceptual technologies that provide the material examples and experimental tools through which a traditional philosophy of consciousness is established and sustained. Whether traced to the artisan engineers of Renaissance Italy, the invention of the printing press, Galileo's physics, Kepler's optics, or the devices for stabilizing and rectifying the visible relations in a perceptual field, the distinctive achievement of classical physics was to use the idealized forms and calculate resources of geometry to recodify sensual relations, expose magnitudes, and distinguish "primary" from "secondary" qualities within the observable manifold.[18]

Within the traditions of existential phenomenology and phenomenological social science, Husserl's philosophy of consciousness has been rejected in favor of philosophies of action that presume an irreducibly historical and intersubjective foundation for any coherent characterization of the perceptual subject. Jean-Paul Sartre, Martin Heidegger, Maurice Merleau-Ponty, Aron Gurwitsch, and Schutz – all of whom were variously indebted to Husserl's philosophical initiatives – overthrew the transcendental ego while insisting that any determinate characterization of the perceptual acts of an ego necessarily presumes a world that is thick with historical and social relevancies.

These critiques retain Husserl's idea that rules of method and calculative techniques obtain their efficacy and adequacy on the basis of an unexplicated

[17] Gibson, *The Ecological Approach to Visual Perception*, pp. 59–60.

[18] For an illuminating discussion of the distinction between primary and secondary qualities, see P. M. S. Hacker, *Appearance and Reality* (Oxford: Blackwell Publisher, 1987). The distinction is conspicuously featured in the Newtonian conception of color, in which Newton defines color as a secondary effect based on sensory stimulation by colorless "rays" traveling at different speeds and obeying mathematical laws.

foundation in the embodied and socially organized praxis of a discipline, but contrary to Husserl, they no longer treat this "foundation" as a unitary source of intuition and practical certainty. Instead, Husserl's centralized consciousness dissolves into the discursive and embodied activities of situated social praxis, and there is no longer a transcendental ego reposing outside the world to endow it with meaning. The role of the ego is taken over by assemblages of acts situated in discursive and embodied articulations of a world that is always and already shot through with meaning. For ethnomethodologists and other inheritors of the Husserlian problematic, the Husserlian *Lebenswelt* is no longer coordinated with the acts of a transcendental consciousness. It is not founded on a unitary domain of experiential acts but is instead treated as a locally organized order of social activities.

Locally organized activities

The term *local organization* (or local production) enjoys currency in ethnomethodology as well as related areas in the social sciences and philosophy. Unfortunately, to speak of local organization or local production is often understood to imply a kind of nominalism or, worse, a kind of spatial particularism. In ethnomethodology, the adjective *local* has little to do with subjectivity, perspectival viewpoints, particular interests, or small acts in restricted places. Instead, it refers to the heterogeneous grammars of activity through which familiar social objects are constituted. Instead of trying to overcome heterogeneity by theoretically postulating an homogeneous domain (e.g., of panlinguistic dispositions, cognitive structures, *doxa,* or historical discourses), ethnomethodologists attempt to investigate a patchwork of "orderlinesses" without assuming that any single orderly arrangement reflects or exemplifies a determinate set of organizational laws, historical stages, norms, or paradigmatic orders of meaning. They do not deny the historical and social "contexts" in which social action and interaction take place; rather, they insist that specifications of such contexts are invariably bound to a local contexture of relevancies.

The transition from a phenomenological account based on the philosophy of consciousness to an ethnomethodological treatment of the local organization of social activities can be reconstructed by reviewing developments in post-Husserlian phenomenology. For expository purposes, I will draw a line of theoretical development from Gurwitsch through Merleau-Ponty before suggesting a way to supplement ethnomethodology's program with some of Michel Foucault's initiatives. I will try to show how Husserl's phenomenology of "experience" can be progressively transformed into a study of heterogeneous fields of practical actions. This reconstructed line of development is not meant to trace an actual historical lineage from existential phenomenology to ethnomethodology. Although it is clear that Garfinkel was influenced strongly by Aron Gurwitsch's teachings and Alfred Schutz's writings when he wrote his thesis and began to conduct his ethnomethodological studies, only later did he incorporate the antifoundationalist views of scientific praxis attributable to Heidegger, Merleau-Ponty, and in a different way, Wittgenstein. I discuss Wittgenstein in the next chapter, and later in this chapter I discuss Schutz's phenomenological studies and their importance for the development of ethnomethodology.

Contextures of activity

Gurwitsch's discussion of *Gestalt contextures* provides an exceedingly simple, though elegant, way to demonstrate the field of phenomenal relations investigated in ethnomethodological studies.[19] He begins the demonstration with a figure of two points set against a homogeneous background:

• •

Figure 4.2

Gurwitsch observes that when the points are in close proximity to each other, we conventionally see them as members of a pair or "dyad": "In this mode of perception, one does not see one point *plus* the other located a short distance away. Rather one point appears as the *right member,* and the other as the *left member* of the pair" (p. 106, emphasis in original).

Gurwitsch adds that "the interval between the two points presents specific phenomenal features which are altogether absent from that part of the field beyond the points" (p. 106). The interval between the points is "closed" and delimited by the "terminal" points, whereas the field outside "extends indefinitely." An entirely different order of predicates can be assigned to what Gurwitsch calls "a row of pairs."

• •   • •   • •   • •

Figure 4.3

[19] See Aron Gurwitsch, *The Field of Consciousness* (Pittsburgh: Duquesne University Press, 1964), pp. 106 ff. Garfinkel and Wieder credit Gurwitsch with a "radical and seminal" contribution to ethnomethodological research. They add that Gurwitsch's demonstrations of contextures of functional significations in the stream of perception were "among ethnomethodology's earliest appropriations from phenomenological studies." See Harold Garfinkel and D. Lawrence Wieder, "Evidence for locally produced, naturally accountable phenomena of order*, logic, reason, meaning, method, etc., in and as of the essentially unavoidable and irremediable haecceity of immortal ordinary society: IV two incommensurable, asymmetrically alternate technologies of social analysis," pp. 175–206, in G. Watson and R. Seiler, eds., *Text in Context: Contributions to Ethnomethodology* (London: Sage, 1992).

The intervals between the pairs now become significant: "Consequently, if in this case the external intervals are relevant, they have significance for the phenomenal structure of the row as a whole consisting of the groups of its 'natural' parts. As to the groups, the external intervals merely function to the internal structure of the groups" (p. 109).

This simple device serves to demonstrate that spatial predicates such as "next to," "a row," and "left/right member" are called into play in accord with the different groupings, but they are nowhere to be found in an inspection of an isolated point in space. The juxtaposition of elements in the figure constitutes an order of spatial predicates like the interval enclosed by a pair of points and the open space outside that interval. Furthermore, if the points are presented successively, their phenomenal properties can take on temporal characteristics, such as rhythmic patterns, gaps, and interruptions. The spatial predicates elucidated by these demonstrations cohere with one another and with the elements in the figure; they emerge from a *contexture* of mutually supportive details. Gurwitsch argues that the unifying basis for such predicates is neither an essence nor an objective property, as the predicates do not reflect an invariant form or identity "behind" the appearances or "in" the material points, as such: "By implying, modifying, and qualifying each other, the several appearances of a perceived thing are given as coordinated by virtue of their mutual intrinsic reference to one another" (p. 296).

Gurwitsch's demonstration is severely limited. The spatial predicates elucidated by the demonstration have a certain existential flavor, but by holding constant the relations between the text and its embodied reader the demonstration too easily allows us to suppose that we are seeing a set of relations in a disembodied space.

Merleau-Ponty's discussion of embodied spatiality offers an antidote to such an intellectualization of space.[20] His strange inventory of clinical observations on brain-injured and bodily disabled patients and his reviews of experiments in perceptual psychology together provide a comparative basis for explicating the "place" of the body in a phenomenology of perception. His reading of accounts of, for example, the "phantom limb" experienced by amputees and experimental subjects' apprehension of tilted visual fields enables him to specify how the lived body, with its perceptual and motile capacities, constitutively reaches into space-time to establish the terms under which "it" is appropriated: "It is never our objective body that we move, but our phenomenal body, and there is no mystery in that, since our body, as the potentiality of this or that part of the world, surges towards objects to be grasped and perceives them."[21]

[20] Maurice Merleau-Ponty, *Phenomenology of Perception*, trans. Colin Smith (London: Routledge & Kegan Paul, 1962).

[21] Ibid., p. 106.

For Merleau-Ponty, embodied spatiality is not simply a "subjective" gloss describing a transcendental space or a set of indexical descriptors to be negated by generalizing beyond a particular perceptual "viewpoint." Nor is it an "ideal" space emanating from a deep intellectual reserve and imposed on a formless chaos.

> There must be, as Kant conceded, a "motion which generates space" which is our intentional motion, distinct from "motion in space," which is that of things and of our passive body. But there is more to be said: if motion is productive of space, we must rule out the possibility that the body's motility is a mere "instrument" for the constituting consciousness. . . . The "motion which generates space" does not deploy the trajectory from some metaphysical point with no position in the real world, but from a certain here towards a certain yonder, which are necessarily interchangeable. The project towards motion is an act, which means that it traces the spatio-temporal distance by actually covering it.[22]

Movement establishes the predicates under which things take form *in* space. These include the standard modes of orientation, typical facets and fronts, distinguishable surfaces and points of entry, boundaries, and synesthetic ensembles that identify a recognizable object or spatial environment. For Merleau-Ponty, "objective" phenomena are intertwined with the many ways in which things present themselves in accordance with our practical activities. "If bodily space and external space form a practical system, the first being the background against which the object as the goal of our action may stand out or the void in front of which it may *come to light,* it is brought into being, and an analysis of one's own movement should enable us to arrive at a better understanding of it."[23]

From Gurwitsch's demonstration we learn that spatial relations are topically bound in a contexture of elements in a visible field. The pair of points establishes a local spatiality, with its lateral relations of "next-to," "left-right," and "interval." Merleau-Ponty allows us to see that the "field" of spatial relations is constituted in reference to our bodily capacities and practical actions. He draws a contrast between the spatiality of *situation* and the spatiality of *position.* The former is the lived space through which we operate prereflectively, and the latter is what is commonly called *physical* space, a space whose coordinates are abstracted from situated perception.

Merleau-Ponty's definition of embodied action is limited to the primordial possibilities inherent in "naked perception." Although he does not separate the body from the scenic "spatiality of situation" accessed through its orientational modalities, he treats those modalities as a set of "equipmental" relations that come with the territory of the naked subject. Consequently, his

[22] Ibid., p. 387.
[23] Ibid., p. 102.

philosophy remains in the tradition of transcendental phenomenology even while he struggles to replace Husserl's sublime "Ego" with a thoroughly embodied historical subject. Merleau-Ponty's philosophy of the naked subject still glosses over how the perceptible world is itself a historical architecture, constructed for use by contemporaneous subjects and their predecessors. The very psychology experiments that he cites while developing his phenomenology of perception themselves were architectonic arrangements constructed for the purpose of exposing the "perceptual" subject. These arrangements become forgotten items of furniture once the bodily capacities of the subject take on a stable configuration.[24] As Gibson points out, the typical design of the visual psychology experiment circumscribes the spontaneous operations of the body:

> The textbooks and handbooks assume that vision is simplest when the eye is held still, as a camera has to be, so that a picture is formed that can be transmitted to the brain. Vision is studied by first requiring the subject to fixate a point and then exposing momentarily a stimulus or a pattern of stimuli around the fixation point. I call this *snapshot vision.* If the exposure period is made longer, the eye will scan the pattern to which it is exposed, fixating the parts in succession, unless the subject is prohibited from doing so. I call this *aperture vision,* for it is a little like looking at the environment through a knothole in a fence. The investigator assumes that each fixation of the eye is analogous to an exposure of the film in a camera, so that what the brain gets is something like a sequence of snapshots. (p. 1)

The laboratory setup inhibits head and body movement, so that the subject is precluded from using what Gibson calls "ambient" and "ambulatory" vision. These latter concepts include embodied practices of turning an object around in one's hands and walking around in a field to disclose temporal and relational properties of the objects in the field. In other words, "perception" is itself a product of a disciplinary field in which a "subject" is constituted.

As Foucault's many-faceted research demonstrates, the spatiality of situation is subject to various historical–material transformations within a public order of discourse and technology. To account for the transformations of embodied spatiality brought about in technologically (and textually) mediated

[24] Merleau-Ponty does recognize the irreducibly historical foundation of embodied action when in a famous passage, he says, "Because we are in the world, we are *condemned to meaning,* and we cannot do or say anything without its acquiring a name in history" (ibid., p. xix). He further recognizes that the body is not a constant physiological mechanism but that its active and reactive capacities are shaped and defined by its historical situation. Nevertheless, the relevant relationships he discusses are the inseparable nexus of "psychic" and "physiological" potentialities inherent in the lived body of the historical subject, and he does not take account of the housing provided by the architectural and technological complexes in which such bodies are situated.

action, we need to go beyond the perceptual "technology" of the naked subject.[25]

The idea that "readable technologies"[26] extend embodied perception is, of course, a familiar one and dates back at least to Francis Bacon. It is particularly well developed in Michael Polanyi's discussion of the primitive case of the "probe,"[27] in which he describes how the blind man's stick provides a transparent "dwelling" from within which he gains access to *what* he "feels" with the end of the probe. This is more than a matter of saying that the instrument extends the body's perceptual sensitivities, since the blind man's use of the probe transforms the entire nexus of spatiotemporal relations within which he acts. His lived body acquires an "ergonomic" mode of being consonant with the instrument and its competent use. The apprehended order of things does not harken back to a metaphysical ghost in the instrument, because the contexture of relations that the probing brings into play is encountered at, and in terms of, the probed surface it transacts. This complex of instrumental predications – the "here" and the "just this" of the probed surface – characterize an environment, a set of relevant identities and actions, and the terms of a relevant knowledge.

Foucault's regional analyses, although indebted to Merleau-Ponty's researches, are explicitly divorced from the existential–phenomenological tradition.[28] Foucault emphasizes the discontinuities between the spaces brought into play in historically specific discursive formations, and he problematizes any notion of a naked existential grounding of action or perception. He forcibly opposes any inference (such as might be drawn from the Gurwitsch example) that a discursive formation is an organization of thoughts, a network of concepts, or a structuring of experience.

"Panopticism" – epitomized by Jeremy Bentham's plan for an inverted amphitheater through which a central guard station surveys the mass of captive bodies in a penitentiary – is more than a network of words, concepts, and embodied experiences; it is an architecture for systems of activities concordant with (though not strictly determined by) a set of concrete positions and lines of sight, an order of perceptual asymmetries, classificatory *tableaux vivantes,* and hierarchical relations.

> Slowly, in the course of the classical age, we see the construction of those "observatories" of human multiplicity for which the history of science has so

[25] See Dorothy E. Smith, "Textually mediated social organization," *International Social Sciences Journal* 34 (1984): 59–75.

[26] The term *readable technologies* is taken from Patrick Heelan, *Space Perception and the Philosophy of Science* (Berkeley and Los Angeles: University of California Press, 1983). For another account using phenomenology and Foucault to explicate scientific praxis, see Joseph Rouse, *Knowledge and Power: Toward a Political Philosophy of Science* (Ithaca, NY: Cornell University Press, 1987).

[27] Michael Polanyi, *Personal Knowledge* (London: Routledge & Kegan Paul, 1958), p. 59.

[28] See Michel Foucault, *The Archeology of Knowledge,* trans. A. M. Sheridan Smith (New York: Pantheon Books, 1972), and *Discipline and Punish: The Birth of the Prison,* trans. Alan Sheridan (New York: Random House, 1979).

> little good to say. Side by side with the major technology of the telescope, the lens and the light beam, which were an integral part of the new physics and cosmology, there were the minor techniques of multiple and intersecting observations, of eyes that must see without being seen; using techniques of subjection and methods of exploitation, an obscure art of light and the visible was secretly preparing a new knowledge of man.[29]

Foucault's historical studies are relevant to ethnomethodology's investigations in a restricted and "literal" way. Ethnomethodology developed independently of Foucault's research, and there was little commerce between the two lines of postphenomenological research.[30] Ethnomethodologists treat technological complexes not as metaphors for a "dominant discourse" characteristic of an historical *épistême*. Instead, they investigate the varieties of contemporaneous complexes of technology and human actions without linking them to an overall master plan. The massive congruencies among diverse representational modalities, architectures, and regimes that Foucault discusses are simply not validated by ethnomethodology's investigations of the local–historical production of practical actions. Although it might be said that ethnomethodological investigations of contemporary orders of "ordinary" and "professionalized" activities all take place during a "modern" (or "postmodern") epoch, the orders of actions, entitlements, and relational symmetries and asymmetries that these studies describe do not carry over from one coherent language game to another. It makes every difference in the world whether the "game" takes place as part of a family dinner conversation, a diagnostic encounter, or a courtroom tribunal.

Foucault's descriptions nevertheless can be exemplary for ethnomethodological investigations, because they so clearly identify how material architectures, machineries, bodily techniques, and disciplinary routines make up coherent phenomenal fields. Whereas Foucault problematizes the diachronic continuity of historical discourses, ethnomethodology explodes the contemporaneous landscape of language games into distinctive orders of practice, which are neither hermetically sealed from one another nor expressive of a single historical narrative.[31]

[29] Foucault, *Discipline and Punish*, p. 171.

[30] A few ethnomethodologists have used Foucault: Alec McHoul, "The getting of sexuality: Foucault, Garfinkel and the analysis of sexual discourse," *Theory, Culture and Society* 3 (1986): 65–79, and "Why there are no guarantees for interrogators," *Journal of Pragmatics* 11 (1987): 455–71; Michael Lynch, "Discipline and the material form of images: an analysis of scientific visibility," *Social Studies of Science* 15 (1985): 37–66; David Bogen and Michael Lynch, "Taking account of the hostile native: plausible deniability and the production of conventional history at the Iran-contra hearings," *Social Problems* 36 (1989): 197–224; and Lucy Suchman, "Speech act: a counter-revolutionary category," paper presented at a meeting of the American Anthropological Association, Chicago, November 1991.

[31] Although Jean-François Lyotard (*The Postmodern Condition: A Report on Knowledge* [Minneapolis: University of Minnesota Press, 1984]) speaks of "language games" in a very loose way, what he says about the heterogeneity of contemporary language games is instructive. It is possible, however, that the fragmentation that Lyotard attributes to the postmodern condition can be found in any historical period investigated in sufficient detail.

This distinctive picture of sociotechnical fields can be illustrated by sketching a description of freeway traffic (see the Appendix to this chapter). The "world" of freeway traffic may seem to have little relevance to a discussion of scientific practices.[32] However, the example enables us to see that organized assemblages of actions, engineered spaces, equipment, techniques, and "rules of the road" can provide distinctive matrices for the production and recognition of intentions, rights, obligations, courtesies, conventions, violations, and identities. Scientific laboratories, observatories, linear accelerators, mainframe computers, and other equipmental complexes can be treated similarly as matrices for human conduct that do not simply provide places where human beings work but instead provide distinctive phenomenal fields in which organizations of "work" are established and exhibited.[33]

The phenomenon of "observation" in science is particularly sensitive to these considerations. Although observation is often treated as a systematic application of human perceptual capacities, the example of traffic suggests that "observation" in, as well as of, traffic is not simply an equipmentally mediated form of perception and cognition; it is part of an elaborate system of signals, displays, and concerted movements in an archi-textual environment.[34] If, like drivers, laboratory technicians are situated in machinic assemblages and disciplinary labor processes, their actions are not precisely characterized by reference to generic structures of perception and cognition. An individualistic phenomenology, or a cognitive sociology based on a "naked" conceptual apparatus, will not be adequate to the descriptive task, nor will generalized conceptions of power, prestige, and gender do the job,

[32] The term *world* (or *social world*) has become identified with an approach to organizations, occupations, and scientific work associated with the pragmatist tradition in American sociology. A point of convergence between pragmatist and phenomenological research was established when Alfred Schutz appropriated William James's concept of "finite provinces of meaning" and developed his well-known analysis of "multiple realities." In contemporary sociological research, the James–Schutz emphasis on "worlds" in consciousness has been transformed into an emphasis on the organizational production and reproduction of fields of activity, including distinctive equipment, skills, entitlements, identities, and the like. See, for example, Anselm Strauss, "A social worlds perspective," *Studies of Symbolic Interaction* 1 (1978): 119–28; Elihu Gerson, "Scientific work and social worlds," *Knowledge* 4 (1983): 357–77; Adele Clarke, "A social worlds research adventure," pp. 15–42, in S. Cozzens and T. Gieryn, eds., *Theories of Science in Society* (Bloomington: Indiana University Press, 1990).

[33] See Sharon Traweek, *Beam Times and Life Times: The World of High Energy Physics* (Cambridge, MA: Harvard University Press, 1988); and Steven Shapin and Simon Schaffer, *Leviathan and the Air Pump: Hobbes, Boyle, and the Experimental Life* (Princeton, NJ: Princeton University Press, 1985).

[34] Consider the engineers' and other scientists' modes of observing traffic, in contrast with "observation" from within traffic. The helicopter hovering above the traffic jam in its limited way embodies a pragmatic "transcendence" of the lived situation of the drivers "stuck" in the clotted flow of traffic.

although they certainly can be found locally relevant. For this reason, the variant of ethnomethodology that I recommend requires a break from the predominantly cognitive approach to the phenomenology of the social world represented in Alfred Schutz's writings.

Kaufmann, Schutz, and protoethnomethodology

The many failings attributed to Husserl's phenomenological investigations should not diminish the importance of his having raised the topic of the praxiological foundations of science. As I discussed earlier, subsequent developments have displaced Husserl's emphasis on pre-predicative modes of first-person experience and have focused on fields of communicative action and readable technology that cannot be enclosed in structures of individual consciousness. Nevertheless, two initiatives from Husserl remain significant for ethnomethodological studies of science: (1) his claim that an historical–praxiological genealogy of scientific objectivity begins with "ordinary" modes of reckoning and (2) his proposal that the question of how lawlike expressions correspond to objective properties should be addressed by investigating the practical and contextual production of observable phenomena.

Although with hindsight we can see that these Husserlian initiatives were available well before ethnomethodology arrived on the scene, until recently they have not been taken up in ethnomethodological research. In part, this is because ethnomethodology's phenomenological initiatives were initially drawn from Alfred Schutz's writings. Schutz, an Austrian banker and scholar who emigrated to the United States before World War II, transformed Husserl's phenomenology of the life-world into an explicit sociological approach, and his research was immensely important to the early development of ethnomethodology and to the innovative approach to the sociology of knowledge developed by Peter Berger and Thomas Luckmann.[35]

Garfinkel's debt to Schutz is evident in his dissertation, in which he uses Schutz's phenomenology of the social world as a basis for a critical elaboration of Talcott Parsons's theory of social action. Garfinkel's writings in the late 1950s and early 1960s also strongly rely on Schutz. Schutz's influence is especially prominent in Garfinkel's well-known investigation of "trust"[36] and also in a paper, which later appeared as a chapter in *Studies in Ethno-*

[35] P. Berger and T. Luckmann, *The Social Construction of Reality* (Garden City, NY: Doubleday, 1966).

[36] H. Garfinkel, "A conception of, and experiments with 'trust' as a condition of stable concerted actions," pp. 187–238, in O. J. Harvey, ed., *Motivation and Social Interaction* (New York: Ronald Press, 1963).

methodology, on "the rational properties of scientific and common sense activities."[37]

Aaron Cicourel's penetrating critique of methods in sociology[38] and his more recent program of cognitive sociology[39] also draw on Schutzian themes. Although Schutz's research had great influence on early ethnomethodology, in certain respects Schutz de-radicalized Husserl's praxiology of science, and consequently ethnomethodology's Schutzian inheritance was expressed in what now appears to be a particularly "weak" set of proposals about the relationship between "scientific" and "ordinary" practical actions. Sociologists of scientific knowledge have been less indebted to the Schutzian problematic, and their criticisms of Schutz and ethnomethodology provide some leverage for reexamining some of the assumptions about science that remain prevalent in much ethnomethodological research.[40]

Unlike Husserl, Schutz did not write extensively about the natural sciences. For the most part, his references to the natural sciences provided a backdrop for his investigations of practical inquiry in the human sciences. Although, as I argue, Schutz acknowledged that science was a pragmatic activity performed in specific social circumstances, he drew strict demarcations between scientific theory and scientific practice and between scientific and commonsense rationality. Similar demarcations have been attacked by post-Kuhnian philosophers, historians, and sociologists of science but while recognizing this, Schutz's readers should not overlook the historical context of his writings. Like Mannheim, Schutz tried to define a basis for the circumstantial adequacy of allegedly "loose" or uncertain modes of practical understanding. His efforts implicated the adequacy of both the "practical" modes of everyday knowledge studied by sociologists and the interpretive methods used in the human sciences. As I argued in Chapter 2, Mannheim's distinction between the "exact" sciences and "existentially determined" modes of thought was less of an attempt to valorize science than it was to specify distinct modes of practical validation that do not accord with exalted standards of scientific and mathematical proof. Like Mannheim, Schutz did not question the internal rationality of the natural sciences, since he sought mainly to establish a distinctive foundation for other practical and interpre-

[37] Harold Garfinkel, *Studies in Ethnomethodology* (Englewood Cliffs, NJ: Prentice-Hall, 1967), chap. 8, pp. 262–83.

[38] Aaron Cicourel, *Method and Measurement in Sociology* (New York: Free Press, 1964).

[39] Aaron Cicourel, *Cognitive Sociology: Language and Meaning in Social Interaction* (New York: Free Press, 1974).

[40] For a more elaborate discussion of these criticisms of Schutz, see my "Alfred Schutz and the sociology of science," pp. 71–100, in L. Embree, ed., *Worldly Phenomenology: The Continuing Influence of Alfred Schutz on North American Human Science* (Washington, DC: Center for Advanced Research in Phenomenology and University Press of America, 1988).

tive modes of reasoning. But just as Mannheim was criticized for "exempting" the exact sciences and mathematics from the sociology of knowledge's explanatory program, Schutz has been accused of exempting natural science research from a thoroughgoing ethnography of practical actions and practical relations.

Many of Schutz's views on natural and social science inquiry were influenced by the philosophy of social science developed by his close colleague and friend, Felix Kaufmann.[41] Both Schutz and Kaufmann were proponents of Husserl's phenomenological program, but they also adopted aspects of the Vienna circle's philosophical version of a unified science. Kaufmann was an occasional and marginal participant in the Vienna circle's discussions, and although he remained critical of the philosophy promulgated by the followers of Whitehead and Russell, he did not question the legitimacy of their overall aim to articulate a logical basis for the unity of science. Kaufmann's conceptions of language and rule-governed action antedated Wittgenstein's devastating criticisms of the logical positivistic conceptions of language, meaning, and action.[42] Even though Kaufmann did not accept the "correspondence theory" of truth he attributed to Russell, he did accept the overall picture of linguistic representation and rule-governed action that was fundamental to Vienna circle's unity-of-science movement.

For Kaufmann, science is defined by a set of "basic" procedural rules, analogous to the rules that define the pieces, legitimate moves, and goals of a game like chess. He distinguishes these from "preference rules" that define more or less effective moves and strategies in the course of a game. Kaufmann asserts that the "basic elements of empirical procedure are common to pre-scientific and scientific thinking, and there is no sharp line of demarcation between them" (p. 39). These rules pertain to the methods for accepting or rejecting propositions and linking them together in logical arguments. Propositions are "meanings" expressed by sentences that are amenable to judgments about their truth or falsity, or in a more pragmatic view, they are "statements" that can be empirically verified or falsified. Kaufmann summarizes this as follows:

> From the point of view of the logician, the procedure of an empirical science consists in the acceptance or elimination of propositions in accordance with given rules. Whatever else the scientist may do, whether he

[41] Felix Kaufmann, *Methodology of the Social Sciences* (New York: Humanities Press, 1944). Schutz also made critical use of Weber's theoretical writings on social action and rationality, as well as the writings of the American pragmatists.

[42] I am referring here to Wittgenstein's later writings and not to his *Tractatus logico-philosophicus*, which, as Wittgenstein later acknowledged, adopted the "picture theory" of language fundamental to the classic tradition of logic carried forward by Russell, Frege, Whitehead, and participants in the Vienna circle like Reichenbach, Popper, and Carnap.

> looks through microscopes or telescopes, vaccinates guinea pigs, deciphers hieroglyphics, or studies market reports, his activities will result in changing the corpus of his science either by incorporating propositions that did not previously belong to it or by eliminating propositions that previously did. Such a change in the corpus of a science may be called a *scientific decision*. (p. 48)

For Kaufmann, the "corpus" of a science is a hierarchically organized system of propositions accepted by members of a scientific field in accordance with the procedural rules of the discipline. It is a dynamic system, because the constituent propositions are not simply derived through deduction but are subjected to observation and testing (themselves defined in accordance with basic rules of method). This system is unified and yet dynamic in a way analogous to a legal order (p. 45), in which substantive laws and procedural rules can change within the framework of a relatively stable system. Propositions in a corpus can be added, modified, or removed in the course of a discipline's historical development. Such "scientific decisions" do not occur haphazardly, since any proposed change must be justified by reasons acceptable to the disciplinary community in accordance with the existing corpus and the rules of logic.

Kaufmann's emphasis on procedural rules and the accumulation of a corpus of knowledge was congruent with his more general theory of social action. In a way roughly akin to Habermas's more complex theory of communicative action,[43] Kaufmann proposed that actions carried out in accordance with rules were the foundation for organized systems of conduct.

> A norm is a maxim that governs the behavior of the person who seeks to comply with it. However, for the person who appraises human behavior in terms of the norm, it is a criterion for the correctness of this behavior. In other words, it is for him a definition, or part of a definition, of "correct behavior of a particular type." Correct thinking is defined in terms of agreement with the rules of logic, just as correct speech is defined in terms of agreement with the rules of grammar, or legal behavior in terms of agreement with given norms of positive law. (p. 49)

Kaufmann's account of scientific procedure later was appropriated by Schutz and ethnomethodology. Whereas Kaufmann tried to clarify the procedural rationality of science, Schutz and Garfinkel attempted to clarify the "rules of the game" not only for science but for all domains of social action. According to Kaufmann, such clarifications can produce nonobvious knowledge, since, according to the policy of *Docta ignorantia* (p. 15), "one does not 'really' – that is, not quite clearly – know what one knows." Taken-

[43] Jürgen Habermas, *The Theory of Communicative Action*, vol. 1: *Reason and the Rationalization of Society*, trans. Thomas McCarthy (Boston: Beacon Press, 1984).

for-granted assumptions can be brought to light, and ambiguities and conflated usages can be sorted out through such efforts at clarification.

Schutz (and later, Garfinkel, Cicourel, and Sacks) relied on Kaufmann's writings, especially his conception of the structure and development of the "corpus" of knowledge in a scientific discipline. Schutz's writings on the problem of rationality in the social world retained some important features of Kaufmann's picture.[44] Although like Kaufmann, Schutz argued that the human sciences could not proceed by treating social life as a domain of natural entities and forces, he posited an order of procedural rules shared by the methods of natural and social science.[45] Moreover, for Schutz, not only was it the case that the unity of science could be characterized by reference to a corpus of knowledge and a set of procedural rules, but also the ordinary social world "at large" could be characterized by referring to the "stock of knowledge at hand" and a set of cognitive norms for deploying such knowledge in situations of practical action and social interaction. Kaufmann's image of science thus became an image for describing everyday reasoning, just as his conception of a corpus of knowledge and a set of procedural rules for deploying it later became the dominant model of "methodology" for ethnomethodology's initial investigations of commonsense "methods"[46] (and as I argue in Chapter 6, it is a cornerstone of the conversation analytic program).

Schutz's conceptions of the "worlds" of science and everyday life were cast in cognitive terms, as domains of "thought" located in an individual consciousness.[47] Consequently, phenomenological sociology and proto-ethnomethodology did not emphasize the concrete embodiment of local action in the way developed by Merleau-Ponty, Foucault, and other inheritors of the historical–materialist tradition in Continental philosophy. Schutz, of course, paid attention to social interaction and practical engagement in

[44] See A. Schutz, "Common-sense and scientific interpretation of human action," p. 347 of his *Collected Papers 1: The Problem of Social Reality* (The Hague: Nijhoff, 1962); "Concept and theory formation in the social sciences," pp. 48–66 of *Collected Papers 1;* "On multiple realities," pp. 207–59 of *Collected Papers 1;* and "The problem of rationality in the social world," pp. 64–88 of his *Collected Papers 2: Studies in Social Reality* (The Hague: Nijhoff, 1964).

[45] See Schutz, "Common-sense and scientific interpretation of human action," p. 6.

[46] Don Zimmerman and Melvin Pollner use the expression "occasioned corpus" as a synonym for Schutz's stock of knowledge at hand ("The everyday world as a phenomenon," pp. 80–103, in Jack Douglas, ed., *Understanding Everyday Life: Toward the Reconstruction of Sociological Knowledge* [Chicago: Aldine, 1970]).

[47] Schutz borrows his "postulate of adequacy" ("The problem of rationality in the social world," p. 85) from Weber and explicitly retains its reference point of methodological individualism. He formulates this postulate as follows: "Each term used in a scientific system referring to human action must be so constructed that a human act performed within the life-world by an individual actor in the way indicated by the typical construction would be reasonable and understandable for the actor himself, as well as for his fellow-men."

systems of action, but he traced these back to the constitutive center provided by an "ego" situated in an expansive field of associations and consociations.[48] This is especially clear in his distinction between practical action and the social scientist's contemplative "attitude":

> This world is not the theatre of his activities, but the object of his contemplation on which he looks with detached equanimity. As a scientist (not as a human being dealing with science) the observer is essentially solitary. He has no companion, and we can say that he has placed himself outside the social world with its manifold relations and its system of interests. Everyone, to become a social scientist, must make up his mind to put somebody else instead of himself as the center of the world, namely, the observed person. But with the shift in the central point, the whole system has been transformed, and, if I may use this metaphor, all the equations proved as valid in the former system now have to be expressed in terms of the new one. If the social system in question had reached an ideal perfection, it would be possible to establish a universal transformation formula such as Einstein has succeeded in establishing for translating propositions in terms of the Newtonian System of Mechanics into those of the theory of Relativity.
>
> The first and fundamental consequence of this shift in the point of view is that the scientist replaces the human beings he observes as actors on the social stage by puppets created by himself and manipulated by himself. What I call "puppets" corresponds to the technical term "ideal types" which Weber has introduced into social science.[49]

Accordingly, the "scientist" performs a kind of transcendental reduction of the everyday practical attitude in order to construct a simulacrum of the actor's practical orientation. As Habermas observed, the professional "interpreter" becomes a "virtual participant" who acts on "a different plane" than that of the actors in the social field and "pursues goals that are not related to the given context but to *another* system of action."[50] By the same token, the actor becomes a virtual agent whose motives are connected by the professional interpreter to a generalized representation of the "social context."

In contrast with Garfinkel's later discussion of the "cultural dope" of classic social theory, Schutz's conception of the ideal–typical puppet explicitly subscribes to the legitimate grounds for its construction. Although the puppet incorporates only and entirely what the social theorist puts into it, Schutz does not repudiate the project of constructing "personal ideal types";

[48] Ibid., p. 80.

[49] Ibid., p. 81.

[50] Habermas, *The Theory of Communicative Action,* vol. 1, p. 113. Latour and Woolgar's "stranger" discussed in the last chapter is fashioned along the lines of such a "virtual participant" who is in, but not of, the laboratory life world.

instead, he demands that any such type be checked against the "mind of the individual actor" described by it.[51]

Although Garfinkel radically transformed Schutz's cognitive approach, ethnomethodology never entirely discarded some aspects of it. It is important to keep in mind that Kaufmann, Schutz, Garfinkel, Sacks, Cicourel, and their followers developed a general sociology of knowledge that retained Kaufmann's conception of scientific methodology, if only to compare it with "members' methods." In an early paper on the "rationalities" of scientific and commonsense actions, Garfinkel expanded on Schutz's discussions of common sense and scientific rationality by compiling a list of norms of rationality that he separated into those shared by common sense and scientific actions and those distinctive to science. The former include standards and procedures for categorizing and comparing, assessing degree of error, searching for adequate means, devising effective strategy, following procedural rules, and making predictions. The exclusively "scientific" rationalities included the use of principles of formal logic to guide investigation, an orientation to semantic clarity and distinctness for their "own sake," and the use of specifically "scientific" knowledge as a background for judgment.[52] Garfinkel's pluralization of "rationality" and his argument that "commonsense" rationalities were included in the principles and operative practices of science provided a radical antidote to the idea that common sense was a domain of "prescientific" notions.

Sociologists of scientific knowledge have criticized Garfinkel for defining the "scientific" rationalities tautologically (because among other things, he adopts Kaufmann's idea that a corpus of scientific knowledge provides the stock of knowledge that "scientists" take into account) and for presuming that scientists act in accordance with rules of logic when they conduct their experiments.[53] Such critiques tend to miss Garfinkel's point that "the model [of scientific rationality] furnishes a way of stating the ways in which a person would act were he conceived to be acting as an ideal scientist."[54] This is different from saying that scientists actually live up to such ideals.

[51] Schutz ("The problem of rationality," p. 84) raises the question "But why form personal ideal types at all?" Rather than eschew such constructions, he then goes on (p. 85) to formulate a postulate of subjective interpretation that regulates the analytic construction of such ideal types by reference to "what happens in the mind of an individual actor whose act has led to the phenomenon in question." The only alternative Schutz considers is simply to collect empirical facts, and he argues that one cannot do this without taking account of subjective categories.

[52] The latter are taken directly from Schutz's postulates of scientific rationality (ibid., p. 86).

[53] Karin Knorr-Cetina, *The Manufacture of Knowledge: An Essay on the Constructivist and Contextual Nature of Science* (Oxford: Pergamon Press, 1981), p. 21; Bruno Latour and Steve Woolgar, *Laboratory Life: The Social Construction of Scientific Facts* (London: Sage, 1979; 2nd ed., Princeton, NJ: Princeton University Press, 1986), pp. 152–53.

[54] *Studies in Ethnomethodology*, p. 280.

Nevertheless, Garfinkel did not explicitly discount Schutz's proposal that an "attitude of scientific theorizing" defined a cognitive "world" remote from the "world of daily life" (the latter including everyday actions in the laboratory). And although he made clear that everyday "rationalities" were distinctive phenomena in their own right, and not unorganized precursors of scientific rationality, he did not challenge the adequacy of "rules that govern the use of propositions as grounds of further inference and action" within the restricted world of scientific theorizing.[55]

Garfinkel's early studies retained Schutz's and Kaufmann's conception of knowledge as propositional and (pre)suppositional, as well as their view of scientific procedure as an enactment of the procedural rules and norms of correct judgment defined by traditional philosophy of science. Only later did Garfinkel and other ethnomethodologists move away from a view of science and other practical activities as cognitive domains defined by distinctive constellations of norms.[56]

In many of his early investigations, Garfinkel used a heuristic method for disturbing the apparent adequacy and apparent objectivity of particular social activities and social scenes. These interventions included his well-known "breaching" experiments, a set of exercises that he and his students performed for disrupting ordinary scenes. In one case, students pretended to be strangers in their family households. In other exercises they treated customers as salespersons and patrons as waiters or disrupted intimate conversations by asking their partners to explain in detail the commonplace expressions they used. These "experiments" were designed as "aids to a sluggish imagination"[57] rather than as tests of hypotheses, and they resembled practical jokes more than the more familiar variety of social–psychological experiment. Their point was to disclose "seen but unnoticed" background expectancies operating in everyday settings as well as to exhibit the "bewilderment" produced when the subjects were unable to restore the disrupted scenes or flee from them. In addition to deliberately inducing trouble for analytic purposes, Garfinkel used the troubled life-situations of persons like "Agnes," an "intersexed person," to elucidate the practical production and practical management of taken-for-granted social identities. It is commonly believed that these experiments exposed the tacit *rules* or cognitive *norms* operating in everyday scenes. Garfinkel suggested this in his "trust" paper, and more selectively in his book, but he also discounted the

[55] Ibid., p. 281. Garfinkel cites Kaufmann, *Methodology of the Social Sciences*, pp. 48–66.

[56] As Marek Czyzewski points out, Heritage's influential exposition of *Garfinkel and Ethnomethodology* retains the early ethnomethodological emphasis on "cognitive norms." See Marek Czyzewski, "Reflexivity of actors vs. reflexivity of accounts," to appear in *Theory, Culture, and Society* (in press).

[57] Garfinkel, *Studies in Ethnomethodology*, p. 38, attributes the quoted phrase to Herbert Spiegelberg.

unilateral (or "bicameral") cognitivism implied by such analytic expressions as "background expectancies," "common understandings," and "commonsense knowledge of social structures."[58] Garfinkel's discussion of indexicality and reflexivity in the first chapter of *Studies* and his subsequent paper written with Sacks ("On Formal Structures of Practical Action") evidenced a move away from the earlier emphasis on norms.[59] As I discuss in Chapter 5, the later studies began to develop a more radically situated version of how rules and other formal expressions come into play in the course of embodied action.

In the past two decades, ethnomethodologists and conversation analysts have not made extensive use of Schutz's writings. It may be fair to say that Garfinkel's and Cicourel's reliance on Schutz was part of a "protoethnomethodological" development that has been superseded by contemporary research. To bury Schutz in history, however, would be to dishonor the memory of his achievements, and just as significantly it would fail to bring under review those aspects of the Schutzian conception of science that continue to be presumed in much ethnomethodological research. This applies particularly to the still-prevalent conception that ethnomethodology is a research program for describing the "rules" of everyday action and social interaction. It also applies to the implication many ethnomethodologists take from Schutz, that academic "analysis" can somehow be separated from the social involvements, local judgments, and embodied actions that ethnomethodologists study. All too often, Garfinkel's policy of ethnomethodological indifference and his related distinction between "topic and resource" are taken to imply that ethnomethodology can remain aloof from "merely practical" concerns.

Ethnomethodological indifference

Garfinkel coined the expression *ethnomethodological indifference* to distinguish ethnomethodology's approach from the project of analytic sociology:

[58] Garfinkel's studies were never congruent with "cognitive science" as it is currently conceived. Although he does speak of "background expectancies" and "common understandings" in his discussion of "Studies of the routine grounds of everyday activities" (pp. 35–75 of *Studies in Ethnomethodology*), he demonstrates how these are intertwined with the "scenic" features of commonplace settings rather than being founded in a normative or cognitive space. See Jeff Coulter, "Cognition in the ethnomethodological mode," pp. 176–95, in G. Button, ed., *Ethnomethodology and the Human Sciences* (Cambridge University Press, 1991).

[59] The fact that *Studies in Ethnomethodology* includes essays written at different times can lead to a confusing impression that the different chapters express a coherent research program rather than a series of efforts to come to terms with what ethnomethodology might be about.

> Ethnomethodological studies of formal structures . . . [seek] to describe members' accounts of formal structures wherever and by whomever they are done, while abstaining from all judgements of their adequacy, value, importance, necessity, practicality, success, or consequentiality. We refer to this procedural policy as "ethnomethodological indifference" . . . our "indifference" is to the whole of practical sociological reasoning, and *that* reasoning involves for us, in whatever form of development, with whatever error or adequacy, in whatever forms, inseparably and unavoidably, the mastery of natural language. Professional sociological reasoning is in no way singled out as a phenomenon for our research attention. Persons doing ethnomethodological studies can "care" no more or less about professional sociological reasoning than they can "care" about the practices of legal reasoning, conversational reasoning, divinational reasoning, psychiatric reasoning, and the rest.[60]

Rather than addressing whether sociologists ever can achieve adequate or acceptable accounts of the phenomena they study, the policy of indifference opens up the alternative topic of how members conduct their "methodological" activities by pragmatically establishing what counts as adequacy, accuracy, and appropriateness. Sociologists' methodological troubles and remedies are thus placed in a vast field of practical activities in which methods are generated and used.

The policy of indifference not only applies to questions about the "ultimate" validity and reliability of sociologists' descriptions, explanations, and measurements; it also covers Schutz's normative proposals concerning the "special" character of scientific cognition, including his theoretical contrast between the natural and social sciences. Since indifference is not equivalent to denial or opposition, the policy does not imply that social scientists' methods have a "merely" commonsense basis. What other kind of basis could they have? Nor does it imply that there are no distinctions to be drawn among sociologists', coroners', physicists', or any other lay or professional methods. Rather, it states that any such distinction is contingent, locally organized, and in a peculiar way discoverable.

Garfinkel's study of "following coding instructions" provides an early case in point.[61] Coding is often a preliminary step in quantifying social science data. In Garfinkel's study of selection criteria in a psychiatric outpatient clinic, two sociology graduate students were given the task of coding standardized information from a large collection of case folders. Each folder contained a "clinic career form" on which clinic personnel were

[60] Garfinkel and Sacks, "On formal structures of practical action," pp. 345–46. For a more recent account, see Benetta Jules-Rosette, "Conversation avec Harold Garfinkel," *Sociétés: revue des sciences humaines et sociales* 1 (1985): 35–39.
[61] Garfinkel, *Studies in Ethnomethodology*, pp. 18 ff.

supposed to record information about their initial contact with the patient, to specify any tests they administered and treatments they recommended, and to note when the case was "terminated." The two research assistants were then given a set of instructions for extracting standardized information from the folders and recording it on a "coding sheet," and a reliability procedure was used to assess the amount of agreement between the coders' judgments. Rather than simply relying on the coders' training and skills to produce adequate data for the study, Garfinkel (p. 20) investigated how the coders managed to accomplish this mundane research task.

> A procedure was designed that yielded conventional reliability information so that the original interests of the study were preserved. At the same time the procedure permitted the study of how any amount of agreement or disagreement had been produced by the actual ways that the two coders had gone about treating folder contents as answers to the questions formulated by the Coding Sheet. But, instead of assuming that coders, proceeding in whatever ways they did, might have been in error, in greater or lesser amount, the assumption was made that *whatever* they did could be counted correct procedure in *some* coding "game." The question was, what were these "games"?

To answer this question, Garfinkel formulated a list of "*ad hoc* considerations" the coders used to decide "the fit between what could be read from the clinic folders and what the coder inserted into the coding sheet" (p. 21). He designated these with a short list of rhetorical terms, including "et cetera," "unless," "let it pass," and "factum valet" (an action that is otherwise prohibited is counted as correct once it is done). The coders used these practices to assess the substantive and "reasonable" fit between the folder contents and the categories on the coding sheet, without getting bogged down by a "literal" assessment of what was or was not in any folder. That is, the coders relied on what they "knew" about the clinic and the staff, including the exigencies of patient presentation and clinic record keeping, to discern what each folder "said" in more than so many words. Their competence thus presupposed an understanding of the state of affairs that the coding sheet categories formulated; indeed, what they recorded on any coding sheet was essentially tied to what they "knew" the respective clinic folder must have contained in addition to and despite its literal contents.

As Garfinkel points out (pp. 21–22), such reliance on ad hoc practices is the very sort of "commonsense" practice that sociological methods seek to replace with disinterested and objectively defensible judgments. He adds, however, that every attempt to upgrade the coding procedure in order to restrain or eliminate such practices itself relied on and reproduced them.

Later in his discussion (pp. 66 ff.) Garfinkel makes a more general point about the relation between "common understandings" and social science "models of man." He clearly does not offer methodological advice for evaluation researchers or survey analysts, since he explicitly argues that ad hoc considerations are an irremediable part of routine social science research practices, as well as more "ordinary" modes of practical sociological reasoning. He argues instead that coding instructions, along with the ad hoc practices used in following them, "furnish a 'social science' way of talking so as to persuade consensus and action within the practical circumstances of the clinic's organized daily activities, a grasp of which members are expected to have as a matter of course" (p. 24). This does not imply that social science discourse amounts to nothing more than a fancied-up version of common sense; instead, it recommends a *constitutive* rather than a descriptive characterization of the relationship between sociological methods and the social activities studied.

Although Garfinkel's discussion of coding practices may raise questions about the validity and reliability of the relation between clinic folders and coding sheets, its main objective is to describe how that very relationship – along with the methodological considerations that accompany it – is an aggregate product of coders' ad hoc procedures for handling the singular contents of their "raw data."

At first glance, the policy of ethnomethodological indifference suggests little more than a reorientation to the detailed practices through which methods are employed in sociology and other fields of practical action. Presumably such a program of study could coexist with sociology and perhaps be of some technical use for sociologists' continuing efforts to improve on their methodologies. To leave it at that, however, would be to miss a more subversive implication of the policy. Rather than confronting sociological methods in terms of an immanent concern with validity and reliability, ethnomethodological indifference turns away from the *foundationalist* approach to methodology that gives rise to principled discussions of validity, reliability, rules of evidence, and decision criteria. The implications of this move can be threatening and even incomprehensible to sociologists.

A vivid indication of sociologists' consternation over questions of "method" in ethnomethodology can be found in the transcribed dialogue between ethnomethodologists and sociologists presented in the *Proceedings of the Purdue Symposium on Ethnomethodology*.[62] The sociologists' questions and complaints about "method" persistently punctuate the dialogue. While Garfinkel, Sacks, and others present a series of examples and demonstrations

[62] Richard J. Hill and Kathleen Stones Crittenden, eds., *Proceedings of the Purdue Symposium on Ethnomethodology* (Purdue, IN: Institute for the Study of Social Change, Department of Sociology, Purdue University, 1968).

of ethnomethodological studies, the sociologists withhold judgment while waiting for an a priori warrant, decision rule, or criterion of correctness, relevance, or acceptability:

Hill: Hal [Garfinkel], you have not told us yet what rules of evidence you accept or employ. (p. 27)

Hill: You have to be able to tell us how you make such distinctions in terms of decision rules. I believe this illustrates a kind of question many of us have with regard to how one presents a warrant for the evidence that one uses to reach a decision. (p. 28)

DeFleur: . . . What are the rules by which you unravel who is right? We have been asking for methodological information and you have been giving us subject matter. A moment ago, Hal said, "Well, we don't have any new science up our sleeve." How about some old science? (p. 39)

DeFleur: How do you reject a thing? What are the rules of evidence on which you reject or accept an explanation? (p. 40)

Questions like these persist throughout the symposium and are never satisfactorily resolved in their own terms. The questions presuppose a set of methodological standards independent of what happens to be under investigation, and they imply that a description or demonstration cannot be sensible or plausible until compared with such standards. In effect, the sociologists immobilize the dialogue by deferring acknowledgment of the sense and intelligibility of the ethnomethodologists' descriptions until given a set of general methodological assurances. The sociologists demand extrinsic criteria of truth and intelligibility before accepting, or even "hearing," what the ethnomethodologists tell them. Their questions and complaints embody Lord Kelvin's memorialized dictum "If you cannot measure, your knowledge is meager and unsatisfactory," which in this instance might be translated into "If you cannot tell us what rules of evidence and decision criteria you respect, your claims are unfounded." By demanding such epistemic guarantees, the sociologists become apt targets for a distinctive form of rejoinder (p. 34):

McGinnis: What criteria would you accept as grounds for arguing that it [a conversationalists' rule for identifying persons, which Sacks had just discussed] is false? What criteria would you require from me to assess my assertion that your claim is false?
Garfinkel: Why don't you just state your objection?

Garfinkel's rejoinder casts McGinnis's academic question into a "vulgar" conversational frame. McGinnis's question proposes that a particular observation about an everyday phenomenon should be tested in reference to a criterion of falsification. By sequentially treating the question as a roundabout "objection," Garfinkel's rejoinder cuts through McGinnis's hypothetical voice, disregarding the question's deference to a criterion that would justify subsequent belief and implicating it as an allusion to what McGinnis already is prepared to argue with or without "criteria." Garfinkel's "vulgar" move situates his interlocutor in a conversational competency that requires no extrinsic criterion.

The entire edifice of "method" is thus challenged, not through an explicit argument, but in the way it is submerged into a "vulgar" competency. This is ethnomethodology's indifference; it is a move that simply leaves the scene of sociology's methodist discourse. Such a move does not leave "knowledge" behind, nor does it place ethnomethodology in a realm without sense or reason; instead, the move pivots on the demonstrable fact that both McGinnis and Garfinkel are acting, have acted, and will continue to act, with no time out, in a dialogue that is already intelligible, mutually recognizable, and characterizable. It is not that Garfinkel's "professional expertise" as an ethnomethodologist enables him to recognize what McGinnis has just said and to identify it as a "preobjection"; rather, his rejoinder acts contentiously within while making an issue of the ordinary grounds of McGinnis's demand for a criterion. McGinnis's privileging of criteria is cast into ironic relief as an "academic" posture in an unremitting and already intelligible conversation. A related implication is produced by Sacks's response to a similar demand by a sociologist:

> Hill: . . . Could you tell us without reference to the subject matter what the structure of [an ethnomethodological] demonstration would be?
>
> Sacks: Do you know what that is asking? You are asking, "Could you tell me, without knowing what kind of world we are in, what a theory would look like?" . . . I do not know in the first instance what it is that sociology should look like to be satisfactory. That is not an available phenomenon.[63]

Sacks's reply undercuts Hill's distinction between "method" and "subject matter" and places sociology in the substantive field investigated. He is not advocating an inductive procedure; rather, he is questioning Hill's programmatic separation between a unitary method of scientific inquiry and the particular subject matter investigated in any science. Sacks's refusal to go

[63] Ibid., p. 41.

along with this picture implicates an alternative view of science, in which "methods" are put into distinctive constellations of activity, equipment, investigative sites, and investigated phenomena.[64] In 1968 this was a radical view of method for a sociologist to espouse, and although it is now familiar to students of the sociology of scientific knowledge, it has yet to be included in textbook accounts of sociological method.

The policy of indifference takes questions of method off the table, except insofar as they provide the subject matter for ethnomethodological studies. The methods used by persons who call themselves ethnomethodologists are implicated in what hey say about diverse lay and professional practices, but these methods are not placed under a distinct heading of "scientific methodology." "Methods" (whether avowedly scientific or not) do not provide a priori guarantees, and the initial requirement for an ethnomethodological investigator is to find ways to elucidate methods from within the relevant competence systems to which they are bound.

Topic and resource

The policy of ethnomethodological indifference is sometimes summarized by placing ethnomethodology's analytic concerns entirely outside the field of lay and professional sociology. This is expressed in theoretical discussions of ethnomethodology by proposals to the effect that investigators should not confuse topic and resource when conducting studies of practical action.[65] According to this policy, classic methods for discerning structures of social action should be (re)formulated as members' practices to be studied. Richard Hilbert, for instance, argues: "While members may view such constructions [of structure] as objectively 'out there' and invoke them in explanations,

[64] This rebuttal anticipates the sort of "anarchistic" attack on the unity of scientific method launched by Paul Feyerabend in *Against Method* (London: New Left Books, 1975). Whether Sacks consistently espoused such a view of science is another matter (see Chapter 6).

[65] The "confusion" is said to consist of a variety of procedures through which an "unexamined" reservoir of tacit knowledge deriving from ordinary judgments about everyday phenomena is allowed to inform sociological analysis. This is more than a matter of the analysts' reliance on unexamined personal knowledge because the "confusion" is procedural as well as cognitive; it includes the way that questionnaire respondents are invited to employ their "natural theorizing" to categorize, assess, predict, and estimate matters of social fact (e.g., when ranking occupational categories on sociological prestige scales, entering "father's occupation" on a questionnaire form, and supplying lexical descriptors for kinship relations). It also applies to methods of coding and analyzing aggregate responses. Sociologists worry and argue about such matters, and their worries and arguments make sense only by reference to the possibility of what Cicourel calls "literal description." But once it is recognized that literal description is impossible, it follows that there can be no secure methodological position from which "methods" can be deployed that are purified of all connection to subject matter.

sociologists cannot so orient themselves without 'going native' and reifying social structure."[66]

Ethnomethodologists, in contrast, are said to study the "members' methods" or "ethnomethods" through which structures are produced and reproduced, but they are not supposed to employ conceptions of social structure as explanatory resources. In terms reminiscent of Husserl's transcendental reduction, ethnomethodologists are encouraged to "bracket" or suspend naturalistic beliefs in "the stuff with which structural studies are normally concerned, such things as institutions, classes, organizations (on the macro end) and persons, individuals, subjective content, interaction processes and patterns (on the micro end)."[67] Instead of "reifying" structures or "going native," ethnomethodologists are enjoined to investigate the lay and professional methods by which structures are constructed.

This understanding of the aim and task of analysis is also prominent in constructivist programs in the study of social problems, in which the sociologist's analytic task of reconstructing social problems discourse is distinguished from the naturalistic claims and counterclaims made by participants in that discourse.[68] Accordingly, to study social problems in order to "solve" them or to aid the cause of one or another faction in the popular disputes concerning them is to slip out of the "analytic" perspective.

Nothing is more emblematic of what I have called protoethnomethodology than such a conception of the topic/resource distinction. This conception follows directly from Schutz's definition of the "world of both the natural and the social scientist" as a "world of thought" that differs radically from "the world within which we act and within which we are born and die."[69] Protoethnomethodology is not only an historical precursor to ethnomethodology – a residue of transcendental phenomenology expressed in the policies of a "radical" research program – it includes a persistent tendency in contemporary ethnomethodological research to define predominantly in cognitive terms the current situation of inquiry for investigators and participants alike. Protoethnomethodology pauses at the threshold of ethnomethodology, and perhaps it can be said that no self-avowed ethnomethodologist can avoid it, just as no "deconstructionist" can entirely avoid the aporias of classical philosophy. Simply put, this threshold is constituted by an understanding that there can be no intelligible theoretical

[66] Richard Hilbert, "Ethnomethodology and the micro–macro order," *American Sociological Review* 55 (1990): 794–808. The best-known exposition on the "topic–resource" distinction is by Zimmerman and Pollner, "The everyday world as a phenomenon."

[67] Hilbert, "Ethnomethodology and the micro–macro order," p. 796.

[68] For instance, Peter Ibarra and John Kitsuse, "Vernacular constituents of moral discourse: an interactionist proposal for the study of social problems," in G. Miller and J. Holstein, eds., *Reconsidering Social Constructionism* (Hawthorne, NY: Aldine de Guyter, in press).

[69] Schutz, "The problem of rationality in the social world," p. 88.

position "outside" the fields of practical action studied in sociology. Although this is an easy phrase to memorize and repeat, it expresses a lesson that is exceedingly difficult to take to heart. Indeed, the lesson is continually subverted by one after another move into transcendental analysis. Often, and even characteristically, the lesson is subverted in the service of implementing it.

The lesson is suggested in the opening lines of Garfinkel and Sacks's paper: "The fact that natural language serves persons doing sociology, laymen or professionals, as circumstances, as topics, and as resources of their inquiries furnishes to the technology of their inquiries and to their practical sociological reasoning *its* circumstances, *its* topics, and *its* resources."[70] Far from recommending a transcendental reduction or a similarly heroic cognitive maneuver, Garfinkel and Sacks can be read as saying that there can be no "methodological" transcendence from the fields of language and practical action inhabited by sociologists and "members" alike.[71] Rather than concerning themselves with the structure (or structures) of social action or, by the same token, altogether abandoning the question of structure, Garfinkel and Sacks address what Derrida at one time called "the structurality of structure," and by doing so they "displace" the relevance of structural description and explanation in the affairs of the human sciences.[72] Such a displacement is not, and cannot be in its own terms, a matter of "stepping back" from the fields of action studied. It is not a transcendence.

To make a topic of the structurality of structure (or of the reflexivity of inquiry and the indexical properties of analytic language) may seem to divest an "investigator" of even the most elementary resources for "taking account" of human actions. But this problem can arise only when one presupposes the possibility of a position or standpoint outside the *topos* constituting the structurality (or structuration) of structure. As Derrida warns those who would attack Western metaphysics as though from a position outside its history:

> *There is no sense* in doing without the concepts of metaphysics in order to attack metaphysics. We have no language – no syntax and no lexicon – which is alien to this history [of metaphysics]; we cannot utter a single destructive

[70] Garfinkel and Sacks, "On formal structures of practical actions," p. 337.

[71] Other uses of this distinction perhaps imply the possibility of analytical transcendence. Note the use of "only and exclusively" in the following passage: "The 'rediscovery' of common sense is possible perhaps because professional sociologists, like members, have had too much to do with common sense knowledge of social structures as both a topic and resource for their inquiries and not enough to do with it only and exclusively as sociology's programmatic topic" (Garfinkel, *Studies in Ethnomethodology*, p. 75).

[72] Jacques Derrida, "Structure, sign, and play in the discourse of the human sciences," pp. 247–72, in R. Macksey and E. Donato, eds., *The Structuralist Controversy: The Languages of Criticism and the Sciences of Man* (Baltimore: Johns Hopkins University Press, 1970).

proposition which has not already slipped into the form, the logic, and the implicit postulations of precisely what it seeks to contest.[73]

Accordingly, we "confuse" topic and resource every time we speak; we "reify" structure every time we write; and we "go native" every time we act. To insist that all sociological methods are substantive "ethnomethods" may seem to imply the strongest and most comprehensive of analytic positions, and yet at the same time it is the weakest, the most marginal, and the most negative position imaginable in the discourse of a human science. Just as the strength of the strong program in the sociology of knowledge resides in its dubious presumption of an ability to give sociological explanations of actions in all the other sciences, the apparent strength of ethnomethodology inheres in a presumption to comprehend the objectifying practices accomplished by all manner of sociological "natives." This position weakens at the height of its strength, however, as soon as the elementary lesson is driven home: There can be no intelligible theoretical position outside the fields of practical action studied in sociology.

The confusing and even paradoxical implications of the topic/resource distinction can be sorted out, first, by recalling that ethnomethodological indifference does not imply that members' methods invariably lack precision, efficacy, rigor, and predictability. Consequently, the ethnomethodological claim that sociology relies on commonsense methods in its day-to-day procedures does not necessarily carry critical implications. Such critical implications would make sense only if we retain the Schutzian contrast between the attitude of daily life and the attitude of scientific theorizing (which, as Schutz defines it, oddly resembles the "attitude" implied by Husserl's transcendental reduction) or if we take seriously Cicourel's rhetorical contrast between actual methods in the social sciences and the "straw" position of literal description.[74] These contrasts are akin to Lévi-Strauss's distinction between the *bricoleur* and the engineer,[75] a distinction that has been used (and undermined) by sociologists of science and ethnomethodologists who have described the *bricolage* of laboratory shop practice.[76]

The *bricoleur* is a jack-of-all-trades who adapts "the means at hand" – a collection of tools, scraps of material, and heterogeneous skills – in trial-and-error fashion to contend with the contingencies arising in an open series of applications. In Schutz's terms, the *bricoleur* uses "cookbook knowledge," a kind of "know-how" using approximate and typical relations, judgments of likelihood, and relatively free substitutions of ingredients and

[73] Ibid., p. 250.
[74] Cicourel, *Method and Measurement in Sociology,* p. 2.
[75] Claude Lévi-Strauss, *The Savage Mind* (Chicago: University of Chicago Press, 1966).
[76] Knorr-Cetina, *The Manufacture of Knowledge;* Garfinkel, Lynch, and Livingston, "The work of a discovering science."

materials.[77] Lévi-Strauss contrasts the *bricoleur* with the engineer, whose tools and skills are dedicated precisely to specific projects of action, in means–end fashion. But as Derrida points out, when it is realized that the engineer "is a myth," the contrast between *bricoleur* and engineer ultimately collapses:

> The notion of the engineer who had supposedly broken with all forms of *bricolage* is therefore a theological idea; and since Lévi-Strauss tells us elsewhere that *bricolage* is mythopoetic, the odds are that the engineer is a myth produced by the *bricoleur.* From the moment that we cease to believe in such an engineer and in a discourse breaking with the received historical discourse, as soon as it is admitted that every finite discourse is bound by a certain *bricolage* and that the engineer and the scientist are also species of *bricoleurs* then the very idea of *bricolage* is menaced and the difference in which it took on its meaning decomposes.[78]

Schutz's distinctions between ordinary and scientific methods, and ethnomethodology's much abused programmatic contrasts between objective and indexical expressions and reflexive and unreflexive accounts, are similarly "menaced." Following Derrida, the ideals of objectivity and scientific method can be identified as mythopoetic constructs made in the service of actions that in their own domain are ordinary. Again, this does not imply that methods are necessarily faulty or that it makes no sense to speak of objective states of affairs. Although no transcendental grounding may warrant its efficacy and certainty for all time, nothing precludes scientific conduct from being orderly, stable, reproducible, reliable, and ordinary.

To say that sociology takes ordinary methods for granted carries no necessary critical implication for the orderliness, stability, reproducibility, and reliability of sociological studies and findings. Sociological topics and findings are, of course, often accompanied by practical uncertainties, endless disputes about methodological considerations, and political controversies. In the course of such disputes, the multitudes of ordinary judgments incorporated into any study's indices and interpretive procedures are selectively brought under critical scrutiny. On the other hand, the most reliable sociological knowledge is, almost by definition, "trivial": widely distributed, widely understood, and taken for granted by professional sociologists as well as by those whom they study.

Although the Derridian "menace" leaves everything as it was, it cuts very

[77] Schutz, "The problem of rationality in the social world," p. 73.

[78] Derrida, "Structure, sign, and play," p. 256. The emphasis on tinkering, negotiation, contingencies, and so on in laboratory studies gives detailed support to Derrida's assertion about the "myth" of the engineer (or, likewise, of a "purely" rational–purposive scientific method). A key point that Derrida makes, however, is not only that Lévi-Strauss's portrayal of the engineer is idealized but also that it is generated by the *bricolage* practices of writing. In Garfinkel's terms, this writing uses a "documentary method" through which an account of the *bricoleur*/engineer distinction is constructed and used.

deeply. Thus far, I have drawn a contrast between protoethnomethodology and ethnomethodology, but readers may fairly ask at this point, Where are we to find this ethnomethodology of which you speak? I have located it with an understanding that there can be no intelligible theoretical position outside the fields of practical action studied in sociology. And yet the entire literature in ethnomethodology implicates such an "outside." Can we find any studies in that literature that do not describe members' practical actions, instances of indexical expressions, and idealizations of the natural attitude as though there might be an alternative mode of action, expression, or attitude? Schutz, of course, explicitly defines the "attitude of scientific theorizing" as such an alternative. But if an authentic "ethnomethodology" begins with a denial of the outside and a corresponding affirmation of the "irremediably" indexical and reflexive properties of language use and practical action, its logical conclusion would be the annihilation of the professional form of life in which "studies" can be produced, published, read, compared, and organized into a coherent literature. Ethnomethodology's ground zero would thus be an affirmation of the organized and intelligible character of a social world untouched by academic hands.

Beyond protoethnomethodology?

Given the deep angst raised by the idea that the very distinctions most often associated with the academic field of ethnomethodology are "menaced" by its primary lesson, it is understandable that protoethnomethodology necessarily provides the default form of the discipline. It does so by establishing a zone of professional "everydayness" in which respectable academic work can be produced and presented. A "reasonable" ethnomethodology – that is, a recognizable program of study comfortably situated in the social sciences – can be sustained only by erecting a classic set of distinctions as though they framed a viable "position" in a set of academic disputes.

Given the primary lesson from ethnomethodology, the most crucial of these distinctions is also the most contradictory: the distinction between "professional analysis" and "members' methods." According to this distinction, "vernacular" or "commonsense" accounts (whether attributed to professional scientists, lay persons, or both) are placed on one side of a divide, and (ethnomethodological) "analyses" are opposed to them. At times, vernacular accounts are characterized as though they made up a dim version of "positivism" or "naive realism."[79] A variant of *Docta ignorantia* then comes into play: The "actor" does not become

[79] Melvin Pollner (*Mundane Reason* [Cambridge University Press, 1987]) speaks of "positivistic common sense," and James Holstein and Gale Miller ("Rethinking victimization: an interactional approach to victimology," *Symbolic Interaction* 13 [1990]: 103–22) characterize an "everyday life" orientation to "a reality that is objectively 'out there,' existing apart from the acts of observation and description through which it is known" (p. 104). This imputation of a coherent philosophical view is similar to the tendency in the sociology of scientific knowledge to attribute a "positivistic" or "realist" orientation to "the scientist" as

the infamous "cultural dope" but instead becomes a philosophically naive agent who takes for granted a "mundane world" that analysis recasts into a product of taken-for-granted "social" practices.

Not only does this distinction create endless work for analysts, it also creates a coherent stance in the ubiquitous realist–constructivist debate. "Realist" opponents can now be accused of taking for granted an unanalyzed member's sense of the objective facticity of social structure. The ethnomethodological analyst then shows that this sense of facticity is interactively constructed and retained: It is a "reality" that is "talked into being"[80] or constituted through "mundane reason."[81] Analysis undermines the claim that members (whether sociologists or lay actors) are merely reporting about an objective reality. Consequently, social, rhetorical, and interactional agencies – constructive practices and ethnomethods – occupy the grammatical role of ideas in an idealist rebuttal to realism.

What I have called protoethnomethodology is very inclusive. If it is a failing, it is one that is always immanent in the attempt to do ethnomethodological studies. Even some of the most often cited writings by Garfinkel and Sacks do not consistently move "beyond" protoethnomethodology. Although Garfinkel and Sacks attacked the idea of transcendental analysis very early in the game, they did not discard that idea simply by foreswearing a programmatic objective and asserting a rejection of a "correspondence theory" of representation. In numerous ethnomethodological studies, the topic/resource distinction has become an effective rhetorical device when presented to readers who presume that sociology ought not to be contaminated by its "subject matter."

Even though ethnomethodologists disavow any interest in purging their analyses of the "methods" they study, this disavowal itself can defuse the critical import of saying that sociologists confuse topic and resource. The distinction cannot be maintained once we allow that every move to distinguish analytic from vernacular methods invariably borrows from the immanent sensibility of ordinary language. To envision a postanalytic ethnomethodology in which professional analysis no longer provides a stable resource for superseding the limitations attributed to its vernacular counterpart is to do something akin to contemplating suicide, since it calls into question the rhetorical scaffolding that has enabled ethnomethodology to gain a tenuous foothold in the social sciences. In lieu of suicide, perhaps "therapy" can help us discover a way to emancipate analysis from its opposition to common sense. In the next chapter I examine the possibility that Wittgenstein's later philosophy offers just the kind of therapy we need at this point.

well as the philosopher of science. See D. Bogen and M. Lynch, "Do we need a general theory of social problems?" pp. 213-37, in G. Miller and J. Holstein, eds., *Reconsidering Social Constructionism* (Hawthorne, NY: Aldine de Gruyter, 1993).

[80] John Heritage, *Garfinkel and Ethnomethodology* (Oxford: Polity Press, 1984), p. 290.

[81] Pollner, *Mundane Reason*.

Appendix: The linear society of traffic

The following description intends to demonstrate what an account might look like that treats "methods" as substantive actions situated in distinctive constellations of activity, equipment, investigative sites, and investigated phenomena. Rather than choosing an example from science or mathematics, I will describe a domain of action that is disgustingly familiar to most of us, and so there should be no need for a preliminary tutorial to enlist readers' recognitions of the assemblage of technological devices, built environments, observable events, and communicative actions that make up a distinctive field of action. Obviously, there are endless differences between, for example, "navigating" in an electron-microscopic field and driving in traffic, but a description of traffic will serve my purpose of laying out an exemplary description of local actions in a distinctive historical–material context. Moreover, it is a domain of action that was frequently addressed by Garfinkel and his students, and so I can develop previous work on the topic.[82]

Traffic is perhaps the most public of spaces on the modern urban landscape: Persons whose lives otherwise are segregated encounter one another in a public space, and when acting within that space they trust (while sometimes testing) one another's competency and accountability. Driving is a game in which instant and violent death can result from a lapse in mutual attentiveness, a misbegotten gesture, or any of a variety of asymmetries in the communicational order.

Despite its public and highly organized character, surprisingly little has been written about the social system of traffic. Sociologists and social psychologists occasionally discuss it as a domain of application for concepts drawn from studies of face-to-face interaction, and the automobile has received some attention as a technology whose use and symbolic value has transformed preexisting forms of community.[83] Erving Goffman formulated a concept of "vehicular unit" that applies specifically to traffic, but his

[82] Garfinkel often used the example of traffic in lectures I attended in the Department of Sociology at UCLA in 1973–78 and 1980–82. He occasionally assigned observational exercises for students on the topic and in discussions with me referred to unpublished papers and personal communications by his students Chris Pack, Stacy Burns, and Britt Robillard. Garfinkel has expounded on the example in numerous public lectures, including a seminar given at Boston University in December 1989 entitled "The curious seriousness of professional sociology." He also gives a brief account of traffic in a draft of his paper "Two incommensurable, asymmetrically alternate technologies of social analysis" (Department of Sociology, UCLA, 1990), an early draft of Garfinkel and Wieder, "Evidence for locally produced, naturally accountable phenomena of order." My discussion of traffic in the following section is indebted to Garfinkel's many discussions of the topic, although I would not want it to be read as an exposition of "Garfinkel's views" or as a report of an ethnomethodological investigation. For the most part it is a loosely constructed example, and among other things, it develops "Foucaultian" themes that do not derive from Garfinkel's treatment.

[83] See, for instance, the classic "community study" by Robert and Helen Lynd (*Middletown: A Study in Modern American Culture* [New York: Harcourt Brace, 1929], pp. 251 ff.), in which they discuss the introduction of the automobile to small-town life.

generalized treatment of the concept does little to illuminate the singular orderliness of driving in traffic.[84] It is perhaps difficult to conceive of traffic as anything more than a simplified communicational environment in which interactional relations are reduced to fleeting and impersonal modes of "contact" (or, better, avoidance of contact), and the highly stereotypical forms of signaling may be too obvious to be worth serious analytic attention by students of human communication. Despite the relative lack of sociological interest in traffic, the immense literatures on traffic engineering and accident research can be reread as accounts of how vehicular units and their built environments produce a distinctive social space.[85]

For ethnomethodology, traffic is an example of social order sui generis, a perspicuous instance of a Durkheimian "social fact." As highway engineers recognize, freeway traffic is a standardized, predictable, and repetitive order of things, and its order is independent of the particular cohorts of drivers whose actions compose it. From the "panoptic" point of view of a helicopter hovering above a jammed freeway, the traffic might as well be a physical system, and in Foucault's terms, its order is inscribed within a coherent, centrally designed, and materially constrictive geometry that both enables and restricts the cellular movements of the constituent vehicles. From the engineering point of view, traffic occurs within a semiotically dense field of signals and relays, which enable a regulated "flow" of vehicles and facilitate numerous points of surveillance.

Garfinkel performs something of a gestalt switch on the engineering account by raising the question of how the social facts of traffic are recognized by drivers situated in mobile platoons of vehicles. Consequently, the traffic is no longer a text to be read panoptically from "outside" (or from "above"); it is a field within which the cellular units together achieve an intelligible order.[86] Garfinkel points to the significance of "gaps" between the car in front and the car behind as situationally organized phenomena.

In contrast with the "interval" between the pair of dots in Gurwitsch's demonstration (Figures 4.2 and 4.3), these "gaps" between cars are temporally composed and modified by a complex assemblage of drivers' actions mediated by and expressed in the traffic. Recognizable social relations are established in reference to gaps between cars, as each driver adjusts to the relative speeds, temporal relations of leading and following, and the common forward-moving directionality of the local traffic.

[84] Erving Goffman, *Relations in Public* (New York: Harper & Row, 1971).

[85] This possibility of rereading the engineering literature was suggested by Harold Garfinkel (personal communication).

[86] This gestalt switch can also be appreciated by comparing Foucault's accounts of the prison and asylum with Goffman's descriptions of inmate life in the "total institution"; see *Asylums* (Garden City, NY: Doubleday, 1961). Although Goffman's account may presuppose the asylum's historical design, it describes an array of practices and strategies through which inmates maintain activities that elude, resist, subvert, and remain indifferent to institutional surveillance.

The topological order of parallel lanes, gaps, directionality, and speed is strikingly linear, although a driver's view of this topology is analogous to that of the fantastic inhabitants of Edwin Abbott's "Flatland" and "Lineland."[87] The topological field is not viewed panoptically from above but is "lived in" from the standpoint (or "movepoint") of mobile inhabitants on a flat surface. The spatial restriction of the field is especially evident when driving at night. In the surrounding darkness, the visibility of the vehicular *umwelt* is largely confined to a frontal and rearward view of a stream of lights, in which headlights illuminate a linear foreground, mirrors enable a limited rearward vision, and the fore-and-aft placement of lights and signals provides for the visible presence of intentional actions within the line of traffic. This topological linearity is, needless to say, concrete, as it provides a systematically designed environment that circumscribes the orderliness of embodied actions taking place in traffic.

The social category of "driver" is topically bound[88] to actions within this built environment, in which space is already fashioned for use, in which language is already inscribed on and in the field in an ubiquitous and impersonal way, and in which a world that is mediated by technology is known in no less immediate a fashion than is any other experiential lifeworld.[89] A "driver" is not a ghost in the machine, since his or her very identity is topically bound to the linear society of traffic. For Merleau-Ponty's naked embodied actor, we can substitute "driver in traffic," an agent who perceives and acts through the medium of a vehicle and whose "intentional" actions are temporally disclosed by movements and positional relations in an engineered environment constructed for drivers' use.This concrete environment is also a graphic "text," as it is composed of a grid of lines and intersections along with inscribed notations and directional signs, all of which encompass and inform an actively developing field of other vehicles. It is a world thick with signs and signals, which are placed and formatted in standardized reference to a linear flow of readers situated in the traffic. Drivers know how fast they are going not only by reading

[87] Edwin Abbott, *Flatland: A Romance of Many Dimensions* (New York: Dover, 1952).

[88] Category-bound activities are discussed by Harvey Sacks in his "On the analysability of stories by children," pp. 216–32, in R. Turner, ed., *Ethnomethodology* (Harmondsworth: Penguin Books, 1974), p. 225; also see Jeff Coulter, *Mind in Action* (Oxford: Polity Press, 1989), p. 39.

[89] Bruno Latour (*Science in Action* [1987, Cambridge, MA: Harvard University Press, p. 254]) provides a lucid example of reading a map in which both map and engineered environment are constituted as texts. "When we use a map we rarely compare what is written on the map with the landscape . . . we most often *compare* the readings on the map with the road *signs* written in the *same* language. The outside world is fit for an application of the map only when all its relevant features have been written and marked by beacons, landmarks, boards, arrows, street names and so on."

speedometers but also by observing the local context of the traffic as it moves along the lanes of a linear society.[90]

Perceptual space for the driver is neither a variation on nor an extension of the phenomenological space of a naked subject. For one thing, it is a space meticulously engineered for an aggregation of standardized vehicular units. The singularity of the driver's perceptual space is defined by a place in the traffic more than by any invariant perceptual capacities of the naked subject. It is circumscribed by a vehicular enclosure with a "forward" orientation supplemented by "reflective" rearward vision. Headlights, signals, and the simple sets of codes for using them are embedded in the linear matrix of traffic. Although the driver's body in the machine is still in play – as both an actual and presumed agent and a source of significations – its actions in the field of traffic are circumscribed by the conventional modes of perception, gesture, and communication within a speeding and nomadic assemblage of vehicular units.

The world of perception and action for drivers in traffic may seem to be relatively impoverished, but it is not as impoverished as one might imagine. The flux of events in traffic makes possible different orders of visibility and varying modes of bodily expression. Without actually seeing anything more than another car's swerve in traffic or its relative articulation of speed and following distance, a driver can compose remarkably precise complaints about the "kind of guy" (or, stereotypically, "the woman") another driver is.

Moreover, a simple movement, position, or communicational gesture in traffic can take on a precise specificity in relation to the flow of scenic events. So, for instance, a honking horn can be heard variously as a "greeting" to another car or a pedestrian, a "sexist come-on," an insult, a complaint, a warning, or (in New Delhi) an audible marker of a vehicle's presence in relation to its immediate neighbors. In each instance, the mere blast of the horn can be assigned an elaborate intentional structure ("Who is he honking at? Is it me? What did I do?"), and it can be absorbed into a complex set of responsive actions. The horn's enunciation can be poetically and intonationally modulated by an articulation of relative amplitude, duration, repetition, and pace, as well as by "conversational" relations to the beeps, honks, and other gestures in the local environment.[91]

[90] Harvey Sacks points out ("On members' measurement systems," edited by G. Jefferson from unpublished transcribed lectures, in *Research on Language and Social Interaction* 22: [1988/89]: 45–60) that driving "with the traffic" provides a kind of metrical basis for determining relative speed. Perhaps facetiously, Sacks notes that many drivers drive as though they were intentionally trying to constitute the background against which fast or slow driving becomes visible.

[91] The example of honking has also been elaborated in unpublished work by Stacy Burns (Garfinkel, personal communication).

The coherence of traffic provides a basis for various "measures" of social potency or power that do not readily transfer to other spaces. So, for instance, the concerted relations that become visible in a temporary platoon of cars speeding down a freeway provide a metrical basis for distinguishing acts of getting ahead, passing, catching up, and impeding mobility. Drivers' epithets can associate these expressions with vocabularies of motive, competitive significances, and typified properties of car and driver ("Get over to the right, slowpoke!"). Mutually visible positional relations in traffic can quickly escalate into explicit games, races, offenses given and offenses taken, and other fleeting or relatively extended modes of encounter.

Although generic vocabularies of social action certainly apply to traffic, they apply distinctively. For example, egoistic expressions of "power" in traffic cannot readily be cashed into other sociotechnical currencies. Accordingly, it would be dubious to treat the social order of traffic as a projection of a general set of power relations. At the same time, however, the structures of sensibility in the economy of traffic are not derived from the possibilities of action ushering from a naked (or "free") subject, and they are not derived from a simple formula for translating "ordinary" modes of action and communication into drivers' modes. Consequently, there is a specificity to the accountable configurations of action in traffic that requires a localized articulation of a historical–materialist understanding.

CHAPTER 5

Wittgenstein, rules, and epistemology's topics

Perhaps the distinguishing mark of recent studies in the sociology of knowledge is their attempt to transform the traditional concerns of epistemology into topics for empirical investigation. Although proponents of the new sociology of scientific knowledge do not follow a single research program, many of them express interest in the philosophy of science. David Bloor paraphrases Wittgenstein by saying that sociology of scientific knowledge is "heir to the field that used to be called philosophy," and he and Barry Barnes propose treating the "contents" of scientific knowledge as an appropriate topic for sociological investigation.

Some sociologists of knowledge use established philosophical positions as springboards for sociological research. Harry Collins, for instance, undertakes what he calls an "empirical relativist programme,"[1] and Karin Knorr-Cetina suggests an empirical sociology supporting a constructivist philosophy of science.[2] Followers of Anselm Strauss's "social worlds" approach, like Elihu Gerson, Susan Leigh Star, Adele Clark, and Joan Fujimura, use ethnographic and historical research to develop some of the epistemic initiatives raised by the American pragmatists.[3] Bruno Latour and Michel Callon go somewhat further with their "actor-network" approach, deconstructing many of the basic conceptual distinctions in sociology and philosophy and placing those distinctions in a unique ontology in which human and nonhuman agencies emerge from a primordial semiotic ooze.[4] Other scholars, like Michael Mulkay, embrace phenomenological and liter-

[1] H. M. Collins, "An empirical relativist programme in the sociology of scientific knowledge," pp. 83–113, in K. Knorr-Cetina and M. Mulkay, eds., *Science Observed: Perspectives on the Social Study of Science* (London: Sage, 1983).

[2] Karin Knorr-Cetina, "The ethnographic study of scientific work: towards a constructivist interpretation of science," pp. 115–40, in Knorr-Cetina and Mulkay, eds., *Science Observed.*

[3] Susan Leigh Star, "Simplification in scientific work," *Social Studies of Science* 13 (1983): 205–28; Elihu Gerson and Susan Leigh Star, "Representation and rerepresentation in scientific work," unpublished paper, Tremont Research Institute, San Francisco, 1987; Adele Clark, "Controversy and the development of reproductive science," *Social Problems* 36 (1990):18–37; Joan Fujimura, "Constructing 'do-able' problems in cancer research: articulating alignment," *Social Studies of Science* 17 (1987): 257–93.

[4] B. Latour, *Science in Action* (Cambridge, MA: Harvard University Press, 1987); M. Callon, "Some elements of a sociology of translation: domestication of the scallops and the fishermen in St. Brieuc bay," pp. 196–223, in John Law, ed., *Power, Action, and Belief: A New Sociology of Knowledge?* (London: Routledge & Kegan Paul, 1986).

ary theoretical initiatives by insisting that epistemological and ontological positions should be treated as discursive registers that come into play in the scientific fields that sociologists of knowledge study.[5]

The more radical approaches in the sociology of science do not simply use philosophy as a general source of assumptions and conceptual themes to be "fleshed out" by empirical study. Instead, they intend to rewrite the philosophy of science in accordance with historical and ethnographic investigations of actual cases.[6] This has led to a sustained and sometimes lively engagement with the philosophy of science. Although Bloor, Barnes, and Collins temper their relativistic proposals with firm commitments to empirical social science, their studies are often treated by philosophers of science as relativistic attacks on the naturalistic and logical foundations of scientific inquiry.

As far as many critics are concerned, the familiar argument that epistemological relativism becomes absurd when turned on itself applies no less strongly to the cultural and historical relativism promoted by sociologists of scientific knowledge.[7] Such criticisms can be justified to some extent by reference to an immanent movement in the sociology of science itself, Steve Woolgar's and Malcolm Ashmore's "reflexive" examinations of the literary rhetoric and empiricist claims in the sociology of science.[8] Their studies demonstrate that proposals for a naturalistic sociology and history of science

[5] Michael Mulkay, *The Word and the World: Explorations in the Form of Sociological Analysis* (London: Allen & Unwin, 1985).

[6] Sociologists were not the first to propose such a sociohistorical turn in the philosophy of science. Kuhn's *Structure of Scientific Revolutions* opened the door, and some philosophers of science who variously aim to preserve a general and normative (if not a foundationalist) philosophy of science also advocate such a turn. See, for instance, Gerald Doppelt, "Kuhn's epistemological relativism: an interpretation and defense," *Inquiry* 21 (1978): 33–86; Larry Laudan, *Progress and Its Problems: Towards a Theory of Scientific Growth* (Berkeley and Los Angeles: University of California Press, 1977); David Stump, "Fallibilism, naturalism and the traditional requirements for knowledge," *Studies in History and Philosophy of Science* 22 (1991): 451–69.

[7] See, for instance, Allan Franklin, *Experiment Right or Wrong* (Cambridge University Press, 1990); Laudan, *Progress and Its Problems,* chap. 7, "Rationality and the sociology of knowledge," pp. 196–222.

[8] Steve Woolgar, ed., *Knowledge and Reflexivity: New Frontiers in the Sociology of Knowledge* (London: Sage, 1988); Malcolm Ashmore, *A Question of Reflexivity: Wrighting the Sociology of Scientific Knowledge* (Chicago: University of Chicago Press, 1989). In a criticism of Woolgar, Collins and Yearley argue that the "reflexive" program in the sociology of science is inherently conservative because it threatens to undermine the empirical support for a social constructivist antidote to the reigning mythologies of positive science. See H. Collins and S. Yearley, "Epistemological chicken," pp. 301-26, in Andrew Pickering, ed., *Science as Practice and Culture* (Chicago: University of Chicago Press, 1992). Although it may be unfair to brand Woolgar's arguments as inherently conservative, they can be appropriated by critics who do not share Woolgar's commitments. Franklin (*Experiment Right or Wrong,* p. 163), for instance, cites Woolgar's argument in "Interests and explanations in the social study of science" (*Social Studies of Science* 11 [1981]: 365–94) while defending the rationality (or, in Franklin's somewhat weaker formulation, the reasonability) of experimental practices against the claims of the strong programme in the sociology of knowledge.

are no less subject to skeptical treatment than are the objective claims of the natural scientists they study. Given the lack of consensus in sociology on the most fundamental theoretical and methodological questions, the programmatic initiatives and explanatory claims in the sociology of scientific knowledge provide especially ripe targets for skeptical criticism.

Such criticisms recall a familiar question that has dogged the sociology of knowledge from its inception: How can a program of explanation that undermines the "internal" rationality and naturalistic supports for other systems of knowledge prevent others from doing the same to its own claims? As discussed in Chapter 2, Mannheim addressed this question by arguing that the unique historical and institutional situation of the sociology of knowledge gave it pragmatic independence from the more familiar ideological positions in religion, politics, and the human sciences.

In their proposals for a "strong program," Bloor and Barnes take a somewhat different tack by attempting to supplement Mannheim's method of demonstration with explanatory strategies that do not necessarily undermine "internal" commitments to the truth and justifiability of particular scientific theories and experimental results (see Chapters 2 and 3). According to their arguments, the fact that sociology of knowledge explanations can be given for even the most elementary propositions in arithmetic does not imply that those propositions are somehow mistaken or arbitrary. Consequently, a reflexive application of the sociology of knowledge would not necessarily show that its own modes of explanation are unfounded, and it could even be used to suggest analogies between the sociology of science and other "strong" modes of argument in the sciences and mathematics. The question comes down to whether "reflexivity" implies skepticism, and more generally whether sociology of knowledge explanations necessarily imply a skeptical regard for the "beliefs" explained.[9]

Wittgenstein and rule skepticism

As described in Chapter 2, Bloor uses Wittgenstein's writings on mathematics to strengthen Mannheim's program. Barnes, Collins, Trevor Pinch, Woolgar, and other sociologists of scientific knowledge also cite Wittgenstein as a key figure in philosophy who initiated a "sociological turn" by showing that the compulsive force of logical and mathematical rules is inseparable from a communal consensus on how such rules are to be applied in particular circumstances of action.[10] In line with their use of the Duhem – Quine

[9] See Wes Sharrock and Bob Anderson, "Epistemology: professional scepticism," pp. 51–76, in G. Button, ed., *Ethnomethodology and the Human Sciences* (Cambridge University Press, 1991).

[10] Barry Barnes, *Scientific Knowledge and Sociological Theory* (London: Routledge & Kegan

underdetermination thesis, these sociologists invariably treat Wittgenstein's writings about rules in mathematics to be pointing beyond philosophy to what Bloor calls a "social theory of knowledge," an essentially sociological account of how stable knowledge is possible.[11]

In this chapter I examine how Bloor and other sociologists of knowledge read Wittgenstein, and I argue that like Saul Kripke, they interpret Wittgenstein to be raising a skeptical challenge and advancing a skeptical solution to the problem of how rules determine actions.[12] Kripke's interpretation of Wittgenstein has been disputed in Wittgensteinian circles, and some of the rebuttals to Kripke apply to Bloor's and other sociologists' skeptical arguments. Contrary to many sociologists of scientific knowledge, I argue that Wittgenstein's discussion of actions in accord with rules can be read as a rejection of epistemological skepticism.

I contend further that an antiskepticist reading of Wittgenstein is compatible with an alternative ethnomethodological program for studying the reflexive relations between rules and practical actions, a version of reflexivity that differs significantly from the theme of self-reflection incorporated into the program for studying reflexivity espoused by Woolgar and Ashmore. Like the sociologists of scientific knowledge, ethnomethodologists try to transform the traditional themes in epistemology into topics for empirical research. But instead of advocating a "sociological turn" in which philosophy's problems are given sociological explanations, ethnomethodologists initiate a "praxiological turn" through which they turn the sociological aim to explain social facts into a situated phenomenon to be described. Sociology's loss becomes society's accomplishment. This "praxiological turn" has far-reaching implications that I hope to spell out in Chapters 6 and 7.

Ethnomethodology and the sociology of scientific knowledge investigate such traditional epistemological topics as representation, observation, experimentation, measurement, and logical determinacy, and proponents of both approaches believe that Wittgenstein's philosophy supports their appropriation of epistemology's topics. As Barnes observes, "There are interesting parallels between [ethnomethodology and the strong program], which derive from their reliance on the late work of Ludwig Wittgenstein."[13] Proponents of ethnomethodology and sociology of science are less concerned about deliv-

Paul, 1974), pp. 163–64, n. 17; Steve Woolgar, *Science: The Very Idea* (Chichester: Ellis Horwood; London: Tavistock, 1988), p. 45; Harry M. Collins, *Changing Order: Replication and Induction in Scientific Practice* (London: Sage, 1985), pp. 12 ff.

[11] David Bloor, *Wittgenstein: A Social Theory of Knowledge* (New York: Columbia University Press, 1983).

[12] Saul Kripke, *Wittgenstein on Rules and Private Language* (Cambridge, MA: Harvard University Press, 1982).

[13] Barry Barnes, *Interests and the Growth of Knowledge* (London: Routledge & Kegan Paul, 1977), p. 24.

ering "faithful" readings of Wittgenstein's texts, as their main interest is to exploit the Wittgensteinian corpus, along with any other suggestive materials, to inspire and guide empirical research.[14]

Despite this common interest in Wittgenstein, sociologists of science and ethnomethodologists develop sharply different readings of his later writings,[15] and their differences recall a familiar debate in philosophy on Wittgenstein's discussion of rules and conduct. Some of Wittgenstein's interpreters read him to be saying that orderly actions are determined not by rules but by social conventions and learned dispositions that circumvent a potential interpretive regress. Others hold that he treats rules inseparably from practical conduct and that his writings give little support for sociological, conventionalist, and related forms of explanation. When the various empirical programs in social studies of science are read in light of these philosophical arguments, they implicate entirely different views of what is "empirical" and how to study it. Whereas sociologists of knowledge give a skepticist reading of Wittgenstein, ethnomethodologists – contrary to what is often said about their program – develop a nonskepticist but not a realist or rationalist extension of Wittgenstein. Although both can cite Wittgenstein's writings to support their positions, the problem for the sociology of scientific knowledge is not that Wittgenstein's writings suggest a path out of philosophy and into sociology. As Peter Winch argues, Wittgenstein problematizes the very possibility of giving general social explanations of epistemologically relevant matters.[16]

Wittgenstein is by no means the only significant philosopher for social studies of science, but he is widely regarded as the pivotal figure for a "sociological turn" in epistemology. Bloor's *Wittgenstein: A Social Theory*

[14] Garfinkel explicitly renounces any attempt to tag ethnomethodology to philosophical predecessors, although he has suggested a practice of "ethnomethodologically misreading" the philosophers. His preference is to "misread" Husserl, Merleau-Ponty, and Heidegger, and unlike Sharrock, Anderson, and Coulter, he has been less explicit about possible resonances with Wittgenstein. The point here is not to show that ethnomethodology is best regarded as an offshoot of Wittgenstein's philosophy but to bring out some strong arguments from Wittgensteinian philosophy in support of research policies in ethnomethodology. To do this is not to imply that those research policies developed in an effort to "follow" Wittgenstein.

[15] Not all ethnomethodologists take the same line on Wittgenstein. Although I propose to speak on behalf of ethnomethodological studies of work, references to Wittgenstein in Garfinkel's and Livingston's writings are scant. In addition, I now view my own discussion of Wittgenstein in my *Art and Artifact in Laboratory Science: A Study of Shop Work and Shop Talk in a Research Laboratory* (London: Routledge & Kegan Paul, 1985), pp. 179 ff., to be inadequate. The view I am currently espousing is most clearly argued by W. W. Sharrock and R. J. Anderson, "The Wittgenstein connection," *Human Studies* 7 (1984): 375–86; R. J. Anderson, J. A. Hughes, and W. W. Sharrock, "Some initial difficulties with the sociology of knowledge: a preliminary examination of 'the strong programme,'" *Manchester Polytechnic Occasional Papers*, no. 1, 1987; Jeff Coulter, *Mind in Action* (Oxford: Polity Press, 1989), pp. 30 ff.

[16] Peter Winch, *The Idea of a Social Science and Its Relation to Philosophy* (London: Routledge & Kegan Paul, 1958; Second Edition, Atlantic Highlands, NJ: Humanities Press, 1990).

of Knowledge has been the most influential treatment of Wittgenstein's later work in social studies of science and mathematics.[17] Wittgenstein's influence is also filtered through many of the "Kuhnian" themes, such as "seeing-as," "incommensurability," and "paradigms," so often discussed in sociology of science. An indication of Wittgenstein's importance is the fact that the concepts of "forms of life," "language games," and "family resemblances" have become common currency in the social studies of science literature, often without much attention to how Wittgenstein used them.

Bloor's central proposal is that Wittgenstein is a pivotal figure in the transformation of epistemology's topics into a set of empirical problems for social science research. Although Wittgenstein made no mention of Durkheim's sociology and explicitly distinguished his approach from behaviorism,[18] Bloor argues that in certain respects Wittgenstein's treatment is compatible with those programs in empirical social science. When faced with glaring discrepancies between Wittgenstein's and Durkheim's writings, Bloor resolves these by repudiating some of Wittgenstein's central proposals.[19]

Bloor makes clear that he is trying to supplement Wittgenstein with an empirical program, and he is willing to creatively misread Wittgenstein to suit this purpose. I have no objection to this, since there is no reason that fidelity to a particular philosophical tradition should sidetrack an attempt to do original sociological research.[20] As Richard Rorty states, there may be no end to efforts to represent correctly the "thought" of a complex body of writings like Wittgenstein's.[21] A creative misreading may serve better to carry forward the conversation on the questions Wittgenstein raises. Unfortunately, Bloor goes well beyond this, since he also claims that sociological research is necessary in order to replace Wittgenstein's "fictitious natural

[17] D. Bloor, *Wittgenstein: A Social Theory of Knowledge* (New York: Columbia University Press, 1983). Other extensive treatments include Derek Phillips, *Wittgenstein and Scientific Knowledge: A Sociological Perspective* (London: Macmillan, 1977); Coulter, *Mind in Action,* chap. 2; Collins, *Changing Order,* chap. 1; H. M. Collins, *Artificial Experts: Social Knowledge and Intelligent Machines* (Cambridge, MA: MIT Press, 1990), chaps. 2 and 7; and Trevor Pinch, *Confronting Nature: The Sociology of Solar Neutrino Detection* (Dordrecht: Reidel, 1986).

[18] Wittgenstein, *Philosophical Investigations,* pp. 307–08; C. G. Luckhardt, "Wittgenstein and behaviorism," *Synthese* 56 (1983): 319–38.; J. F. M. Hunter, *Understanding Wittgenstein: Studies of Philosophical Investigations* (Edinburgh: University of Edinburgh Press, 1985).

[19] Bloor accounts for how Wittgenstein seemed so little inclined to embrace behaviorism or Durkheimian sociology (or any other empirical social science of his day) by suggesting that Wittgenstein's antiscientific predilections (perhaps reflecting Spenglerian influences) blinded him to the natural affinities between his account of language and research in the behavioral sciences.

[20] Ian Hacking makes a similar observation in his review of Bloor's book; see "Wittgenstein rules," *Social Studies of Science* 14 (1984): 469–76.

[21] Richard Rorty, *Philosophy and the Mirror of Nature* (Princeton, NJ: Princeton University Press, 1979).

history with a real natural history, and an imaginary ethnography with a real ethnography."[22] This realist proposal treats Wittgenstein's writings as speculations in need of empirical grounding or correction, and it is entirely out of line with Wittgenstein's repudiation of theory and empiricism in favor of "grammatical" investigations.[23] Wittgenstein's writings no doubt serve to inspire Bloor, even if they do not authorize his project, but they can also be turned against many of his programmatic claims.

As I showed in Chapter 3, Bloor's four-point proposal for a strong program in the sociology of knowledge has influenced a large body of research in the social history of science and has provided a target for numerous criticisms.[24] Bloor's causalist assumptions are not widely accepted in social studies of science, but many sociologists who do not agree with them share his skeptical posture regarding scientists' and mathematicians' truth claims. In calling this a "skeptical" posture, I am not saying that Bloor advocates disbelief in scientists' theories and mathematicians' proofs. In line with Mannheim's "nonevaluative general total conception of ideology," Bloor's "symmetry" and "impartiality" postulates require only that all theories, proofs, or facts be treated as beliefs to be explained by social causes. Bloor's skepticism is primarily methodological, as his aim is to relativize the immanent rationality of what he calls "scientific beliefs" in order to set up a social or conventionalist explanation of science and mathematics. Although it has certainly been successful as a sociological research strategy, a similar skepticist posture has attracted a great deal of criticism among Wittgensteinian philosophers.

[22] Bloor, *Wittgenstein: A Social Theory of Knowledge,* p. 5.

[23] Sharrock and Anderson, "The Wittgenstein connection," argue that Bloor's proposals for an empirical science take the immediate form of a philosophical treatise. Although Bloor cites and summarizes numerous historical studies and suggests what an empirical treatment might consist of, his argument is on the face of it programmatic. Eric Livingston makes a similar point about Bloor's writings: "What Bloor seems to mean by claiming that the sociological investigation of 'scientific knowledge' should follow the canons of scientific procedure is that one should adopt a way of speaking that conforms to current, popular, philosophical theories." See Eric Livingston, "Answers to field examination questions in the field of sociology, philosophy, and history of science," unpublished transcript, circulated in Department of Sociology, UCLA, 1979, pp. 15–16. It is therefore appropriate to treat Bloor's arguments as philosophical exercises rather than as a substantive social theory to be evaluated on empirical grounds. I say this not to demean those arguments but to engage them for what they are.

[24] These critiques include Larry Laudan, "The pseudo-science of science?" *Philosophy of the Social Sciences* 11 (1981): 173–98; Stephen Turner, "Interpretive charity, Durkheim, and the 'strong programme' in the sociology of knowledge," *Philosophy of the Social Sciences* 11 (1981): 231–44; Steve Woolgar, "Interests and explanations in the social study of science," *Social Studies of Science* 11 (1981): 365–94; Anderson et al., "Some initial difficulties with the sociology of knowledge"; and Coulter, *Mind in Action.* The collection edited by M. Hollis and S. Lukes, *Rationality and Relativism* (London: Routledge & Kegan Paul, 1982), includes several papers arguing the pros and cons of the approach. See Chapter 3 for further discussion of these critiques.

Rules, actions, and skepticism

In his essay, *Wittgenstein on Rules and Private Language,* Saul Kripke reviews Wittgenstein's discussion of following rules. He regards Wittgenstein as advancing a novel solution to the classic skeptical problem of how rules determine actions. In Kripke's view, Wittgenstein initially accepts the skepticist thesis that actions are underdetermined by rules but then gives a social constructivist solution to the problem of how orderly conduct is possible. Kripke is not the only philosopher to attribute skeptical and conventionalist views to Wittgenstein,[25] but his essay provoked especially heated criticism in Wittgensteinian circles.[26] Wittgenstein discusses rules in several other manuscripts and collections of notes,[27] but the dispute between Kripke and his critics mainly concerns Sections 143 through 242 of the *Philosophical Investigations (Pl),* which presents the famous example of continuing the number series (2, 4, 6, 8 . . .).

As is typical of Wittgenstein's later writings, numerous threads of argument weave through the text, along with a series of partly overlapping or analogous examples. Questions are posed and seemingly left hanging, and it is sometimes difficult to keep track of when Wittgenstein is asserting his own views and when he is speaking in the voice of an interlocutor. In spite of, or perhaps because of, its difficulty, the argument has been reconstructed in numerous secondary and tertiary sources.

As I understand it, the argument runs as follows: Wittgenstein (*PI,* sec. 143) devises a "language game" in which a teacher asks a pupil to write down a series of cardinal numbers according to a certain formation rule. It is clear from the discussion that this language game and its imaginary pitfalls are to be understood as a paradigm for actions in accord with rules, not only in arithmetic, but also in other rule-ordered activities like playing chess and

[25] See Michael Dummett, "Wittgenstein's philosophy of mathematics," pp. 420–47, in G. Pitcher, ed., *Wittgenstein: The Philosophical Investigations* (Notre Dame, IN: University of Notre Dame Press, 1968); and more ambiguously, Stanley Cavell, *The Claim of Reason: Wittgenstein, Scepticism, Morality, and Tragedy* (Oxford: Oxford University Press, 1979).

[26] G. P. Baker and P. M. S. Hacker, *Scepticism, Rules and Language* (Oxford: Blackwell Publisher, 1984); G. P. Baker and P. M. S. Hacker, *Wittgenstein, Rules, Grammar and Necessity. Vol. 2 of an Analytical Commentary on the Philosophical Investigations* (Oxford: Blackwell Publisher, 1985); Oswald Hanfling, "Was Wittgenstein a sceptic?" *Philosophical Investigations* 8 (1985): 1–16; S. G. Shanker, *Wittgenstein and the Turning-Point in the Philosophy of Mathematics* (Albany, NY: State University of New York Press, 1987).

[27] See especially Wittgenstein, *Remarks on the Foundation of Mathematics,* ed. and trans. G. E. M. Anscombe (Oxford: Blackwell Publisher, 1956), *Zettel,* ed. G. E. M. Anscombe and G. H. von Wright (Oxford: Blackwell Publisher, 1967), and also the collection of lecture notes on mathematics edited by Cora Diamond, *Wittgenstein's Lectures on the Foundations of Mathematics* (Ithaca, NY: Cornell University Press, 1976). Norman Malcolm ("Wittgenstein on language and rules," *Philosophy* 64 [1989]: 5–28) discusses material from an unpublished manuscript (Wittgenstein, MS 165, ca. 1941–44).

speaking a natural language. In the main section of his argument, Wittgenstein (*PI*, sec. 185) asks us to assume that the student has mastered the series of natural numbers and that we have given him exercises and tests for the series "$n + 2$" for numbers less than one thousand.

> Now we get the pupil to continue a series (say +2) beyond 1000 – and he writes 1000, 1004, 1008, 1012.
>
> We say to him: "Look what you've done!" – He doesn't understand. We say: "You were meant to add *two:* look how you began the series!" He answers: "Yes, isn't it right? I thought that was how I was *meant* to do it."

In the skepticist reading, the pupil's "mistake" reveals that his present action is logically consistent with an imaginable series: "Add 2 up to 1000, 4 up to 2000, 6 up to 3000." Because the pupil had not been given examples past 1000, his understanding of the rule is consistent with his previous experience. With enough imagination, numerous permutations of this can be generated. Collins, for instance, says that the rule "Add a 2 and then another 2 and then another and so forth . . . doesn't fully specify what we are to do . . . because that instruction can be followed by writing '82, 822, 8222, 82222' or '28, 282, 2282, 22822' or '8^2', etc. Each of these amounts to 'adding a 2' in some sense."[28] Since we can think of an indefinite variety of understandings of the formula "$n + 2$" based on the finite series of examples that the pupil previously calculated, it seems we have arrived at a radically relativistic position:

> This was our paradox: no course of action could be determined by a rule, because every course of action can be made out to accord with the rule. The answer was: if everything can be made out to accord with the rule, then it can also be made to conflict with it. And so there would be neither accord nor conflict here. (*PI*, sec. 201)

But as Wittgenstein then goes on to say, this paradox is based on the assumption that our grasp of the rule is based on an "interpretation"; a private judgment about the rule's meaning separate from any regular practices in a community. He contests the possibility of such an interpretation by adding that the regularities in our common behavior provide the context in which the rule is expressed and understood in the first place. Imaginable variations in counting rarely, if ever, intrude on our practice. Nor do violent disputes break out among mathematicians over the rules of their practice (*PI*, sec. 212). They simply follow the rule "as a matter of course" (sec. 238).

But the question now is why? Or rather, the question is, How do we manage so unproblematically to extend a rule to cover cases to which we haven't previously applied it? The answer seems to appeal to sociology.

[28] Collins, *Changing Order*, p. 13.

Wittgenstein (*PI,* sec. 206 ff.) likens following a rule to obeying an order, and he notes that the concepts of rule, order, and regularity can have a place only in a nexus of common behavior. How is such orderly action established? Through example, guidance, expressions of agreement, drill, and even intimidation: "When someone whom I am afraid of orders me to continue the series, I act quickly, with perfect certainty, and the lack of reasons does not trouble me" (*PI,* sec. 212).

Since we do indeed act in accord with the rules for calculating, the reason for this is not intrinsic to formal mathematics but to our "form of life" (*PI,* sec. 241). What limits our practice, and eventually the pupil's if he learns it, is not the rule alone but the social conventions for following it in a certain way. If it makes sense to say that logic "compels" us, this is so only in the way that we are, as Bloor puts it, "compelled to accept certain behaviour as right and certain behaviour as wrong. It will be because we take a form of life for granted."[29]

Orderly calculation thus depends on our natural proclivities as well as the social conventions we learn through drill, conventions that are inculcated and reinforced by normative practices in the social world around us.[30] If we read expressions like "the common behaviour of mankind" and "form of life" to refer to a much broader domain than the norms of a particular social group, we can invoke our common biological and psychological capacities. If we assume that mathematics (in this case, elementary arithmetic) is among our most rigorously rule-governed activities, then Wittgenstein appears to be making a powerful argument for turning from philosophy to sociology and other empirical sciences to explain order in mathematics.[31]

What holds for rules can also be said to hold for theories in the natural sciences: They are underdetermined by facts, since no theory can be supported unequivocally by a finite collection of experimental results. Therefore, if consensus is reached on a theory, it is not explained by facts alone but by the social conventions and common institutions in a community of scientists. These aspects of communal life greatly restrict the field of possible theoretical accounts to one or a very few socially recognized and approved versions. Collective habit and, at more heated times, vigorous persuasion and even coercion limit the range of sensible theoretical alternatives.

[29] Bloor, *Knowledge and Social Imagery,* p. 125.

[30] Bloor, *Wittgenstein: A Social Theory of Knowledge,* p. 121.

[31] Bloor invokes experimental psychology and biology as well as sociology when he empirically extends Wittgenstein's philosophy. Collins (*Changing Order,* p. 15) invokes Wittgenstein's private language argument to bar psychology (and presumably biology) from such investigations. For a discussion of an "organic account" – but not a strictly biological one – of Wittgenstein's references to "form of life," see J. F. M. Hunter, "'Forms of Life' in Wittgenstein's Philosophical Investigations," *American Philosophical Quarterly* 5 (1968): 233–43.

The appeal to social studies of science should seem obvious at this point. The skepticist reading of Wittgenstein seems to place the contents of mathematics and natural science at the disposal of the sociologist, since the very algorithms of mathematics and theoretical laws of physics can now be seen to express "the common behaviour of mankind" and not the transcendent laws of reason or the intrinsic relations in a platonic realm of pure mathematical forms.

Bloor's argument does not necessarily try to explain the behavior of scientists or mathematicians by reference to norms or ideological influences arising from "outside" the disciplinary community. Although such "external" influences can be included whenever they are relevant, Bloor's argument also permits relatively small and closed disciplinary communities ("core sets" in Collins's terminology) to be held responsible for their members' conventional practices.[32] Controversies in scientific fields take on special significance, since they exhibit fissures in scientific communities on the "internal" relations among theories, facts, and experimental procedures.

As discussed in Chapter 3, an established procedure in social studies of science is to use historical records (and, whenever possible, interviews and ethnographic observations) to document how interpretive possibilities that remain open during a scientific or technical controversy are closed down when a contending innovation gathers support in the community. According to such arguments, the innovation's superior performance in experimental tests only seems to explain why it vanquished its rivals; its technical superiority became obvious only after the fact, when alternative possibilities that were never definitively ruled out are shunted aside and buried in a black box of taken-for-granted assumptions.[33] From then on, so the argument goes, the successful innovation is retrospectively justified, and depending on the case, the justification can invoke a set of experimental facts that corresponds to "nature," a theory congruent with the dictates of "reason," or an invention

[32] The concept of "core set" is developed in H. M. Collins, "The seven sexes: a study in the sociology of a phenomenon, or the replication of experiments in physics," *Sociology* 9 (1975): 205–24.

[33] A similar argument about technological innovation is made by Trevor Pinch and Wiebe Bijker, "The social construction of facts and artefacts: or how the sociology of science and the sociology of technology might benefit each other," *Social Studies of Science* 14 (1984): 399–441. According to their argument, at an early phase in the social history of invention, alternative pathways for innovation remain very much alive. Eventually these alternatives are closed down, and one or a very few models of, for example, the bicycle, refrigerator, or personal computer, prevail. Pinch and Bijker emphasize the role of interest groups in this process, and they contrast their social–constructivist view with a technological rationalism that supposes the particular model that wins the day is the most efficient. For a case study critiquing this and related arguments, see Kathleen Jordan and Michael Lynch, "The sociology of a genetic engineering technique: ritual and rationality in the performance of the plasmid prep," pp. 77–114, in A. Clarke and J. Fujimura, eds., *The Right Tools for the Job: At Work in 20th Century Life Sciences* (Princeton NJ: Princeton University Press, 1992).

that is more "efficient" than its competitors.[34] The difference between normal and revolutionary science comes down to whether some of the open possibilities for developing science, mathematics, or technology are explicitly disputed or whether they remain submerged in the taken-for-granted habitus of "ready-made science."[35]

The Wittgensteinian critique of skepticism

Although it may be compatible with Bloor's and other sociologists' explanatory programs, Kripke's skepticist thesis about the example of rule-following has been charged with being a fundamentally mistaken reading of Wittgenstein. Stuart Shanker, for instance, argues that Kripke misunderstands the key passage quoted earlier from Section 201 of the *Philosophical Investigations:*

> Far from operating as a skeptic, one of Wittgenstein's earliest and most enduring objectives was . . . to undermine the skeptic's position by demonstrating its unintelligibility. "For doubt can exist only where a question exists, a question only where an answer exists, and an answer only where something *can be said.*"[36]

Shanker argues that Kripke fails to take into account that Wittgenstein's passage "is the culmination of a sustained *reductio ad absurdum.*"[37] Whereas Kripke interprets Wittgenstein in the familiar terms of the realist – antirealist debate in epistemology, Shanker contends that Wittgenstein lends support to neither camp in this debate and that considerable misunderstanding results from any attempt to enlist his arguments on either side.

> But if the premise is wrong – if Wittgenstein belongs to neither school of thought, for the very reason that he had embarked on a course which would undermine the very foundation of the Realist/Anti-realist distinction – the "sceptical" interpretation of *Remarks on the Foundations of Mathematics* is itself undermined at a stroke.[38]

As Shanker reconstructs it, the point of Wittgenstein's number-series argument is to demonstrate the absurdity of a "quasi-causal" picture of rule following, a metaphysical treatment that construes a rule as "an abstract

[34] Bachelard notes that although rationalists and realists emphasize different sides of an epistemic divide, their arguments play a similar justificatory role in discussions of science. Both sides subscribe to the same duality: on one side nature, on the other rational procedures for correctly discerning nature's secrets. There are, of course, significant differences among philosophers who put primary emphasis on one or the other, and within realism there are numerous positions, some of which are compatible with the strong program. See Gaston Bachelard, *The New Scientific Spirit,* trans. Arthur Goldhammer (Boston: Beacon Press, 1984).

[35] Latour (*Science in Action,* pp. 4 ff.) contrasts "ready-made science" and "science in the making."

[36] Shanker, *Wittgenstein and the Turning-Point,* p. 14.

[37] Ibid.

[38] Ibid., p. 4. Wittgenstein's writings are notoriously difficult, and vast amounts of academic writing have been devoted to clarifying them. Often as a prelude to mounting a criticism, expositors often relate Wittgenstein's positions to one or another side in familiar debates about realism/antirealism, positivism/idealism, objectivism/constructivism, and structural

object which engages with a mental mechanism." As Shanker reads him, Wittgenstein replaces this deterministic picture with one that emphasizes the practical basis of rule following. The "impression" that the rule guides our behavior reflects "our *inexorability in applying it.*"[39]

Thus far, the argument is fairly consistent with the lesson that Bloor, Barnes, Collins, and other sociologists of scientific knowledge derive from the example. The arguments soon diverge, however. The skeptic follows Wittgenstein's *reductio ad absurdum* to the point that abandonment of the quasi-causal picture is warranted but then concludes that rules provide an insufficient account of actions. Taken into the realm of sociology of knowledge, this conclusion motivates a search for alternative explanations of how orderly actions are possible. Social conventions and interests fill the void vacated by rational compulsion.

The critical move in the skepticist strategy is to isolate the formulation of the rule from the practice it formulates (its extension). Once the rule statement is isolated from the practices that extend it to new cases, the relation between the two becomes problematic: No single rule is determined by the previous practices held to be in accord with it, and no amount of elaboration of the rule can foreclose misinterpretations consistent with the literal form of its statement. Such indeterminacy is then remedied by a skepticist solution, which is to invoke extrinsic sources of influence on the relation between rules and their interpretations. These extrinsic sources include social conventions, community consensus, psychological dispositions, and socialization – a coordination of habits of thinking and action that limits the alternative interpretational possibilities. A battery of questions can then be raised for further research: How are such conventions established and sustained? How is consensus reached in the face of uncertainty and controversy? What are the relative contributions from our biological makeup, cognitive structure, and social affiliations?

Contrary to the skepticist solution, Shanker points out, "The purpose of the *reductio* is certainly not to question the intelligibility or certainty of the practice of rule following" (p. 25). The path out of the skeptical paradox is not through an antirealist epistemological position but through an examination of "grammar." The "foundations crisis" in epistemology (the realist – antirealist debate) arose from questions that can have no answer, and Wittgenstein offered a way to dissolve such questions. The point of the demonstration, therefore, was not to undermine objectivity but to clarify "in what sense mathematical knowledge can be said to be objective," which is not the same as saying that such knowledge has an objective or transcenden-

determinacy/methodological individualism. This is a familiar fate for phenomenological and ethnomethodological writings as well.

[39] Ibid., pp. 17–18.

tal foundation.[40] For Shanker, the internal relation between the rule for counting by twos and the actions carried out in accord with it is by no means an insufficient basis for the rule's extension to new cases. Nor is there any need to search for such a basis in psychological dispositions, biological mechanisms, or extrinsic social conventions.

G. P. Baker and P. M. S. Hacker also contest Kripke's skepticist reading of the number-series example, in their extended exegesis of the *Philosophical Investigations.*[41] Their particular target is what they call "the community view," the position that rule-following behavior is determined by patterns of reasoning sanctioned by community behavior. Their challenge to the community view at times is overly zealous,[42] but their most telling arguments are worth repeating. In their view, the problem begins with the way that the skeptic initially phrases the question. They argue that the skeptic's question, "How can an object like a rule determine the infinite array of acts that accord with it?" is miscast. As Wittgenstein says in regard to a similar question (*PI,* sec. 189), "'But *are* the steps then *not* determined by the algebraic for-

[40] Ibid., p. 62.

[41] Baker and Hacker, *Scepticism, Rules, and Language,* and *Wittgenstein, Rules, Grammar and Necessity.*

[42] For instance, Baker and Hacker, (*Scepticism, Rules, and Language,* p. 74) say that the community thesis "seems to imply that 'human agreement decides what is true and what is false.' But this, of course, is nonsense. It is the world that determines *truth:* human agreement determines meaning." Apparently this is a paraphrase of Wittgenstein (*PI,* sec. 241): "'So you are saying that human agreement decides what is true and what is false?' – It is what people *say* that is true and false; and they agree in the *language* they use. That is not agreement in opinions but in form of life." Wittgenstein makes no mention here of the world, nor does he say anything about what determines truth. Rather, his passage identifies "what is true and false" with what people "say." I read this to be locating "what is true and false" (and not "truth") in the situated grammars of speaking. Perhaps what people say is not a matter of "agreement" in any facile sense, but there seems to be no basis for attributing it to "the world" as such. Wittgenstein uses different terms for "agreement" in the preceding passage. His term for agreement in language is more akin to the English "consonance" or "attunement," as it draws on a musical metaphor suggested in the German *Ubereinstimmung.* See D. Bogen and M. Lynch, "Social critique and the logic of description: a response to McHoul," *Journal of Pragmatics* 14 (1990): 131–47. Much of Baker and Hacker's critique of the community view is worth taking into account, as is their further discussion of "accord with a rule" in their 1985 book. But as Malcolm ("Wittgenstein on language and rules,") incisively argues, their zealous attack on the community view sometimes strays into individualism, denying or ignoring the overwhelming emphasis on concerted human practice in Wittgenstein's writings about rules. Malcolm greatly clarifies Wittgenstein's emphasis on "quiet agreement" and "consensus in action" in the discussion of rules. This differs from agreement in *opinions* but is no less social. "It seems clear to me . . . that Wittgenstein is saying that the concept of following a rule is 'essentially social' – in the sense that it can have its roots only in a setting where there is *a people,* with a common life and a common language" (p. 23). Note that this is far from an endorsement of Kripke's view or of the sort of sociological reading of Wittgenstein that Bloor gives. Hunter and Cavell also elaborate views on rules and skepticism that are not quite so hostile to all "social" readings of Wittgenstein, but their views are not very compatible with the SSK approach. See J. F. M. Hunter, "Logical compulsion," pp. 171–202, in Hunter, *Essays After Wittgenstein* (Toronto: University of Toronto Press, 1973),

mula?' – The question contains a mistake." The question presupposes the independence between the rule and its extension, as though the rule were external to the actions performed in accord with it.

The skepticist interpretation retains the quasi-causal picture of rule following, since it never abandons the search for explanatory factors beyond or beneath the rule-following practice. The skeptic agrees that the formula "*n* + 2" cannot force compliance but keeps looking elsewhere for the cause: the mind, an interpretation, a socialized disposition.[43] But if it is agreed that an "internal" relation holds between rule and extension – that it makes no sense even to speak of the rule for counting by twos aside from the organized practices that extend it to new cases – then the epistemological mystery dissolves. "'*How* does the rule determine this as its application?' makes no more sense than: 'How does this side of the coin determine the other side as its obverse?'"[44]

This analogy may seem puzzling given the fact that formulations of rules are commonly set down on paper and posted on walls, and they are often recited separately from any acts that do or do not follow them. To clarify this further, consider the following passage from an unpublished manuscript by Wittgenstein:

> A rule can lead me to an action only in the same sense as can any direction in words, for example, an order. And if people did not agree in their actions according to rules, and could not come to terms with one another, that would be as if they could not come together about the sense of orders or descriptions. It would be a "confusion of tongues," and one could say that although all of them accompanied their actions with the uttering of sounds, nevertheless there was not language.[45]

As Norman Malcolm reads this, "a rule does not determine anything *except* in a setting of quiet agreement." In the absence of such concerted action, the rule is isolated, as though it were "naked" and the "words that express the rule would be without weight, without life."[46] This means more than that; for example, the rules of the traffic code have little weight in Boston, since drivers routinely ignore them. It refers instead to the practical adherence that supports a rule's intelligibility, that is, the order of concerted activities already in place when a rule is formulated, notably violated, disregarded, or evidently followed. The statement of a rule or order is a constituent part of such activities, and there is no way to contain or determine those activities in even the most elaborate version of the "naked" statement.

pp. 171–202, and *Understanding Wittgenstein: Studies of Philosophical Investigations* (Edinburgh: University of Edinburgh Press, 1985); and Cavell, *The Claim of Reason.*

[43] Baker and Hacker, *Scepticism, Rules, and Language,* p. 95.

[44] Ibid., p. 96.

[45] Wittgenstein, MS 165, ca. 1941–44, p. 78; quoted in Malcolm, "Wittgenstein on language and rules," p. 8.

[46] Malcolm, "Wittgenstein on language and rules," p. 9.

When we follow a rule we do not "interpret" it, as though its meaning were somehow fully contained in an abstract formulation. We act "blindly," and we show our understanding by acting accordingly and not by formulating a discursive interpretation. Of course, it is possible to misinterpret a rule, and we do sometimes wonder what the rules are and how we can apply them in a particular situation. But such occasions do not justify a general position of rule skepticism, nor do they suggest that in the normal case we interpret rules in order to use them in our actions.[47]

It is important to understand that the antiskepticist argument does not revert to a more familiar "internalist" or rationalist view. The distinction between internal and external in Baker and Hacker's treatment should not be confused with the internalist – externalist distinction in explanations of scientific progress. There is a sense in which Baker and Hacker affiliate to an "internalist" position, to the effect that an organized practice (e.g., calculating) demonstrates its rational organization (i.e., that it is orderly, in accord with relevant rules). However, this does not mean that rationality governs the practice or that one can explain the practice by invoking a set of rules. Again, a quotation from Wittgenstein may help clarify the sort of "internal" relation between rule and practice that is involved here:

> Suppose that we make enormous multiplications – numerals with a thousand digits. Suppose that after a certain point, the results people get deviate from each other. There is no way of preventing this deviation: even when we check their results, the results still deviate. What would be the right result? Would anyone have found it? Would there be a right result? – I should say, "This has ceased to be a calculation."[48]

Despite Baker and Hacker's occasional realist assertions, the argument does not provide a blanket endorsement of epistemological realism. Instead, it is a rejection of both variants of externalism: (1) the platonist position that the transcendental objects of mathematics determine mathematicians' practices and (2) the skepticist position that something else (community norms or individual dispositions) accounts for the relation between rules and behavior. I emphasize this point, since it is clear from Bloor's reply to an earlier version of this argument that it is easily misunderstood. It may be instructive to review how Bloor formulates the lesson from the number-series example, since it exemplifies some of the problems I have been discussing.[49]

> In *Philosophical Investigations,* Section 185, Wittgenstein imagined what would happen if a teacher, seeking to convey a rule in arithmetic, were to

[47] Baker and Hacker, *Scepticism, Rules, and Language,* pp. 93–94.

[48] L. Wittgenstein, in Cora Diamond, ed., *Wittgenstein's Lectures on the Foundations of Mathematics.* Lecture notes taken by four people (Ithaca, NY: Cornell University Press, 1976); quoted in Malcolm, "Wittgenstein on language and rules, p. 14.

[49] David Bloor, "Left- and right-Wittgensteinians," pp. 266–82, in Pickering, ed., *Science as*

> confront a pupil who systematically misunderstood the task. All attempts at correction fail because they too are systematically misunderstood. This is an example of the possibility of an endless regress of rules for following rules. It exhibits the limits of "interpretation" and the endlessness of the task of repairing indexicality. But another aspect of the example is what it says about internal relations. It shows that the deviant applications of a rule themselves stand in an internal relation to the rule as the deviant understands it. The teacher and the pupil here fail to make the usual kind of contact because the pupil constructs his own circle of definitions and his own set of internal relations between his signs and his practices. So the phenomenon of internal relations between a rule and its applications – if conceived narrowly – doesn't serve to define the real nature of rule following as we know it as a feature of a shared practice. At most it challenges us to define the difference between the actual rules of arithmetic and their idiosyncratic alternatives. It brings home, in the way that previous discussions of interpretation did, that something more and different is needed to define the accepted institution of arithmetic. Clearly what is required in the Wittgenstein example is something that breaks the deadlock between the competing internal relations. Such a factor would be consensus, the very thing rejected by Baker and Hacker. Ultimately it is collective support for one internal relation rather than another that makes the teacher's rule correct and the other deviant and incorrect.

Bloor's recitation of the number-series argument displays a critical confusion about the central question of "understanding" a rule, a confusion that points to a very un-Wittgensteinian element of psychologism in his "social" theory of knowledge. In this passage Bloor initially says that the pupil in Wittgenstein's example "systematically misunderstood the task." Shortly thereafter he characterizes this as an application of the rule as "the deviant understands it." From that point on, he places the pupil's "idiosyncratic alternative" into a symmetrical relationship with the teacher's conventional treatment of the rule, and he avows that both display "competing internal relations" between the rule and a possible practice, with consensus breaking the deadlock.

There is a degree of plausibility to Bloor's interpretation. Consider, for instance, the following example of a child learning to count according to instructions given by an adult:[50] The child counts on the fingers of his

Practice and Culture, quotation from pp. 273–74. Bloor (p. 273) prefaces his recitation of Wittgenstein's example by saying that it can be read as a "*reductio ad absurdum* of the position advocated by Lynch, Baker, Hacker, Shanker and other antisociological commentators." As I stated earlier, Shanker (p. 14) speaks of the same argument as "a sustained *reductio ad absurdum*" of rule skepticism.

[50] The example was furnished by Ed Parsons, who described it to me after having seen it and another similar example on a television program called "America's Funniest Home Videos."

hand: "One, two, three, four, five." The adult asks him, "Can you count backwards?" The child turns around, and with his back facing his questioner, he counts "one, two, three, four, five."[51] Following Bloor's recommendations, we might say that this example illustrates how the injunction to "count backwards" is an indexical expression whose sense is bound to the practice in which it is used. The child "misunderstands" the adult's injunction, and yet his application of the word *backwards* implicates an understanding of sorts linking the adult's question to others of the form, "Can you face backwards?" There is nothing intrinsic in the form of the statement that signals its "correct" application. In Bloor's terms, the child "constructs his own circle of definitions and his own set of internal relations" for applying the words *count backwards* to a technique for counting. The deadlock between competing internal relations is broken when the child is laughed at, corrected, and shown examples, and he eventually comes to learn what count backwards means as a constituent expression in a conventional practice.

The problem with this description is that if the child "systematically misunderstands" the injunction to count backwards, he has not demonstrated an understanding of the relevant use of the injunction. When he turns around and counts, "one, two, three, four, five," he inadvertently produces a pun on the words *count backwards,* but what he produces is not the technique we call counting backwards, which would be demonstrated by saying "five, four, three, two, one." He does show a "funny" understanding of the injunction in the way that his actions display an ignorance of the techniques of counting. There is no symmetry or deadlock between "competing internal relations," unless we were to assume that the child's actions establish a viable alternative to the technique invoked by the adult's injunction. But if a practice or technique is not an entirely private affair, it would not make sense to say that the child is understanding the words *count backwards* in terms of "his own" technique.[52]

When Baker and Hacker speak of an "internal" relation between a rule and a practice in arithmetic, they are describing a grammatical relation between the expression of a rule and the techniques of arithmetic. This has nothing to do with the "internal relations" that Bloor mentions when he speaks of a pupil's "own set of internal relations between his signs and his practices" or "his own circle of definitions."[53] Bloor here seems to be using

[51] In another example from the same program, a child is asked, "Can you count higher?" and he responds by raising his hand high above his head while counting, "One, two, three, four, five."

[52] See Wittgenstein *PI,* sec. 199.

[53] In "Left- and right-Wittgensteinians," Bloor (p. 271) gives an explanation of "internal" relations that avoids the implication of a private interpretation: "To say that *A* and *B* are internally related means that the definition of *A* involves mention of *B,* while the definition of

the word *internal* as though it referred to the pupil's private conception of the rule's meaning. But the pupil in Wittgenstein's example performs actions that demonstrate that he only thinks he is following the rule. By treating the internal relation between rule and practice as an individual matter, Bloor creates the need to search for "something more and different" in order to "define the accepted institution of arithmetic." The initial characterization of the action as a "misunderstanding" of the rule only makes sense from a standpoint that is already situated in (i.e., internal to) the "accepted institution of arithmetic," so that there is no comparable standpoint from which to characterize what the pupil is doing as a "competing understanding."

There is no suggestion in Wittgenstein's example that the pupil's misunderstanding is to be placed on equal theoretical footing with the correct way of continuing the number series. To say this is not to express a lack of sympathy for the pupil's predicament but to point out that there is no room in the world to place a "systematic misunderstanding" on such a footing without revising the initial terms of the description. The established practices and techniques of arithmetic are inseparable from the terms under which a relevant action is characterized as an understanding, competing understanding, or misunderstanding. Even if the student's practice displays "misunderstanding," however, it does not "relativize" the rule. "Competing internal relations" are precluded, since the student's practice is defined negatively in reference to the established practice of counting by twos.

I am not saying that there can be no such thing as idiosyncratic options to the usual way a practice or technique is carried out. Competition certainly can arise among different internal relations, and sometimes "deviant" usages (such as "ungrammatical" colloquial expressions or variants of a game initially prohibited by official rules) later gain acceptance. The point is that none of these characterizations turns on the "deviant" (or "eccentric" or "mistaken" or "innovative") agent's "own set of internal relations between his signs and his practices." The agent does not own the internal relations that identify his actions as mistakes, legitimate alternatives, or idiosyncratic instances of some practice. Rather, all of these characterizations presume that the agent's actions already take place in relation to some concerted practice.

B involves mention of *A*. In short. Two things are internally related if they are inter-defined, and so described that you can't have one without the other." In my view, this is an entirely different matter from saying that the pupil in Wittgenstein's example enacts his own "internal" understanding of the rule. For a related discussion see Graham Button and Wes Sharrock, "A disagreement over agreement and consensus in constuctionist sociology," *Journal for the Theory of Social Behavior* (forthcoming).

It would be misleading to treat the pupil in Wittgenstein's example according to the analogy of a scientist whose unconventional theory is rejected during a controversy (such as in Collins's case study of Joseph Weber's gravity-wave experiments discussed in Chapter 3).[54] Despite the once common tendency to reduce the history of science to a chronology of "great men's" ideas, no controversy is generated by an individual's "own set of internal relations between his signs and his practices." A controversial theory's very identity – as a theory about which there is controversy – is internally related to the equipment, techniques, literary practices, observation language, accepted concepts, and so forth in a field, even when historians, or even the scientists who originally promulgated the theory, later characterize it as a "misunderstanding" or "mistake." Consequently, not every imaginable alternative to accepted theories in a discipline counts as a controversial theory, nor can an outside analyst presume to apply a policy of symmetry to every unconventional claim that is made on a matter of fundamental significance. There is no room in the world for such a nonjudgmental standpoint.

Although Bloor confidently lays claim to a "sociological" reading of Wittgenstein, his recitation of the number-series example portrays internal relations in a radically individualistic way, as though the pupil could have his own understanding of arithmetic, at odds with that of the teacher but equally valid. "Consensus" then becomes a factor, independently introduced into the equation, that "breaks the deadlock" between the pupil's and the teacher's individual "understandings." Although Wittgenstein implies a kind of consensus in his discussion of agreement, this "quiet agreement" is so thoroughly and ubiquitously a part of the production of social order that it has little value as a discrete explanatory factor.

Wittgenstein (*PI,* sec. 241) distinguishes between agreement "in opinions" and agreement "in form of life." Agreement in form of life is exhibited in and through the coherence of our activities. It is an evident agreement of activities and their results, an orchestration of actions and expressions that enables mistakes, disruptions, and systematic misunderstandings to become noticeable and accountable. There is no time out from such agreement, even for a student whose actions display a misunderstanding or a sociologist who describes the student's misunderstanding. To describe this consensus and to specify its role in the activity is not to isolate a causal factor.

A similar argument applies to the way that sociologists of science commonly employ the Duhem – Quine underdetermination thesis. The problem of underdetermination is created by separating "evidence" from "theory"

[54] Collins, *Changing Order,* chap. 4.

and then arguing that no finite array of data can compel the acceptance of a single theory, because alternative theories (however implausible) can always be imagined that would account for the existing data. The trouble with sociological uses of this classic argument is they ignore that it is created "from a logical point of view," a point of view that in many respects is imcompatible with the empirical auspices of social studies of science.

Sociological descriptions of experimental practices typically portray an initial situation in which data are not (or not yet) isolated from theory and in which theoretical preconceptions, ordinary linguistic concepts, and trust in laboratory equipment and staff all come into play. Although scientists in such situations may face many interpretive problems, those problems do not boil down to a matter of reconciling isolated data with similarly isolated theoretical statements. Of course, many interesting problems can arise from a division of labor among technicians, experimentalists, lab administrators, and theorists, and these may occasion a variety of practical efforts to standardize instruments, coordinate staff, and reconcile diverse orders of records and evidences.[55]

Such solutions are not designed to meet the kinds of stringent standards of logical proof that would satisfy philosophical concerns about underdetermination; indeed, following Shanker's and Baker and Hacker's expositions of Wittgenstein, one can be led to wonder whether the philosophical problem is at all relevant to the practice it seems to describe.[56] Like the skeptical treatments of Wittgenstein's number-series argument, the way that the underdetermination thesis is formulated misleadingly suggests a yet-to-be-explained determination; that is, we are led to suppose that some sort of quasi-causal determination must be involved when scientists bridge the gap between data and theory. The insufficiency of logical determination thus seems to call for another mode of determination. But if there is no such "gap" in the first place, then no such determinative explanation is needed.

Can there be a sociology of science and mathematics?

The most distressing implication of the antiskepticist argument is that the "contents" of knowledge that Bloor's Wittgenstein delivered to sociology have now been taken back and placed firmly in mathematicians' and scientists' practices (although not in connection with an overarching rationality or reality). Following Wittgenstein's *reductio,* the rule for counting by twos stands as an adequate *members'* account. The student in Wittgenstein's

[55] See Peter Galison, *How Experiments End* (Chicago: University of Chicago Press, 1987) for a discussion of the philosophical implications of such divisions within the high-energy physics community.

[56] See Sharrock and Anderson, "Epistemology: professional scepticism," pp. 54 ff.

example does not display a possible interpretation of the rule; rather, his actions fail to obey the rule. For members, his actions demonstrate a failure of understanding and not the relativistic nature of the rule's sense or application.

Likewise, the rule's unproblematic extension calls for no independent justification outside the organized practices of counting. It is a rule in, of, and as counting by twos. The formulation of the rule does not cause its extension, nor does the meaning of the rule somehow cast a shadow over all the actions carried out in accord with it. The indefinite series of actions sustains the rule's intelligibility "blindly" without pause for interpretation, deliberation, or negotiation. Although this is nothing other than a *social* phenomenon, it does not call for an explanation using concepts proper to a particular social science discipline.

The problem for sociology is that the rule for counting by twos is embedded in the practice of counting. Counting is an orderly social phenomenon, but this does not make it an object for a general, causal, explanatory, and scientific sociology. Similarly, for more complex practices in mathematics, the consensual culture of mathematics is expressed and described mathematically; it is available in the actions of doing intelligible mathematics. To say this does not imply that mathematicians' practices are given a complete and determinate representation by mathematical formulas but that no such representation can be constructed and none is missing. To define the contents of mathematics and science as social phenomena turns out to be a very hollow victory for sociology.[57]

It seems we have arrived at an unhappy position for the sociology of

[57] My mention of a "hollow victory" here should not be confused with a recent polemic that purports to show that the sociology of knowledge makes "empty" claims about scientific discovery. See Peter Slezak, "Scientific discovery by computer as empirical refutation of the strong programme," *Social Studies of Science* 19 (1989): 563–600. Slezak claims that computer programs operating on the basis of general principles of problem solving have indeed made scientific discoveries. He argues that because the cognitive heuristics in these programs have been abstracted from the concrete sociohistorical circumstances of the original discoveries, their success provides a "decisive disproof" of the strong program's argument that scientific achievements are inextricable from historically specific constellations of social circumstances and social interests. In his article Slezak (p. 586) challenges Bloor's reading of Wittgenstein by citing some of the antiskepticist arguments in the philosophical literature. He uses Wittgenstein where it suits his argument in favor of cognitive science, but he shows little understanding of Wittgenstein's sustained attack on mentalism. Moreover, Slezak's claim (p. 591) that sociologists of knowledge have not noticed the "significant intellectual enterprise and body of research" in cognitive science is simply wrong because it ignores Coulter's Wittgenstein-inspired critique of cognitivism. See Jeff Coulter, *Rethinking Cognitive Theory* (New York: St. Martin's Press, 1983). Slezak lumps Wittgenstein together with behaviorism and then asserts that Wittgenstein's critiques of psychologism are now a dead issue in philosophy and psychology. He also characterizes some of SSK's claims and achievements as "trivial" and "empty." Slezak exposes his Cartesian commitments by arguing that an interest in the "grammar" of discovery is "trivial," since it applies to the designation of discoveries but not to their production. In my view, Slezak's argument is far more vulnerable to a Wittgensteinian critique than is Bloor's.

science. The neointernalist view expressed by Shanker, and Baker and Hacker seems to provide little basis for sociology to extend Wittgenstein's project. Mathematics and science (not to speak of innumerably other theory-guided or rule-following activities) now seem to have no need for sociologists to show them what they are missing in their realist preoccupations. Bruno Latour (who is partially sympathetic to constructivist sociology of science) acknowledges this problem in a most forceful way:

> But where can we find the concepts, the words, the tools that will make our explanation independent of the science under study? I must admit that there is no established stock of such concepts, especially not in the so-called human sciences, particularly sociology. Invented at the same period and by the same people as scientism, sociology is powerless to understand the skills from which it has so long been separated. Of the sociology of the sciences I can therefore say, "Protect me from my friends; I shall deal with my enemies," for if we set out to explain the sciences, it may well be that the *social* sciences will suffer first.[58]

This passage succinctly identifies a dilemma for any program of "social" explanation that seeks to show that the "contents" of other disciplinary practices are determined by a distinct configuration of sociological "factors." As Latour suggests, if to explain a practice is to deploy concepts that are independent of the discourse and skills that constitute that practice, such explanatory concepts would have to have a home in an independent form of life. But since sociology's analytic language is not divorced from the vernacular terms by means of which scientists (and other competent language users) develop their operative relations to the world in which they act, sociology seems ill-suited for devising the kind of explanation Latour has in mind.

Latour neatly identifies the problem and disavows any possibility of a causal or explanatory sociology of science, but, as noted in Chapter 3, he tries to solve the problem by borrowing a stock of concepts from A. J. Greimas's semiotics that he holds to be analytically independent of both general (i.e., academic) sociology and the situated sociologies in the other disciplines studied. Ultimately, he takes the program of "stepping back" from the field of investigation to an even further extreme than do the sociologists he criticizes.

In contrast, Wittgenstein attempts to make language use perspicuous, but not by distancing an "observer" (or, in Habermas's terms, a "virtual participant") from the concepts used in the fields of action described.[59] He instead draws explicit attention to the in-use (i.e., situated, occasional, indexical)

[58] Bruno Latour, *The Pasteurization of France*, trans. A Sheridan and J. Law (Cambridge, MA: Harvard University Press, 1988), p. 9.

[59] Jürgen Habermas, *Theory of Communicative Action*, vol. 1: *Reason and the Rationalization of Society* (Boston: Beacon Press, 1984), p. 118.

properties of familiar expressions and to the "quiet agreement" that supports their sensibility. In his imaginary "anthropological" examples, Wittgenstein sometimes suggests that a common ground for intelligibility can be given by such primordial language games as greetings, commands and responses, and the giving and receiving of orders.[60] These social conditions for the intelligibility of practical actions are not the conceptual property of an academic discipline but part of a common human legacy:

> If someone came into a foreign country, whose language he did not understand, it would not in general be difficult for him to find out when an order was given. But one can also order oneself to do something. If, however, we observed a Robinson, who gave himself an order in a language unfamiliar to us, this would be much more difficult for us to recognize.[61]

In ethnographic studies of science and other specialized practices, "familiar" activities like giving orders, asking questions, and giving instructions provide an initial, although far from sufficient, basis for grasping the intelligibility of technical actions. To examine the more esoteric language games requires an analysis situated in the settings studied. As I argued in Chapter 3, efforts to step back from the fields studied – whether done in the interests of analytic sociology or semiotics – disengage the "observer" from the epistemic "contents" in the field that are embodied in situated discourse. Consequently, although the new sociology of knowledge's major claim to fame is its attempt to explain the contents of science, the very practices through which its proponents address those contents ensure that they will remain unrecognizable or, at best, contentiously recognizable in the local idioms in the field studied.[62]

To contend that the discipline of sociology has no privileged access to the practices in other fields is not tantamount to defining such practices as asocial. Even if the antiskepticist argument convinces us about the absurdity of regressive attempts to explain rule following, Wittgenstein's clear references to training, drill, custom, common practice, and quiet agreement do comprise a picture of a public (i.e., "social") domain of activities in which a consensus on how to follow one or another rule is established. The problem with Bloor's account is that it treats Wittgenstein's "social theory of knowledge" as licensing an extension of sociology's existing concepts and methods to cover the subject matter of logic, mathematics, and natural science.

[60] For an anthropological case study that illuminates this point, see Brigitte Jordan and Nancy Fuller, "On the non-fatal nature of trouble: sense-making and trouble-managing in *lingua franca* talk," *Semiotica* 13 (1975): 11–31.

[61] This quotation is taken from an unpublished manuscript by Wittgenstein (MS 165, p. 103), quoted in Malcolm, "Wittgenstein on language and rules," p. 24.

[62] Habermas, *Theory of Communicative Action,* vol. 1, p. 119: "As soon as we ascribe to the actors *the same* judgmental competence that we claim for ourselves as interpreters of their utterances, we relinquish an immunity that was until then methodologically guaranteed. . . . We thereby expose our interpretation in principle to the same critique to which communicative agents must mutually expose their interpretations."

Mathematics and logic are collections of norms. The ontological status of logic and mathematics is the same as that of an institution. They are social in nature. An immediate consequence of this idea is that the activities of calculation and inference are amenable to the same processes of investigation, and are illuminated by the same theories, as are any other body of norms.[63]

What Bloor overlooks is that Wittgenstein's arguments apply no less appropriately to realist and rationalist sociology than to mathematical realism and logicism. Winch, and Sharrock and Anderson point out that far from making science and mathematics safe for sociology, Wittgenstein made things entirely unsafe for the analytic social sciences.[64] If sociology is to follow Wittgenstein's lead, a radically different conception of sociology's task needs to be developed. Bloor's attempts to graft Durkheim's or Mary Douglas's schemes to Wittgenstein's arguments simply do not go far enough.

This is where ethnomethodology comes into the picture, but to make the case for it as a program for pursuing Wittgenstein's initiatives will require our clearing away certain confusions both in and about ethnomethodology.[65] Ethnomethodology has become an increasingly incoherent discipline, despite incessant efforts by reviewers and textbook writers to define its theoretical and methodological program. On the one hand, current research in conversational analysis has diverged sharply from the radical program announced in Garfinkel's central writings (see Chapter 6). On the other hand, in the philosophy of the social sciences and the sociology of knowledge, the "older" ethnomethodology retains interest, but often in confusing ways.

Steve Woolgar, for instance, places some of Garfinkel's "key concepts" in the service of a skepticist treatment of science. He lists indexicality and reflexivity among the "methodological horrors" haunting all attempts at scientific representation.[66] By doing so, he treats Garfinkel's writings in much the same way that Bloor interprets Wittgenstein as licensing a theoreti-

[63] Bloor, *Wittgenstein: A Social Theory of Knowledge,* p. 189.

[64] Sharrock and Anderson, "The Wittgenstein connection"; Winch, *The Idea of a Social Science.* This applies not only to sociology's attempts to explain science scientifically but also to its attempts to explain religious beliefs, magical rituals, and ordinary actions. See Peter Winch, "Understanding a primitive society," pp. 78–111, in B. Wilson, ed., *Rationality* (Oxford: Blackwell Publisher, 1970); and W. W. Sharrock and R. J. Anderson, "Magic, witchcraft and the materialist mentality," *Human Studies* 8 (1985): 357–75.

[65] Wittgenstein's importance is downplayed by Garfinkel and other ethnomethodologists, and Schutz and phenomenology are usually accorded greater prominence in ethnomethodology's philosophical ancestry (cf. John Heritage, *Garfinkel and Ethnomethodology* [Oxford: Polity Press, 1984], chap. 3). As I elaborate in Chapter 6, the early development of conversational analysis and Garfinkel's studies of accounting practices and everyday rule use exhibit strong Wittgensteinian overtones. Although I argued in Chapter 4 that Schutz's influence is undermined by much of the work on science in the sociology of scientific knowledge (SSK) and ethnomethodology, the same cannot be said about Wittgenstein. My saying this should not imply that ethnomethodologists have endeavored to be faithful to the Wittgensteinian or any other philosophical tradition.

[66] Woolgar, *Science: The Very Idea,* pp. 32 ff.

cal program for calling into question the taken-for-granted assumptions in every domain of practical action. This is a common enough treatment of ethnomethodology, and in fact it is one that is occasionally promoted by persons who have unquestionable standing in the field.[67] It would be inaccurate (as well as pretentious) for me to argue that ethnomethodology cannot be understood along such lines, but what I will claim is that such an understanding misses what is most original about Garfinkel's "invention."

The antiskepticist reading of Wittgenstein suggests a way to understand what I see to be ethnomethodology's distinctive treatment of language and practical action: a treatment that avoids the twin pitfalls of sociological scientism and epistemological skepticism. To clarify this point, in the next section I describe an argument by Garfinkel and Sacks about the relationship between "formulations" and practical actions, an argument that I believe is compatible with an antiskepticist reading of Wittgenstein. I then outline some of the differences between ethnomethodology's and the strong program's "empirical" approaches by reviewing an ethnomethodological study of mathematics.

Formulations and practical actions

In their difficult and frequently misunderstood paper "On Formal Structures of Practical Actions," Garfinkel and Sacks discuss ethnomethodology's interest in natural language.[68] They mention Wittgenstein only briefly in this paper, but Sacks gives a more elaborate discussion of Wittgenstein's relevance in a transcribed lecture covering some of the themes discussed in his paper with Garfinkel.[69]

In that lecture Sacks speaks of Wittgenstein's having "exploded" the problem of the referential meaning of "indicator terms" (related to what Garfinkel calls "indexical expressions" – see Chapter 1). These terms have traditionally boggled logicians, as their reference changes with each occasion of use. Ordinary language was often held to be defective because it typically did not facilitate strict logical inference. Before Wittgenstein, a

[67] For instance, Melvin Pollner ("Left of ethnomethodology," *American Sociological Review* 56 [1991]: 374, n. 3), approvingly cites Woolgar's version of reflexivity as part of his argument for a revival of a "radical" ethnomethodology that avoids the positivism and professionalism that have crept into the field.

[68] Harold Garfinkel and Harvey Sacks, "On formal structures of practical actions," pp. 337–66, in J. C. McKinney and E. A. Tiryakian, eds., *Theoretical Sociology: Perspectives and Development* (New York: Appleton-Century-Crofts, 1970).

[69] Harvey Sacks, "Omnirelevant devices; settinged activities; indicator terms," transcribed lecture (February 16, 1967), pp. 515–22 in *Lectures on Conversation*, vol. 1, G. Jefferson, ed. (Oxford: Blackwell, 1992). I am assuming that Sacks's lecture expresses themes arising in his collaboration with Garfinkel.

common solution in the philosophy of language was to "remedy" these defects by translating indexical expressions into formulations that more precisely "captured" their referential meaning. Such remedial translations were like social science coding practices, in the way they attempted to substitute a limited set of analytic operators for a polysemous array of natural linguistic expressions.

Garfinkel and Sacks challenge the adequacy of such a translation practice, first by questioning the demand that it places on the ordinary practice of "formulating," and second by suggesting that the attempt to remedy indexical expressions necessarily misses the "rational properties" inherent in their ordinary use.[70] For Garfinkel and Sacks, formulating is not simply a professional analytic procedure; it also includes a wide range of ordinary linguistic actions: naming, identifying, defining, describing, explaining, and, of course, citing a rule. They point out that in both lay and professional discourse, such expressions are used to clarify the unequivocal sense of activities, although they also do many other things.[71]

Formulations are often used in attempts to repair the indexical properties of language by substituting "objective expressions" for "indexical expressions" (see Chapter 1 for a distinction between indexical and objective expressions). For a demonstration of the point that formulations do much more than clarify or correct prior usage, consider the following excerpt from an interrogation:

Mr. Nields: Did you suggest to the Attorney General that maybe the diversion memorandum and the fact that there was a diversion need not ever come out?

Lt. Col. North: Again, I don't recall that specific conversation at all, but I'm not saying it didn't happen.

Mr. Nields: You don't deny it?

[70] Sacks noted that the expression "You like to drive cars fast" – a recorded remark made by one "hot rod" enthusiast to another – loses its precision when it is translated into a particular speedometer reading. As it stands, the expression "fast" is measured by reference to "normal traffic" under different circumstances and is thus "stable" in the face of variations in road conditions, speed laws, police surveillance, and the like: "The stability of the terms, and the conditions under which they're usable, are such that time, place, speed laws, whatever else, are all irrelevant to their use. Changes in speed laws, changes in the capacity of cars, changes in personnel, new generations, new places – this thing can hold." See Harvey Sacks, "Members' measurement systems," *Research on Language and Social Interaction* 22 (1988–89): 45–60, quotation from p. 49; originally in H. Sacks, University of California at Irvine, lecture 24, Spring 1966; pp. 435–40 in Harvey Sacks, *Lectures on Conversation, Vol. 1*, ed. Gail Jefferson (Oxford: Blackwell Publisher, 1992).

[71] John Heritage and D. R. Watson discuss several systematic uses of formulations in conversation. See their "Aspects of the properties of formulations in natural conversations: some instances analyzed," *Semiotica* 30 (1980): 245–62.

Lt. Col. North: No.
Mr. Nields: You don't deny suggesting to the Attorney General of the United States that he just figure out a way of keeping this diversion document secret?
Lt. Col. North: I don't deny that I said it. I'm not saying I remember it either.[72]

In this brief but complicated interchange, one can see numerous interlarded "formulations" at work: formulations of prior conversations (with the attorney general), formulations of the pragmatic implications of "not recalling" that conversation, formulations of what "I said" or might have "said" and what "I'm not saying" now, and formulations that suggest irony and the like. Without going further into this, it should be obvious that these formulations do not simply refer to something; they act as thrusts, parries, feints, and dodges in the interrogatory game.

One especially interesting kind of formulation occurs in ordinary conversation and takes the form of a reflexive inquiry about "what we are doing" in the self-same conversation: "Was that a question?" "Are you inviting me to go along with you?" "I already answered your question, didn't I?" "Would you please get to the point!" What is striking about these is that although they apparently refer to "what we are doing" in conversation, they have distinct intelligibility as conversational acts, because of the way they are positioned in the dialogue. This property of formulations is made perspicuous in the following response to a rather demonic inquiry (the brackets are supplied by Garfinkel and Sacks (p. 350) as a notational device for marking that the formulation is a *doing* in the conversation as well as a referential expression):

HG: I need some exhibits of persons evading questions. Will you do me a favor and evade some questions for me?
NW: [Oh, dear, I'm not very good at evading questions.]

As a recognizable "doing," NW's reply performs the very "evasion" it disavows, so that the referential and performative aspects of the expression stand in a paradoxical relationship to each other. The formulation does not stand outside the temporality of the dialogue, so as to make a "metacomment" about the relationship; rather, it makes sense through the way it can be heard as a substantive move in that dialogue.

After explaining a series of related examples, Garfinkel and Sacks make two major points about formulations: (1) The "work" of doing "accountably

[72] Excerpt of dialogue from *Taking the Stand: The Testimony of Lieutenant Colonel Oliver L. North* (New York: Pocket Books, 1987), p. 33. Also see David Bogen and Michael Lynch, "Taking account of the hostile native: plausible deniability and the production of conventional history in the Iran–contra hearings," *Social Problems* 36 (1989): 197–224.

rational activities" can be accomplished, and recognizably so, by participants in an activity without need for formulating "this fact," and (2) "there is no room in the world *to definitively* propose formulations of activities, identifications, and contexts" (p. 359).

To relate this to our earlier discussion of rules, consider Baker and Hacker's discussion of formulating a rule:

> Typically explanations by examples involve using a series of examples *as a formulation of the rule.* The examples, thus viewed, are no more *applications* of the rule explained than is an ostensive definition of "red" (by pointing to a tomato) an application (predication) of "red." . . . The formulation of a rule must itself be *used* in a certain manner, as a canon of correct use.[73]

The series of examples acts to formulate the rule (i.e., make it evident, clear, relevant), without the rule's being stated in so many words. The appropriateness, sense, intelligibility, and recognizability of the rule are displayed in and through the examples, without the need for additional commentary. Garfinkel and Sacks distinguish between "formulating" (saying in so many words what we are doing) and "doing" (what we are doing), but their point is similar: Formulations have no independent jurisdiction over the activities they formulate, nor are the activities otherwise chaotic or senseless. Far from it, the sense and adequacy of any formulation is inseparable from the order of activities it formulates. It does not act as a substitute, transparent description, or "metalevel" account of what otherwise occurs.

Like Wittgenstein's discussion of rules, Garfinkel and Sacks's discussion of formulating can be misunderstood to imply either of two antithetical positions: (1) a skepticist interpretation of the effect that any attempt to formulate activities is beset by the "problem" of indexicality, so that description, explanation, and the like are essentially indeterminate; and (2) a realist interpretation that recommends empirical study of formulations in order to enable social scientists to attain an objective understanding of members' activities. A close reading of their argument should permit us to see that neither view is adequate.

Garfinkel and Sacks's argument undermines their initial contrast between "objective" and "indexical" expressions (and similarly between "formulations" and "activities").[74] Formulations themselves are used as "indexical expressions," and in so using them, members routinely find that "doing formulating" is itself a source of "complaints, faults, troubles, and recommended remedies, *essentially.*"[75] By the same token, "formulations are *not*

[73] Baker and Hacker, *Wittgenstein, Rules, Grammar and Necessity,* p. 73.

[74] See Paul Filmer, "Garfinkel's gloss: a diachronically dialectical, essential reflexivity of accounts," *Writing Sociology* 1 (1976): 69–84. Filmer closely analyzes Garfinkel and Sacks's argument, particularly the apparent distinction between objective and indexical expressions.

[75] Garfinkel and Sacks, "On formal structures of practical action," p. 353.

the machinery whereby accountably sensible, clear, definite talk is done."[76] "'Saying in so many words what we are doing' can be "recognizably incongruous, or boring, . . . [furnishing] evidence of incompetence, or devious motivation, and so forth."[77] Conversationalists manage to maintain topical coherence, often without naming the topic,[78] and as Garfinkel's breaching exercises demonstrate, attempts to "repair" the indexicality of any text or set of instructions further compound and extend the indexical properties of the text. The conclusion that Garfinkel and Sacks (p. 355) draw from this may initially seem to support a skepticist reading (emphasis and brackets in the original): "*For the member it is not in the work of doing formulations for conversation that the member is doing [the fact that our conversational activities are accountably rational].* The two activities are neither identical nor interchangeable."

But carefully note the passage that follows (p. 355, brackets in original): "In short, doing formulating for conversation itself exhibits for conversationalists an orientation to [the fact that our conversational activities are accountably rational]." This clearly differs from the skepticist conclusion that meaning is indeterminate or that the intelligibility of conversation is an illusion that rests on a basis lying behind the members' apparent sense of it. Also note that realism and rationalism are not recommended either: "The question of what one who is doing formulating is doing – which is a members' question – is not solved by members by consul-ting what the formulation proposes, but by engaging in practices that make up the *essentially* contexted character of the action of formulating" (p. 355).

For the rule "add two," no formulation can provide a complete or determinate account of how the rule is to be extended to new cases (as though the rule

[76] Ibid., pp. 353–54.

[77] Ibid., p. 354. For an example of how formulating can often deepen the misery in which a speaker is enmeshed, consider the following formulation, which was stated during a particularly disastrous public lecture: "I'm going to tell a joke, but it isn't very funny."

[78] Sacks demonstrates that topical coherence is achieved through systematic placement of a second utterance vis-à-vis a first. The placement of an utterance answers such unasked questions as "'Why did you say that?' 'Why did you say that *now*?'." This is done "automatically" and not by any formulation: ". . . that persons come to see your remarks as fitting into the topic at hand, provides for them the answer for how come you said it now. That is, it solves the possible question automatically. Upon hearing the statement a hearer will come to see directly how you come to say that" (Sacks, "Topic: utterance placement; 'activity occupied' phenomena; formulations; euphemisms," transcribed lecture (March 9, 1967), pp. 535–48 in *Lectures on Conversation,* vol. 1, quotation on p. 538. Although resolved on an entirely different historical scale, Sacks's analytic approach is strikingly, if perversely, in line with Foucault's treatment of historical discourse: "The meaning of a statement would be defined not by the treasure of intentions that it might contain, revealing and concealing it at the same time, but by the difference that articulates it upon the other real or possible statements, which are contemporary to it or to which it is opposed in the linear series of time" (Michel Foucault, *The Order of Things* [New York: Vintage, 1975], p. xvii).

"contained" a representation of an endless series of applications). Citing the rule is an activity in its own right (an instruction, warning, correction, reminder, etc.), but the rule's formulation does not say in so many words what is to be done with it. The sense of the rule is "essentially contexted" by the orderly activity in which it is invoked, expressed, applied, and so forth. But this does not imply that the activity has no rational basis or that the participants' understandings of what they are doing are necessarily incomplete or faulty.

In the concluding section of their paper, Garfinkel and Sacks assert that how "members do [the fact that our activities are accountably rational] . . . is done without having to do formulations" (p. 358). They add further that this "work" can be organized as "a machinery, in the way it is specifically used to do [accountably rational activities]" (brackets in the original). They then spell out the critical implications of this for the social sciences:

> That there is no room in the world for formulations as serious solutions to the problem of social order has to do with the prevailing recommendation in the social sciences that formulations can be done for practical purposes to accomplish empirical description, to achieve the justification and test of hypotheses, and the rest. Formulations are recommended thereby as resources with which the social sciences may accomplish rigorous analyses of practical actions that are adequate for all practical purposes . . . insofar as formulations are recommended as descriptive of "meaningful talk" something is amiss because "meaningful talk" cannot have that sense. (p. 359)

Insofar as the formal structures of practical actions (i.e., the "achieved fact" that activities are accountably rational) are not recovered by formulations, these structures elude constructive – analytic attempts to codify and statistically represent them. "The unavailability of formal structures is assured by the practices of constructive analysis for it *consists* of its practices" (p. 361). Garfinkel and Sacks speak of "constructive analysis" in an inclusive way, but a more precise sense of the term can be gained by examining a particular style of functional analysis that flourished some decades ago in North American sociology.

For example, Bernard Barber begins his essay "Trust in Science"[79] by noting that the concept of trust "has very frequently been used ambiguously by past social thinkers, by the man-in-the-street, by journalists, and by contemporary social scientists." He then proposes to remedy this "conceptual morass" by constructing a definition: "To put us on more solid analytical and empirical ground, we need to examine trust in light of our general

[79] Chap. 7 of Bernard Barber, *Social Studies of Science* (New Brunswick, NJ: Transaction Publishers, 1990), pp. 133–49.

understanding of social relationships and social systems. The construction resulting from this examination should, of course, be empirically usable and testable. Very briefly, I offer just such a construction."[80]

Barber goes on to formulate "two essential meanings" of trust in science, which he defines as socially shared expectations of "technically competent role performance" and of "fiduciary obligations" and responsibility. He then describes the latter by referring to the Mertonian norms of science (universalism, disinterestedness, communalism, and organized skepticism). Without going into the question of whether Barber's analysis achieves its stated objective, the point in contention for ethnomethodology is the initial construction that establishes the analysis. By addressing the ordinary concept of trust by considering all of the uses of the term together under a single conceptual heading, Barber finds a "morass" of different meanings that he hopes to remedy by stipulating a more restrictive definition. Both his definition and the theoretical apparatus he uses in his analysis derive from a coherent theoretical source (Talcott Parsons's model of the social system). By subsuming the various uses of the concept of trust under a general definition, Barber never considers that those uses – however diverse and confusing to consider all at once – may be orderly and investigable in their own right.

Ethnomethodology does not solve the epistemological problems arising from classic efforts to substitute theoretical formulations for an unexplicated "morass" of ordinary activities. By remaining indifferent to the aims and achievements of constructive analysis, ethnomethodologists try to characterize the organized uses of indexical expressions, including the various lay and professional uses of formulations. Inevitably, ethnomethodologists engage in formulating, if only to formulate the work of doing formulating, but unlike constructive analysts, they "topicalize" the relationship between formulations and activities in other than truth-conditional terms. That is, they do not treat formulations exclusively as true or false statements; instead, they investigate how they act as pragmatic moves in temporal orders of actions. Two main questions arise from this program: (1) How is it that, in the course of their being done, activities exhibit regularity, order, standardization, and particular cohort independence (i.e., "rationality") in advance of any formulation? and (2) How, in any instance, do members use formulations as part of their activities?

From this we can see the stark contrasts between ethnomethodology and the classic sociology that Bloor invokes when he proposes a scientific study of science. Where Bloor maintains a distinction between sociology's foundation as a science and the sociologically explained contents of the sciences

[80] Ibid., p. 133.

studied, Garfinkel and Sacks place sociology squarely in the ordinary society it studies.

After the "formal structures" paper was written, ethnomethodology's program diverged into two different lines of research. As mentioned in Chapter 1, one line of studies, conversational analysis, sought to elucidate "rational properties of indexical expressions" by investigating sequential structures in "naturally occurring" conversation. These studies described the regular procedures for turn taking, adjacency-pair organization, referential placement and correction, topical organization, story structure, place formulation, and other phenomena. In Wittgenstein's terminology, such phenomena are the "language games" through which order, sense, coherence, and agreement are interactionally achieved.[81] The other development was Garfinkel's ethnomethodological studies of work.

Garfinkel characterizes this program as an approach to the production of social order that breaks with classical conceptions of the problem of order.[82] For Garfinkel, both the detailed methods for producing social order and the conceptual themes under which order becomes analyzable are members' local achievements. There is no room in such a universe for a master theorist to narrate the thematics of an "overall" social structure. Instead, the best that can be done is to study closely the particular *sites* of practical inquiry where participants' actions elucidate the grand themes (e.g., of rationality, agency, structure) as part of the day's work. Of particular interest for our discussion are ethnomethodological studies of scientists' and mathematicians' practices. In this body of research, the questions that Garfinkel and Sacks raise about how formulations arise in practical activities remain much livelier than in conversational analysis (as I elaborate in Chapter 6).

One might figure that formulations such as maps, diagrams, graphs, textual figures, mathematical proofs, and photographic documents differ significantly from the formulations of activity that Garfinkel and Sacks discuss. Maps, after all, represent objective terrain and territory, and mathematical proofs represent functions in mathematics. They are not, in any

[81] Wittgenstein's use of the term *language game* is multifaceted. Conversational analysis develops from the sense of "language game" that Wittgenstein (*PI*, sec. 23) emphasizes when he says the term "is meant to bring into prominence the fact that the *speaking* of a language is part of an activity, or of a form of life." He then provides a list of examples, including giving and obeying orders, describing the appearance of an object, constructing an object from a description, and telling stories and jokes. Wittgenstein (*PI*, sec. 25) characterizes some of these activities ("commanding, questioning, recounting, chatting") as "primitive forms of language," and he observes that these "are as much a part of our natural history as walking, eating, drinking, playing."

[82] Harold Garfinkel, "Evidence for locally produced, naturally accountable phenomena of order, logic, reason, meaning, method, etc., in and as of the essential quiddity of immortal ordinary society (I of IV): an announcement of studies," *Sociological Theory* 6 (1988): 103–6.

precise sense, used as formulations of "what we are doing." But to treat maps and proofs as isolated pictures or statements ignores the activities that compose and use them. To analyze a document's use does not discount its referential functions, but it does demolish suppositions about the essential difference between formulations of "things" and formulations of "our activities."

An example is the following conversation recorded during a session in which two laboratory assistants (J and B) review some electron-microscopic data they prepared while the lab director (H) looks on and comments:

J: . . . if you *look* at this stuff it – things that are degenerating are very definite, and there's no real *question* about it.
B: That's the thing that really blew me *out.* Once I was looking at the three-day stuff, and the terminals were already phagocytized by the uh, by the glia.
J: Oh yeah, there are some like that now.
(silence: 3 seconds)
H: Yeah, *I'm* not worried about that. It's the false positives that worry me.
J: Yeah, yeh.
H: Like this.
J: Oh yeah, well that one – I didn't mark I don't *think* – You know I just put a little "X" there, because that's *marginal,* but *this* one looks like it has a density right there.
H: Yeah, and this one looks pretty good . . .[83]

Roughly characterized, the fragment starts when J assesses the analytic clarity of the data that he and B have just finished preparing. B then supports this assessment with a comparison with other data. H expresses a "worry" that challenges what the two assistants have just said, and J then fends off the challenge by simultaneously explicating details of the document and his method for preparing it. The fragment ends as H begins to accede to J's assessment. (The interchange continues well beyond the transcribed fragment.)

Without analyzing the fragment in detail, let me just mention a few points relevant to the current question about formulations of "things." The participants say "things" about the electron microscopic photographs they inspect together. These references include at least the following:[84]

1. J's initial references to "this stuff" and to "degenerating" organelles of the brain tissue presumably resulting from an experimental lesion.

[83] This is a simplified version of a transcript that originally appeared in M. Lynch, *Art and Artifact in Laboratory Science,* pp. 252–53.

[84] My glosses on what these indexical expressions "refer to" were not generated from the

2. B's comparison of the present materials with "three-day stuff," in which "three" formulates the number of days between the lesion and the sacrifice of the animal.
3. B's reference to phagocytosis, a process by which glial cells are said to "clean up" degenerated tissue following brain injury.
4. H's "worry" about "false positives," which in this instance can be understood as visible profiles of organelles that should appear to be degenerating but look normal in the particular micrographs.
5. J's mention of the "little X" that he says he marked on the surface of the micrograph to denote a "marginal" entity.
6. H's assessment that "this one" looks "pretty good."

Each of these references to things makes a point about the materials being inspected. Some references seem to point to visibly discriminable features of the data – instances of "degenerating" axon profiles (1), of a "marginal" case (5), and of "this one" that looks "pretty good" (6) – and these indicator terms may be accompanied by the characteristic gestures of ostension. Other references invoke temporal and conceptual horizons of the particular case at hand, for example, B's references to other cases and phagocytosis (2, 3), and C's mention of a possible methodological problem (4). Still others, for example, J's reference to "this stuff" (1), point with a rather thick and hazy finger, or rather, they may indicate any of several things. "This stuff" could indicate the micrographic document as a whole, a delimited feature in the frame of the document, a series of comparable micrographs, various analytic indices and markings, a characteristic phenomenon, and so forth. But the parties do not take time out to clarify such references (except when challenged to do so), and this is not because of an occult process that supplies the knowing participant with a mental image of what the indicator term "stands for." Moreover, each of the successive references to things is included in utterances that make a point vis à vis a local context of utterances and activities.

From this example, we can see that references to things act simultaneously as references to (and within) activities. The participants do not act like talking machines emitting nouns that correspond to pictorial details. Their

transcribed text alone but rely also on my ethnography of the lab's common techniques and vernacular usage. Their intelligibility for this analysis hinges on my (rather tenuous, in this case) grasp of the disciplinary specific practices studied. To mention the tenuousness of my glossing practices is not, contrary to Latour's criticisms, a mea culpa regarding my ignorance of technical science so much as a reminder that what I have to say about the practices is – whether adequate, inadequate, or trivial – an extension of the competency described. See B. Latour, "Will the last person to leave the social studies of science please turn on the tape-recorder?" *Social Studies of Science* 16 (1986): 541–48, which is a review of M. Lynch, *Art and Artifact in Laboratory Science*.

references implicate the adequacy of J's and B's work and the success of the project (i.e., the references to "definite" features of the data imply that things are going well, that a discriminable phenomenon seems to be emerging in the data). So, the general argument Garfinkel and Sacks made about formulations of activities is no less pertinent to formulations of things in laboratory shoptalk.

If we recall once again the contrasts between the two readings of Wittgenstein's number-series argument, we can now bring into relief how ethnomethodology's program extends Wittgenstein in very different ways than does the strong program's. The skepticist reading treats the rule as a *representation* of an activity that fails to account uniquely for the actions carried out in accord with it. The skeptical solution invokes psychological dispositions and/or extrinsic social factors to explain how an agent can unproblematically extend the rule to cover new cases. The nonskepticist reading treats the rule as an expression in, of, and as the orderly activity in which it occurs. The rule's formulation contributes to an orderly activity, insofar as order is already inherent in the concerted production of that activity.

As discussed earlier, Garfinkel and Sacks treat "indexicality" as a chronic problem for logicians' and social scientists' attempts to represent objectively linguistic and social activities. This problem disappears for ethnomethodology, not because it is solved or transcended, but because of a shift in the entire conception of language. As Garfinkel and Sacks elaborate in their discussion of "the rational properties of indexical expressions," such expressions are the very stuff of clear, intelligible, understandable activities. From their point of view, indexicality ceases to be a problem except under delimited circumstances. A sense of it as a ubiquitous "methodological horror" only accrues when indexical expressions are treated as tokens isolated from their meanings.[85]

[85] For Woolgar (*Science: The Very Idea*, pp. 32 ff.), the "methodological horrors" are a set of problems raised by skeptical treatments of representation, including the indeterminate relationships between rule and application and between theory and experimental data. Woolgar gives a methodological rationale for his global skepticism about scientists' representational practices. The policy of unrestricted skepticism licenses the sociological "observer" to impute methodological horrors to practices that would otherwise appear unperturbed. This interpretive policy requires us to envision a picture of scientists endlessly laboring to evade or circumvent the problems that a skeptical philosopher could raise. If this looks like a familiar move in the game of ideology critique, it is no accident. Woolgar (p. 101) states that "science is no more than an especially visible manifestation of the ideology of representation." The latter he defines (p. 99) as "the set of beliefs and practices stemming from the notion that objects (meanings, motives, things) underlie or pre-exist the surface signs (documents, appearances) which give rise to them." His critique is squarely aimed at scientific practice as well as a particular metaphysical view of science, and he thus may seem liable to Hacking's (*Representing and Intervening*, p. 30) charge of conflating what specialized scientists do with what philosophers of science would have them do. In Woolgar's defense it can be argued

Insofar as scientists and mathematicians use such expressions as part of a nexus of routine activities, they do not manage to evade indexicality by some rhetorical or interpretive maneuver; rather, the general "horror" never arises in the first place. This is not to say that scientists have no methodological or epistemic problems or that indexical expressions are simply a benign resource, but only that these problems arise and are handled as occasional (and sometimes "diabolical") contingencies in the course of disciplinary specific work.

From the sociology to the praxiology of mathematics

From Garfinkel and Sacks's argument we learn that far from disturbing or forestalling efforts to formulate activities, "the rational properties of indexical expressions" furnish an indispensable basis for understanding the sense, relevance, success, or failure of any formulation. In those cases in which rules or related formulations are regarded as rigorous, invariant, or even transcendental descriptions of activities, the basis for their rigor is provided by the practices in which such formulations are used. The contrast between this proposal and SSK's program becomes clear when we examine issues raised in Bloor's review of Eric Livingston's ethnomethodological study of mathematicians' work.[86]

Livingston introduces a phenomenon he calls the "pair structure" of a mathematical proof.[87] This involves a distinction between a "proof ac-

that practicing scientists often do indeed give realist (whether naive or otherwise) accounts of their results when asked to explain them (for many examples, see G. Nigel Gilbert and Michael Mulkay, *Opening Pandora's Box: A Sociological Analysis of Scientists' Discourse* [Cambridge University Press, 1984]), and it would not be off base to say that scientists' writings are a particularly "realistic" literary genre. But although it may be appropriate to criticize the "ideology of representation," it is not at all clear whether such criticisms implicate the "vulgar competence" of scientists' routine activities (see Garfinkel et al., "The work of a discovering science," p. 139). And Woolgar's statement that science is "no more than" a manifestation of an ideology is particularly difficult to accept when we take into account that the "ideology of representation" is a rather thin and often irrelevant account of scientists' practices.

[86] David Bloor, "The living foundations of mathematics," *Social Studies of Science* 17 (1987): 337–58, which is a review of Eric Livingston's *The Ethnomethodological Foundations of Mathematics* (London: Routledge & Kegan Paul, 1986).

[87] Livingston's *(The Ethnomethodological Foundations of Mathematics)* treatment develops the theme of the "*Lebenswelt* pair" introduced in Garfinkel's recent work (Harold Garfinkel, Eric Livingston, Michael Lynch, Douglas Macbeth, and Albert B. Robillard, "Respecifying the natural sciences as discovering sciences of practical action, I & II: doing so ethnographically by administering a schedule of contingencies in discussions with laboratory scientists and by hanging around their laboratories," unpublished manuscript, Department of Sociology, UCLA, 1989, pp. 123–4). The "pair" consists of a "first segment" (e.g., the proof statement in Livingston's example) and the "'lived' work-site practices - 'the work' - of proving the theorem." Garfinkel and his colleagues and Livingston take pains to point out

count" (the textual statement of a proof's "schedule") and "the lived work of proving" (the course of activities through which a "prover" works out the proof on any particular occasion). In his demonstrations of Gödel's proof and a simpler proof from Euclidean geometry, Livingston emphasizes the internal relation of proof account to the lived work of proving, by which he means the practices through which mathematicians "work out" the proof by sketching figures, using systems of notations, making calculations, discussing and debating what to do next, and so forth.

Livingston's treatment assumes that neither the proof account nor its associated lived work stands alone. For a competent mathematician, acting alone with pencil on paper or together with colleagues at the blackboard, the proof account comes to articulate the lived work of proving. Once worked through, it becomes a "precise description" and "transcendental account" of the work of proving.

> The puzzling and amazing thing about the pair structure of a proof is that neither proof-account nor its associated lived-work stands alone, nor are they ever available in such a dissociated state. The produced social object – the proof – and all of its observed, demonstrable properties, including its transcendental presence independent of the material particulars of its proof-account, are available in and as that pairing. A prover's work is inseparable from its material detail although, as the accomplishment of a proof, that proof is seen to be separable from it.[88]

The relation to the antiskepticist reading of Wittgenstein should be obvious. Livingston avoids the "question contain[ing] a mistake" by insisting that the intelligibility of a proof statement does not stand isolated from the practices of proving. The lived work that the proof formulates, though it is nothing other than mathematicians' work, is at the same time a social phenomenon.

> One of the consequences of the discovered pair structure of proofs is that the proofs of mathematics are recovered as witnessably social objects. This is not because some type of extraneous, non-proof-specific element like a theory of "socialization" needs to be added to a proof, but because the natural accountability of a proof is integrally tied to its production and exhibition *as* a proof.[89]

In his extensive and in some ways trenchant review of Livingston's

that the "pair structure" is not simply another example of formulations and activities. They raise the possibility that the *Lebenswelt* pair occurs only in mathematics and other "discovering sciences of practical action." Although they are not proposing to exempt mathematics and physical science from ethnomethodological study, this policy does seem to imply that these fields are "special." I pursue this matter in Chapter 7.

[88] Eric Livingston, *Making Sense of Ethnomethodology* (London: Routledge & Kegan Paul, 1987), pp. 136–37.

[89] Ibid., p. 126.

volume, Bloor raises a set of objections that clearly expose the differences between his approach and ethnomethodology's. He enlists Wittgenstein on his side of the fray but, as I point out, he does so at great risk to his own position. Bloor chides Livingston for having made no mention of Wittgenstein and then lectures him about what he should have known about Wittgenstein's "social theory of knowledge." While doing so, Bloor fails to grasp how strongly Livingston's treatment accords with an antiskepticist reading of Wittgenstein.[90] Bloor characterizes Livingston's position as follows:

> The amazing feat of creating universally compelling, eternal mathematical truths is managed entirely by what goes on, say, at the blackboard. If we examine the precise details we will see how transcendence is accomplished then and there. We don't need to enquire into the surroundings of the episode, or into the possibility that the feat depends on something imported into a situation from the surroundings. That would be to involve non-local features and circumstances beyond the "worksite."[91]

Livingston can, of course, only fail by Bloor's reckoning, because Bloor demands a general causal explanation, whereas Livingston tries to investigate the practical intelligibility of singular proofs. Bloor points out that Livingston refers to "familiar" aspects of a proof, thereby implying a wider horizon of accepted arguments and common tendencies among mathematicians. But to count this against Livingston is to miss the point of his focus on the internal relation between a proof statement and the lived work of proving.

What Livingston is trying to demonstrate is that the lived work of proving (the public production of mathematics at the blackboard or with pencil and paper) generates the proof statement's "precise description" of that same activity. In retrospect, there is no better formulation than the proof statement itself, although its adequacy is established not by a referential function of the statement alone but through the lived activity of proving. If a better formulation is to be developed, it too will arise from the historicity of mathematicians' activities. This, of course, implicates a communal setting of "quiet agreements" and orderly practices, but it is not enough for Bloor, since there is no sociological explanation in Livingston's demonstration. Bloor contends that the seeds of such an explanation are found in Wittgenstein's later philosophy:

[90] To be sure, Livingston fails to mention Wittgenstein in his volume *(The Ethnomethodological Foundations of Mathematics)*, and in his subsequent book (*Making Sense of Ethnomethodology* [London: Routledge & Kegan Paul, 1987], pp. 126 ff.) he mentions Wittgenstein only in relation to a particular example. Nonetheless, both texts use what I would argue are Wittgensteinian arguments, which Livingston may have drawn from Garfinkel's teachings.

[91] Bloor, "The living foundations of mathematics," p. 341.

> Wittgenstein, despite what is sometimes said, elaborated a *theory.* He argued that constructing mathematical proofs could be understood as a process of reasoning by analogy. It involves patterns of inference that were originally based on our experience of the world around us, and which have come to function as paradigms. They become conventionalized, and begin to take on a special aura as a result. We think that mathematics shows us the *essence* of things but, for Wittgenstein, these essences are conventions (*RFM,* I–74). We might say that in Wittgenstein, Mill's empiricism is combined with Durkheim's theory of the sacred.[92]

In a basic way, Bloor's "Wittgensteinian" critique of Livingston might as well be a critique of Wittgenstein. If Livingston fails to state a social scientific theory and fails to explain mathematical practice causally, so too does Wittgenstein "fail" as a matter of explicit policy!

> It was true to say that our considerations could not be scientific ones. . . . And we may not advance any kind of theory. There must not be anything hypothetical in our considerations. We must do away with all *explanation,* and description alone must take its place. And this description gets its light, that is to say its purpose, from the philosophical problems. These are, of course, not empirical problems; they are solved, rather, by looking into the workings of our language, and that in such a way as to make us recognize those workings: *in despite of* an urge to misunderstand them. The problems are solved, not by giving new information, but by arranging what we have always known. Philosophy is a battle against the bewitchment of our intelligence by means of language.[93]

Far from offering a "social theory of knowledge" in line with the dream of classical sociology, Wittgenstein here disavows the relevance of science, theory, and explanation to his investigations. Ethnomethodology also eschews the most basic elements of scientific sociology: its explanatory aims, its disciplinary corpus, and its definition of society.[94] In that sense, ethnomethodology extends Wittgenstein without having to repudiate his challenge to scientism and foundationalism.

In recommending description rather than explanation, Wittgenstein took into account that a description is not a "word-picture of the facts" and that descriptions "are instruments for particular uses" (*PI,* sec. 291). He did not

[92] Ibid., pp. 353–54.

[93] Wittgenstein, *Philosophical Investigations,* sec. 109.

[94] Bloor's reconstruction of Wittgenstein's "theory" is paralleled by Heritage's version of Garfinkel's "interpretively based theory of action" (*Garfinkel and Ethnomethodology,* p. 130). Like Bloor, Heritage reads Wittgenstein (and Garfinkel as well) as advancing a "finitist" conception of the relationship between rules and practical actions. Here I have argued for a reading of Wittgenstein not as a theorist who addresses this classic problem but as an antitheorist (or atheorist) who systematically investigates ordinary language to demonstrate how the problem arises only through a dubious treatment of linguistic expressions. Like Wittgenstein, Garfinkel also avoids characterizing his studies as a systematic theory.

propose delivering singularly correct descriptions of language use. Instead, he advocated a kind of reflexive investigation, in which philosophy's problems are addressed by "looking into the workings of our language."

Toward an empirical extension of Wittgenstein

When Wittgenstein recommended a descriptive rather than an explanatory approach to language, I take it that he meant neither an empirical sociology of language nor an introspective form of reflection. With regard to the latter, he saw no need to develop a second-order philosophy to "reflexively" comprehend its "unreflexive" counterpart: "One might think: if philosophy speaks of the use of the word 'philosophy' there must be a second-order philosophy. But it is not so: it is, rather, like the case of orthography, which deals with the word 'orthography' among others without then being second-order."[95]

How, then, are we to "look" into the workings of our language? Wittgenstein remarks that "we do not *command a clear view* of the use of our words. – Our grammar is lacking in this sort of perspicuity" (*PI,* sec. 122). In the reflective attitude of traditional philosophy, we are easily led to ascribe essential or core meanings to such resonant terms as *know, represent, reason,* and *true* and to develop hypostatized concepts of *Knowledge, Representation, Reason,* and *Truth.* By citing intuitively recognizable examples from ordinary usage and constructing imaginary "tribes" and "language games" systematically different from our customary usage, Wittgenstein is able to problematize epistemology by showing the variations, systematic ambiguities, and yet clear sensibilities in the everyday usage of "epistemological" expressions.

As Bloor points out, Wittgenstein develops an imaginary ethnography and not an empirical ethnography of language. This is not necessarily a failing, however, since Wittgenstein (*PI,* sec. 122) devises his cases as "perspicuous representations" – examples that are arranged systematically to show "connections" in our grammar. Wittgenstein's project may create a role for empirical cases, but not, as Bloor suggests, to transform a speculative method into an explanatory one. Instead, as Garfinkel advises, empirical investigations can be devised primarily as "aids to a sluggish imagination."[96] Garfinkel's well-known troublemaking exercises can be viewed as methods of perspicuous representation – interventions that disrupt ordinary scenes in order to make visible their practical organization. For the more recent studies

[95] Wittgenstein, *Philosophical Investigations,* sec. 121.
[96] Harold Garfinkel, *Studies in Ethnomethodology* (Englewood Cliffs, NJ: Prentice-Hall, 1967), p. 38.

on scientific work, Garfinkel devised systematic interventions for turning the central terms in epistemology (rationality, rules, agency, etc.) into "perspicuous phenomena."[97]

The idea of perspicuous representation also applies to early conversation-analytic investigations. Sacks initially took up the analysis of tape-recorded conversation to supply examples of commonplace language use that elude the reflective method of grammatical analysis used by ordinary language philosophers and speech-act theorists. Many of Sacks's early lectures were discourses inspired by one or another excerpt from his collection of transcribed conversations. In the course of one such discussion, Sacks remarks that "what I'm trying to do here is make my transcript noticeable to me."[98] Although, as I point out in Chapter 6, Sacks also expressed scientistic ambitions in his early lectures, his treatment of tape-recorded conversations contrasted with his and his colleagues' later development of rule-governed models of conversational systems. In the earlier lectures, issues in logic and philosophy of language were never far from the surface.[99] Sacks used particular fragments of conversation to critique logical – grammatical investigations based on intuitive examples.

The extension of Wittgenstein's later philosophy produced in ethnomethodology is therefore not a move into empirical sociology so much as an attempt to rediscover the sense of epistemology's central concepts and themes. The word *rediscover* is used here in a particular way. Although as speakers of a natural language, we already know what rules are and what it means to explain, agree, give reasons, or follow instructions, this does not mean that our understanding can be expressed in definitions, logical formulas, or even ideal–typical examples. Ethnomethodology's descriptions of the mundane and situated activities of "observing," "explaining," or "proving" enable a kind of rediscovery and respecification of how these central terms become relevant to particular contextures of activity. Descriptions of the

[97] An example of this is Friedrich Schrecker's study of experimental practice in which Schrecker (a graduate student in Garfinkel's seminars) assisted a disabled chemistry student in his laboratory work. Schrecker acted in effect as the student's "body" at the bench during lab exercises. The interaction between the two was videotaped. The verbal instructions from the chemistry student to Schrecker was a clear instance of the work of moving and arranging equipment into a "sensible" display of the current state of the experiment. See Friedrich Schrecker, "Doing a chemical experiment: the practices of chemistry students in a student laboratory in quantitative analysis," unpublished paper, Department of Sociology, UCLA, 1980. Schrecker's paper is discussed in M. Lynch, E. Livingston, and H. Garfinkel, "Temporal order in laboratory work," pp. 205–38, in Knorr-Cetina and Mulkay, eds., *Science Observed.*

[98] Harvey Sacks, "Omnirelevant devices...," transcribed lecture (March 9, 1967), pp. 515–22 in *Lectures on Conversation,* vol. 1.

[99] See G. Jefferson, ed., *Harvey Sacks – Lectures 1964–1965,* a special double issue of *Human Studies* 12 (1989); republished under the same title by Kluwer Academic Publishers, Dordrecht, 1989.

situated production of observations, explanations, proofs, and the like provide a more differentiated and subtle picture of epistemic activities than can be given by the generic definitions and familiar debates in epistemology. This involves less a substitution of "real" ethnographies for "imaginary" investigations of language use than a movement from definitions of key concepts to investigations of the production of the activities glossed by such concepts. In the remainder of this volume I address some of the problems attending such an approach in order to substantiate its outlines.

CHAPTER 6

Molecular sociology

Conversation analysis (CA) is the most sustained and coherent research program that has developed out of ethnomethodology. Since the late 1960s, researchers in the field have produced a steady accumulation of technical studies that have built on one another's findings. These studies have been published in numerous edited collections and professional journals in sociology, linguistics, communication studies, and anthropology.[1] Although CA is by no means a dominant program in any of these disciplines, it is moderately well established,[2] and it has been praised as a rare example of a "normal science" research program in sociology.[3]

In this chapter, I critically examine some programmatic claims for CA's status as a scientific discipline, and I argue that a "mythological" conception of natural science has become entrenched in the CA's observation language and in its conventions for presenting data and disseminating analytic reports. Like many other social scientists, conversation analysts often espouse a conception of unified scientific method that is now widely criticized in the history, philosophy, and sociology of science. Just as the new sociology of scientific knowledge provided a basis for criticizing the Kaufmann–Schutz version of science in Chapter 4, so can it enable a critical review of some of CA's assumptions about how a natural observational science can be implemented in the human sciences. My aim in this chapter, however, is not just to criticize yet another scientistic research program but to suggest how some of CA's programmatic initiatives and exemplary studies can be reincorporated

[1] See, for instance, the collections by G. Psathas, ed., *Everyday Language: Studies in Ethnomethodology* (New York: Irvington Press, 1979); J. M. Atkinson and J. Heritage, eds., *Structures of Social Action* (Cambridge University Press, 1984); and G. Button and J. R. E. Lee, eds., *Talk and Social Organization* (Clevedon: Multilingual Matters, 1987). Hundreds of other studies are listed in B. J. Fehr, J. Stetson, and Y. Mizukawa, "A bibliography for ethnomethodology," pp. 473–559 in Jeff Coulter, ed., *Ethnomethodological Sociology* (London: Edward Elgar, 1990).

[2] In his article "Left of ethnomethodology" (*American Sociological Review* 56 (1991): 370–80) Melvin Pollner speaks of conversation analysis as having moved to the "suburbs" of sociology, as compared with the time when ethnomethodology was very much at the margin of the discipline.

[3] See John Law and Peter Lodge, *Science for Social Scientists* (London: Macmillan, 1984, p. 283, n. 15). Law and Lodge point out that conversation analysts can apparently use one another's findings to accumulate a body of results. This is rare in sociology, as most fields are locked into endless debates about fundamental theoretical and methodological issues.

into a "postanalytic" line of ethnomethodological research that remains indifferent to the allure of science and the pitfalls of scientism.

A natural observational science of human behavior

If conversation analysis were simply another in the series of attempts to show that a social science can be built along the lines of the established natural sciences, it would not be especially interesting. What makes it significant for my purposes is the proposal, initially articulated by Harvey Sacks, that the natural sciences had already achieved a natural observational science of human behavior. Sacks set out to construct a nascent behavioral science that was based substantively, and not only analogically, on the existing natural sciences. In essence, he suggested that an incipient sociology of practical actions, existing in the natural sciences, could be grafted onto an independent disciplinary root in the social sciences, a root that would nurture the existing scientific sociology to a more complete fruition.

For Sacks, a scientific sociology was not to be constructed by adopting an abstract "scientific" method; instead, it was already present as a "sociology" in the existing natural sciences. He made this point explicit in his earliest lectures and writings by elaborating on his aim to build a scientific sociology that could produce formal descriptions based on observations of "the details of actual events."[4]

Primitive natural science

Sacks observed that early in their history, natural sciences like ancient astronomy and nineteenth-century biology included a "primitive" structure of accountability, in which virtually "anyone" in a nonspecialized community could go into the field, look at what there is to be seen, and describe it in vernacular terms.[5] "If you read a biological paper it will say, for example, 'I used such-and-such which I bought at Joe's drugstore.' And they tell you just what they do, and you can pick it up and see whether it holds. You can re-do the observations."[6] He added that such observers "could see it with their eyes; they didn't need a lot of equipment, and they knew what an account would look like." In his lectures Sacks occasionally drew analogies between his own studies of natural conversation and such primitive natural sciences,

[4] Harvey Sacks, "Notes on methodology," in Atkinson and Heritage, eds., *Structures of Social Action*, p. 26.

[5] Harvey Sacks, "On sampling and subjectivity," transcribed lecture (spring 1966, lecture 33), p. 983–8 in Harvey Sacks, *Lectures on Conversation*, vol.1, G. Jefferson, ed. (Oxford: Blackwell, 1992). See especially pp. 487–8.

[6] Sacks, "Lecture 4: An impromptu survey of the literature," pp. 26–32 in Harvey Sacks, *Lectures on Conversation*, G. Jefferson, ed., the quotation is on p. 27.

and he told his students that the opportunity to study conversation in such a primitive fashion was "probably [a] very short term possibility, so you'd better look while you can."[7]

Sacks's disarmingly simple version of science provided a refreshing contrast with the Byzantine methodologies so often constructed in contemporary sociology. Rather than starting his investigations with a complicated theory of action, he began with descriptions of observable social activities: simple sequences of conversation, proverbs, and various other recurrent expressions and gestures. He assumed that the surface of the social world was already well ordered and that its scattered and heterogeneous facts could be observed, collected, described, and analyzed without a great deal of preparation. He suggested further that there was no need to search for "big issues" to begin a study of social order. Order was visible "at all points," even in the least interesting and most accessible of places.[8] Thus, an intensively focused analysis of the most mundane and unremarkable events could yield enormous understanding, analogous to the way that intensive analysis of the humble intestinal bacterium *E. coli* produced revolutionary breakthroughs in genetics and molecular biology.[9] By starting with "simple" and "observable" social objects rather than obviously significant historical episodes and massive social institutions, Sacks tried to develop a grammar for describing the social production of communicative actions.

In a posthumously published argument, Sacks outlined how "the fact of science's existence" could provide "a foundation for a natural observational science of sociology."[10] He did not say that the exact sciences offer a general method for sociology to emulate. Instead, he proposed that natural scientists "naively" and routinely produce "scientific descriptions of human actions" when they make observations, report them to colleagues, and try to replicate the observations from the reports. He made it clear that the linkages among observations, reports, and replications were essentially and irreducibly communicative. In this he was consistent with Karl Popper, who identified the practical and communicational process of reproducing observations as a "social aspect of scientific method" that the sociology of knowledge has

[7] Sacks, "On sampling and subjectivity," p. 488.

[8] Ibid. See also Sacks, "Notes on methodology," p. 22.

[9] Sacks, "An impromptu survery of the literature," p. 28. Sacks recommended that his students read James Watson's *Molecular Biology of the Gene* (New York: Benjamin, 1965) in order to appreciate how a domain of intricate phenomena could be built from simple recurrent structures (Alene Terasaki, personal communication).

[10] This argument is presented in "Introduction," in G. Jefferson, ed., *Harvey Sacks – Lectures 1964–1965*, a special issue of *Human Studies* 12 (1989), pp. 211–15. In this same book, E. A. Schegloff, "An introduction/memoir for Harvey Sacks – lectures 1964–1965," p. 207, n. 5, states that Sacks wrote the introduction in 1965 and that it was intended for a book that he never published, entitled *The Search for Help*. Schegloff (p. 202) mentions that Sacks was working on the argument as early as 1961–62.

ignored.[11] In a formulation that largely resembled Sacks's account of primitive natural science, Popper stated, "An empirical scientific statement can be presented (by describing experimental arrangements, etc.) in such a way that anyone who has learned the relevant technique can test it."[12] Note, however, that for this to describe a primitive natural science, as Sacks defined it, the "relevant techniques" must be ordinary and nonspecialized.

Like Popper, Sacks treated natural scientific methods as formal analytic structures of practical action, that is, as organized complexes of action, reproduced again and again at different times and places by different production cohorts, which would include techniques for producing, certifying, and distributing descriptions of observable phenomena. However, in a move more reminiscent of Durkheim than of Popper, Sacks argued that scientific practices not only are means for getting access to natural facts; they also are social facts in themselves. And finally, following Garfinkel rather than Durkheim, Sacks noted that *members'* descriptions of those practices would be sociological descriptions.[13] His sociological program would not simply be modeled after a successful natural science; it would exploit a feature that he understood to be inherent in the production of scientific facts.

Although it is often supposed that descriptions of human actions are truly scientific only when they are based on the results of neurological or biological researches, Sacks turned the tables by making a brilliantly simple observation: "The doing of natural science, indeed the doing of biological inquiries, was something which was reportable, first, and second, the reports of the activities of doing science did not take the form that the reports of the phenomena under investigation took."[14] Neurologists' instructional texts and research reports include vernacular instructions on how to replicate observations and experiments, but the reliable use of these descriptions cannot be explained by any substantive neurological findings about human perception and brain activity.[15] Like other natural scientists, neurologists rely on stable

[11] Karl Popper, "The sociology of knowledge," pp. 649–60, in J. E. Curtis and J. W. Petras, eds., *The Sociology of Knowledge* (New York: Praeger, 1970). Jürgen Habermas (*The Theory of Communicative Action,* vol. 1: *Reason and the Rationalization of Society,* trans. Thomas McCarthy [Boston: Beacon Press, 1984], p. 111) also speaks of this "'forgotten theme' in the analytic theory of science: the intersubjectivity that is established between ego and alter ego in communicative action."

[12] Karl Popper, *The Logic of Scientific Discovery* (New York: Harper & Row, 1959), p. 99.

[13] "Members" in this case would be masters of the relevant scientific techniques. This conception of natural and social scientific descriptions is outlined in Sacks's early paper "Sociological description," *Berkeley Journal of Sociology* 1 (1963): 1–16.

[14] Sacks, "Introduction," in *Harvey Sacks – Lectures 1964–1965,* p. 213.

[15] Enthusiasts for cognitive science may argue that they can (or soon will be able to) model the behavior of scientists, but regardless of such claims scientists were able to reproduce their methods accountably and reliably long before the days of artificial intelligence.

modes of description, instruction, and demonstration that are not based on the particular findings of a natural science discipline. Sacks insisted that scientists' reports of both their own activities and the phenomena they observed were necessary features of science.[16] A "stable" (i.e., reproducible, replicable) science would be impossible without both kinds of descriptions.[17]

Sacks went on to ask, "What is it . . . that made scientists' descriptions of their own activities adequate?" The answer, he said, was obvious: "Scientists' reports of their own activities are adequate, i.e., they provide for the reproducibility of their actions on the part of themselves or others, by the use of methods."[18] And since science is far from the only social activity that is reproducible, it seemed "obvious enough" to Sacks that "whatever activities of humans could be adequately described as methodical could be then said to be adequately scientifically described."[19]

In his "Introduction/memoir" to the volume of Sacks's lectures in which the argument is included, Emanuel Schegloff concisely enumerates the key points in the argument:

> So, Sacks concluded, from the fact of the existence of natural science there is evidence that it is possible to have (1) accounts of human courses of action, (2) which are not neurophysiological, biological, etc., (3) which are reproducible and hence scientifically adequate, (4) the latter two features amounting to the finding that they may be stable, and (5) a way (perhaps *the* way) to have such stable accounts of human behavior is by producing accounts of the methods and procedures for producing it. The grounding for the possibility of a stable social-scientific account of human behavior of a non-reductionist sort was at least as deep as the grounding of the natural sciences. Perhaps that is deep enough.[20]

Rather than proposing a science that had yet to be born, Sacks suggested that scientific sociology was already on the scene, embodied in what Garfinkel has called the "instructable reproducibility" of natural science

[16] Sacks, "Introduction," in *Harvey Sacks – Lectures 1964–1965*, p. 213.

[17] A point of contrast can be made between Sacks's emphasis on "descriptions" in science and Lyotard's assertion that the continued existence of "science" requires "metanarratives of legitimation" (e.g., in the form of utilitarian justifications and promises) while at the same time it rejects "narrativity" as prescientific. See Jean-François Lyotard, *The Postmodern Condition: A Report on Knowledge* (Minneapolis: University of Minnesota Press, 1984). Far from acknowledging such a "crisis," Sacks identified descriptive "narrative" as a productive machinery inseparable from the local organization of "science" in laboratories and texts. Sacks did not express "incredulity" toward metanarratives; instead, he was indifferent to the necessity of overarching legitimations for a practice that is sustained by a "molecular" narrativity. Of course, Sacks's argument itself provides a partly qualified legitimation for science (and, as I elaborate below, for conversation analysis).

[18] Sacks, "Introduction," in *Harvey Sacks – Lectures 1964–1965*, p. 214.

[19] Ibid.

[20] Schegloff, "An introduction/memoir," p. 203. See also Schegloff, "Introduction" to Harvey Sacks, *Lectures on Conversation*, vol. 1, pp. ix–lxii. See especially pp. xxxi–xxxii.

observation. Sacks added that an activity can be adequately described as methodical, regardless of whether those who do it methodically generate their own descriptions. "Indeed many great scientists do not make adequate reports of their procedures; others do it for them." The key to the technical development of a molecular sociology was that methodical activities are describe-*able* and adequate descriptions are use-*able* as instructions for (re)generating those activities. As he envisioned it, the task ahead for sociology was to extend and technically elaborate "the body of reports of scientific activities" by producing formal descriptions of the full range of methodical human actions.[21]

Primitive science rewritten as scientific mythology

Sacks's proposals about the "primitive natural sciences" can be reexamined in light of subsequent developments in the sociology of scientific knowledge. Using the advantages of hindsight, it is easy to fault Sacks for holding a quaint, and even mythological, view of science in which observation, description, and replication provide the "grounding" of publicly verified knowledge. Although I have no intention of diminishing the brilliance of his and his colleagues' achievements, I do think that it is worth giving some critical attention to Sacks's assumptions about science.

Sacks explicitly stated that his argument presumes that "science exists,"[22] and although it is not entirely clear what he meant by saying this, I take it that he was alluding to the historical "fact" that a remarkably fertile program for certifying knowledge emerged in Europe a few centuries ago. When he elaborated on this "fact" of science's existence, he emphasized the following essential elements of scientific methodology:

1. Science is based on naturalistic observations.
2. Such observations are describable as methods.
3. Adequate methods descriptions enable anyone to replicate the observations described.
4. Adequate methods descriptions include two analytically distinct components:
    a. Accounts of specialized findings about, for example, chemical, biological, and astronomical phenomena.
    b. Vernacular accounts of methodic human behavior.

[21] Sacks, "Introduction," p. 214. Sacks cites L. S. Vygotsky, *Thought and Language* (Cambridge, MA: MIT Press, 1962), chap. 6, as a related discussion of science as a basic human activity. Although Sacks's argument was written before the English translation of Claude Lévi-Strauss's *The Savage Mind* (Chicago: University of Chicago Press, 1966), Lévi-Strauss's discussion of "the sciences of the concrete" provides another relevant basis of comparison. Unlike Sacks, however, Lévi-Strauss ultimately contrasts the primitives' *bricolage* with a principled version of the rationality in modern science and engineering.

[22] Sacks, "Introduction," p. 212.

5. The existence of adequate vernacular accounts of scientific method provides the grounding for the possibility of a stable science of human behavior.

This is by no means an unfamiliar picture. Three centuries ago, Robert Boyle devised what Steven Shapin and Simon Schaffer describe as a "language game" for producing experimental *matters of fact.* A matter of fact in Boyle's experimental program was "the outcome of the process of having an empirical experience, warranting it to oneself, and assuring others that grounds for their belief were adequate. In that process a multiplication of witnessing experience was fundamental."[23] Boyle treated matters of fact as "both an epistemological and social category."[24] A matter of fact had to be relayed through an ordered series of communications, and its very identity as a fact was a product of that communicational circuit. Boyle described his "new experiments" in letters to other experimenters, carefully instructing them on how to replicate them without error. He also expressed a desire to instruct "young gentlemen" on how to perform some of the simpler experiments. Some of them, he remarked, "require but little time, or charge, or trouble in the making" and could even be tried "by ladies."[25]

This program for multiplying witnessing experiences recalls the structure of accountability (observation–report–replication) that Sacks attributed to primitive natural science. Moreover, at least some of Boyle's descriptions were written so that "anyone" could redo the experiments.[26] Shapin and Schaffer add, however, that Boyle's efforts to get others to replicate his experiments did not often succeed. Eight years after performing his famous air-pump experiments, he "admitted that, despite his care in communicating details of the engine and his procedures, there had been few successful replications." Still later, "Boyle . . . expressed despair that these experiments would ever be replicated. He said that he was now even more willing to set down divers things with their minute circumstances 'because' probably many of these experiments would never be either re-examined by others, or re-iterated by myself."[27]

Despite these difficulties, however, Boyle was far from unsuccessful in promoting his experimental program. But rather than devising a method by

[23] Steven Shapin and Simon Schaffer, *Leviathan and the Air Pump: Hobbes, Boyle, and the Experimental Life* (Princeton, NJ: Princeton University Press, 1985), p. 25.

[24] Ibid.

[25] Robert Boyle, "The experimental history of colours," pp. 662–778, in Thomas Birch, ed., *The Works of the Honourable Robert Boyle,* 2nd ed., vol. 1 (London: J.& F. Rivington, 1772). Quoted in Shapin and Schaffer, *Leviathan and the Air Pump,* p. 59.

[26] As Shapin and Schaffer observe, "anyone" was neither any member of the "scientific community" nor any person but something like the classical concept of "citizen" or perhaps the equivalent of today's "average intelligent reader."

[27] Shapin and Schaffer, *Leviathan and Air Pump,* pp. 59–60; Boyle quotation from "Continuation of new experiments. The second part," p. 505, in Birch, ed., *The Works of the Honourable Robert Boyle.*

which anyone could directly witness what he observed, Boyle constructed what Shapin and Schaffer call a "technology of virtual witnessing . . . a technology of trust and assurance that things had been done and done in the way claimed." This "technology" included a set of material, textual and organizational practices for (1) laboriously producing a rare and privileged "space" for experimental observation (i.e., the laboratory and the air-pump apparatus), (2) utilizing prolix descriptions and detailed engravings to convey a sense of the circumstantial details of the experiment and equipment, and (3) displaying the modest virtues of a credible gentleman of the Royal Society. The repetition of the experimental experience was a reproduction of an original observation, not in the canonical sense of "replication," but as a displacement of verisimilitudinous renderings.[28] Boyle's evidently sincere exasperation over his inability to persuade others to reproduce his experiments itself contributed to his credibility, so that for all practical purposes his experiments might just as well have been replicated.

For Boyle, as Shapin and Schaffer reconstruct his program, the ordered ensemble of observation–report–replication was a mythological description of the work of an experiment that helped proselytize an entire experimental way of life.[29] The air pump was the centerpiece of this way of life, because the care, management, description, reproduction, and standardization of its mechanisms were intertwined with the prospects of Boyle's experimental matters of fact. "The capacity of this machine to produce matters of fact crucially depended upon its physical integrity, or, more precisely, upon collective agreement that it was air-tight for all practical purposes."[30] Consequently, the technical competencies associated with building and managing the machinery of the air pump came to authorize claims about experimental facts that, only in principle, were verifiable by "anybody."

Here it might be objected that owing to its use of fairly complex instrumentation, Boyle's experimental program was not an entirely "primitive" science in Sacks's terms. Perhaps a more apt case would be provided by a field science, such as ornithology. In a study of eighteenth- and nineteenth-century ornithology, Paul Farber shows that democratic procedures for looking and telling were only occasionally featured in the genealogy of that science. The story is again one of controlled access to observations, disciplined observational spaces, and literary technologies. According to Farber,

[28] The appropriate sense of "reproduction" in this instance can be drawn from Walter Benjamin, "The work of art in the age of mechanical reproduction," pp. 217–51, in W. Benjamin, *Illuminations,* ed. Hannah Arendt, trans. Harry Zohn (New York: Schocken Books, 1969). For a detailed account of the circulation of scientific texts, see Bruno Latour, "Drawing things together," pp. 19–68, in M. Lynch and S. Woolgar, eds., *Representation in Scientific Practice* (Cambridge, MA: MIT Press, 1990).

[29] This is a somewhat strained paraphrase of Wittgenstein, *Philosophical Investigations,* ed. G. E. M. Anscombe (Oxford: Blackwell Publisher, 1958), section 221.

[30] Shapin and Schaffer, *Leviathan and the Air Pump,* p. 29.

the eighteenth-century naturalist Pierre-Raymond de Brisson developed his taxonomy of birds by meticulously drawing the specimens in his jealously guarded museum collection.[31]

For Brisson the museum was a privileged observational site. Specimens for his and other museum collections were gathered from diverse places, sometimes with the aid of market hunters who had their own methods of "field study." Carcasses were stuffed and preserved (often very badly) and then juxtaposed as tabular "entries" in the cellular compartments of the museum drawers.[32] The *tableau mort* of the museum drawer provided a preliterary organizational field for systematic inspection, reinspection, and comparison. Field study and amateur bird watching developed only later, after the distribution of portable field manuals and field glasses and the emergence of ornithological societies and socially instituted canons of proper description.[33] The natural science that emerged was inseparable from the organized methods for collecting, preserving, circulating, and arranging materials, measurements, and communal activities, along with the literary conventions for composing and juxtaposing pictures and descriptions. Similar themes also appear in recent accounts of the origins of microbiology, geology, and meteorology.[34]

Sacks's account of a primitive natural science seems problematic in its focus on direct observation, adequate description, and replication. Numerous studies in the sociology of scientific knowledge (see Chapters 2 and 3) open up questions about just what observation, adequate description, and replication involve as socially organized epistemic practices.

1. As Ian Hacking puts it, observation has been overrated in the history of science: "Often the experimental task, and the test of ingenuity or even greatness, is less to observe and report, than to get some bit of equipment to exhibit phenomena in reliable way."[35] And as Shapin and Schaffer's study illustrates, an entire disciplinary program can

[31] Paul Farber, *The Emergence of Ornithology as a Scientific Discipline: 1760–1850* (Dordrecht: Reidel, 1982).

[32] Susan Leigh Star and James Griesemer, "'Translations' and boundary objects: amateurs and professionals in Berkeley's Museum of Vertebrate Zoology, 1907–39," *Social Studies of Science* 19 (1989): 387–420.

[33] For an account of some of the complexities in the "novice's literary language game" of amateur bird-watching, see John Law and Michael Lynch, "Lists, field guides, and the descriptive organization of seeing: birdwatching as an exemplary observational activity," *Human Studies* 11 (1988): 271–304; reprinted in M. Lynch and S. Woolgar, eds., *Representation in Scientific Practice,* pp. 267–99.

[34] Bruno Latour, *The Pasteurization of France,* trans. Alan Sheridan and John Law (Cambridge, MA: Harvard University Press, 1988); Martin Rudwick, *The Great Devonian Controversy: The Shaping of Scientific Knowledge Among Gentlemanly Specialists* (Chicago: University of Chicago Press, 1985); Robert Marc Friedman, *Appropriating the Weather: Vilhelm Bjerknes and the Construction of a Modern Meteorology* (Ithaca, NY: Cornell University Press, 1989).

[35] Ian Hacking, *Representing and Intervening* (Cambridge University Press, 1983), p. 167.

hinge on an ability to invent, standardize, and legitimate the use of such equipment.

2. Numerous ethnographic and historical studies emphasize that descriptions of observations do not reproduce what an observer originally witnesses in the field; rather, they constitute literary orders and graphic renderings that have textual and pragmatic organization independent of any observational experience. Moreover, the adequacy of scientists' methods reports is inseparable from the ability to produce the described procedure "as a matter of course."[36]
3. The concept of replication is problematic in several respects. As studies by Harry Collins and numerous others have demonstrated (see Chapter 3), the question of what counts as a replication of an experiment is bound together with local inquiries and arguments about what counts as "the same" equipment, "competent" use of that equipment, and "comparable" results.[37]
4. Studies of how scientists communicate findings to other practitioners indicate that descriptions of methods are intertwined with descriptions of particular phenomena. Consider a case described by Garfinkel, Lynch, and Livingston.[38] On the night of January 16, 1969, three astronomers using a telescope and electronic apparatus observed what appeared to be an "optical pulsar," although what they were observing and whether it was a pulsar were subject to the vicissitudes of a "first time through" course of action.[39] After repeating the observation several times under different conditions, while checking their equipment for sources of electronic "noise"

[36] See Bruno Latour and Steve Woolgar, *Laboratory Life: The Social Construction of Scientific Facts* (London: Sage, 1979; 2nd ed., Princeton, NJ: Princeton University Press, 1986); Star and Griesemer, "Translations and boundary objects"; K. Amann and K. Knorr-Cetina, "The fixation of (visual) evidence," pp. 85–122, in Lynch and Woolgar, eds., *Representation in Scientific Practice;* M. Lynch, "Discipline and the material form of images: an analysis of scientific visibility," *Social Studies of Science* 15 (1985): 37–66.

[37] There are many sources for documenting these and related issues. A good source for the problems of replication is H. M. Collins, *Changing Order: Replication and Induction in Scientific Practice* (London: Sage, 1985). Also see Gerald Holton, *The Scientific Imagination: Case Studies* (Cambridge University Press, 1978). For ethnomethodological accounts, see H. Garfinkel, M. Lynch, and E. Livingston, "The work of a discovering science construed with materials from the optically discovered pulsar," *Philosophy of the Social Sciences* 11 (1981): 131–58; and Kathleen Jordan and Michael Lynch, "The sociology of a genetic engineering technique: ritual and rationality in the performance of the plasmid prep," pp. 77–114 in A. Clarke and J. Fujimura, eds., *The Right Tools for the Job: At Work in 20th Century Life Sciences* (Princeton NJ: Princeton University Press, 1992). For a clear and concise discussion of the relationship between instructions and technical actions, see Lucy Suchman, *Plans and Situated Actions* (Cambridge University Press, 1987).

[38] Garfinkel et al., "The work of a discovering science."

[39] Ibid., pp. 132 ff.

and optical imprecision, they sent out a telegram to major observatories throughout the world. This brief telegram announced their findings simply by formulating the date and time, period of the pulse, celestial coordinates, and identity of the "source" star in the Crab Nebula. During that same night, astronomers at other observatories replicated the observation. In this case, the report of the astronomical object did not include instructions on how to verify it. Or rather, the celestial coordinates and frequency reading were the instructions, but only for "anyone" who was prepared to follow them. In this case "anyone" did not include very many people. Of course, there were no guarantees that the observation would be replicated, but the point is that no separate account of a human course of action was necessary. The relevant human actions were astrophysically accountable.[40]

5. The existence of adequate vernacular accounts of scientific method is less of a "grounding" for the possibility of a stable science of human behavior than an indication of the local achievement of such stability. As Garfinkel expresses it, each natural science can be viewed as "a distinctive science of practical action."[41] Although this does imply that each natural science embodies a "natural science of human behavior," the descriptive adequacy of particular reports and methods recipes cannot be separated from the distinctive analytic culture of that science.

This is not to say that scientists do not or cannot reproduce laboratory methods, but it does problematize the final point enumerated by Schegloff in his summary of Sacks's argument: "A way (perhaps *the* way) to have such stable accounts of human behavior is by producing accounts of the methods and procedures for producing it." Methods and descriptions are certainly not

[40] Ibid., p. 140. The issue of astrophysical accountability allows us to consider a critical remark by Schegloff ("From interview to confrontation: observations of the Bush/Rather Encounter," *Research on Language and Social Interaction* 22 [1988–89]: 215–40), who criticized Garfinkel and his colleagues for studying the astronomers' work without first taking account of the "generic domain" of mundane conversation:. "before addressing what is unique, analysis must specify what is the generic domain within which that uniqueness is located" (p. 218). Given the way that a structure of accountability (observation–report–replication) is communicated through a mere mention of astrophysical features, in this instance an alternative "generic domain" – of mundane astronomy – seems relevant to the analysis of what the parties are doing.

[41] Harold Garfinkel, Eric Livingston, Michael Lynch, Douglas Macbeth, and Albert B. Robillard, "Respecifying the natural sciences as discovering sciences of practical action, I & II: doing so ethnographically by administering a schedule of contingencies in discussions with laboratory scientists and by hanging around their laboratories," unpublished manuscript, Department of Sociology, UCLA, 1989, pp. 3ff.

useless, and learning to compose and use step-by-step instructions is an important part of scientific training, but such accounts do not provide the stable grounds for reproducing a practice. Although it is possible to reproduce an observation from a written description, a text can only allude to what eventually may count as a replication of the observation. Schegloff's phrasing also suggests something of a regress: If reproducible methods depend on reproducible accounts of those methods, what accounts for the reproducibility of those accounts? It might be more advisable to say that methods accounts are part and parcel of the concerted practices that enable them to be descriptive and instructive.[42]

The upshot of much research in the sociology of science is that familiar epistemic themes like observation, description, and replication do not provide a "grounding" for natural or social scientific inquiries. Although as I argued in Chapter 5, efforts to give "social explanations" of epistemic activities run into their own difficulties, the sociology of scientific knowledge has succeeded in transforming "the logic of scientific investigation" into a phenomenon for sociological analysis. To an extent, Sacks's early investigations of ordinary descriptions, categorization devices, measurement terms, and inferential practices provided exemplary studies of the central themes of scientific methodology. At the same time, his occasional references to the natural sciences expressed an aspiration to build a science of human behavior whose methods for multiplying witnessing experiences transcended the limitations of common sense. Whereas on the one hand, he conducted inquiries on the vernacular production of observations, descriptions, and replications in, of, and as ordinary activities, on the other hand, he proposed building an objective science of human behavior.

Like Boyle, Sacks succeeded in constructing a specialized technology of virtual witnessing while laying claim to a universal program of observation, description, and replication. To an extent, he shared Boyle's "alchemical" interest in building a laboratory in order to establish an appropriate setting for examining the epistemic crafts, and like Boyle he set in motion a stable program of objective investigation in which systematic "misunderstandings" of the means of scientific production were incorporated into a promotional and instrumental endeavor. In the case of CA, it did not take very long before a program for investigating mundane practices of observation, description, and replication developed into a professional social science discipline.

[42] See Eric Livingston, *The Ethnomethodological Foundations of Mathematics* (London: Routledge & Kegan Paul, 1986) for a series of demonstrations on how the sense and adequacy of a mathematical proof statement depends on the course of action that the statement "describes."

The professionalization of conversation analysis

The development of a professionalized CA can be traced back to the role of "method" in Sacks's argument about primitive science, and specifically to the special status he assigns to "scientific" method. Although CA's operative version of "method" developed out of an ethnomethodological (or praxiological) understanding of scientific and ordinary actions, it gradually took on a more disciplined and scientistic cast. As Schegloff recently observed, Sacks's early proposals for the possibility of a natural observational science of sociology were "undoubtedly motivated, at least in part, by Sacks's engagement with Garfinkel," but they provided a point of departure for a distinct program of study. "For the tenor at least of Garfinkel's arguments was anti-positivistic and 'anti-scientific' in impulse, whereas Sacks sought to ground the undertaking in which he was engaging in the very fact of the existence of science."[43]

As mentioned in Chapter 1, Garfinkel coined the term *ethnomethodology* to describe the methodical production of practical actions and practical reasoning in "lay" as well as "professional" settings. Sacks's understanding of the link between ordinary social actions and methods in the natural sciences and his conception of the natural accountability of practical actions were directly derived from Garfinkel's programmatic writings and exemplary studies. Although Schegloff misunderstands Garfinkel's arguments when he suggests, with some qualification, that their "tenor" was "anti-scientific," he correctly observes that Sacks set in motion a program of study that left behind some of the distinctive commitments associated with Garfinkel's ethnomethodology. As Schegloff observes, the differences between conversation analysis and ethnomethodology can be traced to their divergent orientations to the work of the sciences, but it would be far too simple to say that former approach aspires to be scientific and that the latter is antiscientific in impulse or action.

Even though many of the case studies in ethnomethodology and the sociology of scientific knowledge may "problematize" the logical empiricist terms that Sacks adopted when he proposed a possible science of human behavior, CA would do quite well if its formal descriptions were to attain the historical significance of Boyle's experimental findings. Consequently, my allegation that CA has lost its original relation to ethnomethodology might simply support Schegloff's point that CA follows a scientific rather than an antiscientific agenda. The problem, however, is that many conversation

[43] Schegloff, "An introduction/memoir," pp. 203–04, in *Harvey Sacks – Lectures 1964–1965*. See also Schegloff, "Introduction" to Harvey Sacks *Lectures on Conversation*, vol. 1, p. xxxii.

analysts, like so many aspiring social scientists, have confused the "fact of science's existence" with a grounding for an empirical program. As CA developed from a natural–philosophical mode of investigation into a professional discipline, the very practice of "analysis" incorporated the terms and postures of a logical–empiricist conception of science, and thus the empirical research remained indebted to a no-longer-acknowledged philosophical starting point.

Phase 1: A natural philosophy of ordinary language

In his early lectures, Sacks's investigations often took an overtly "natural philosophical" form even while he made clear that he was trying to build a science of human behavior.[44] A prominent aspect of his natural philosophical investigations was his reliance on the intuitive recognizability of the conversational objects being analyzed. His investigations enlisted "our" recognition of the orderly details of ordinary actions in a critical and reflexive examination of classic versions of language and social action, and not incidentally, Sacks challenged the system of education that promotes a self-reflective and analytic mastery of the "unreflective" details of common knowledge and ordinary action.

In a typical lecture, Sacks would begin by playing a tape-recorded utterance or conversational sequence, and he would then explicate the critical significance of the fragment in light of traditional analytic concerns with indexical expressions, proverbs, paradoxes, structures of argument, and description.[45] He argued that tape-recorded "data" offered an advantage over "imagined" examples of linguistic usage. Since these data were easily accessible, their details could be studied repeatedly and other investigators could use them as a documentary basis for assessing particular analytic claims.

> I started to work with tape-recorded conversations. Such materials had a single virtue, that I could replay them. I could transcribe them somewhat and

[44] See Michael Lynch, "Review of G. Jefferson, ed., *Harvey Sacks–Lectures 1964–1965.*" *Philosophy of the Social Sciences* 23 (1993), 395–402.

[45] In a way, Sacks was using his tape recordings in order to give an intuitive precision to a way of working akin to ordinary language philosophy. It is important to recognize that at this point in his investigations, "language" or "conversation" was not the object of investigation any more than "language" is the object for the philosophy of ordinary language. As Stephen Turner points out (*Sociological Explanation as Translation* [Cambridge University Press, 1980], p. 4), the label is "misleading in that it suggests that the philosophy of ordinary language is 'about' ordinary language, as the philosophy of science is 'about' science. Instead, it is about everything that ordinary language is about: from activities like atonal music to activities like promising." To this it can be added that Sacks's early investigations, when focused on such topics as ordinary descriptions, accounts, and uses of measurement terms, were "about" the general themes in the philosophy of science.

> study them extendedly – however long it might take. The tape-recorded materials constituted a "good enough" record of what happened. Other things, to be sure, happened, but at least what was on the tape had happened. It was not from any large interest in language or from some theoretical formulation of what should be studied that I started with tape-recorded conversations, but simply because I could get my hands on it and I could study it again and again, and also, consequentially, because others could look at what I had studied and make of it what they could, if, for example, they wanted to be able to disagree with me.[46]

The scholarly claims that he judged in this critical way were established "classical" accounts of the elementary logic of investigation, argument, analysis, observation, description, and reasoning. In other words, Sacks used tape-recorded data to launch reflexive examinations of familiar epistemological themes in established traditions of logic and philosophy of science. In effect, he designed a natural philosophical method for exhibiting and inspecting the material organization of observation, description, and replication, the very modes of accountability he associated with primitive natural science.

At this early point in Sacks's investigations, a program of observation – description – replication did not provide a foundation for a science of conversation analysis; rather, these programmatic themes were featured among the topics of investigation. For Sacks, tape-recorded "materials" presented investigators with an inspectable order of details that vastly surpassed even the most insightful reflections or recollections about "our" language and reasoning. Like Wittgenstein, he attempted to review ordinary linguistic competencies for a kind of "therapeutic" respecification of previous scholastic treatments of action and reasoning. And like Wittgenstein's investigations, his reflexive examinations did not take the form of first-person reflections on "our knowledge"; instead, they described what "we" are able to say about public performances, viewed from a third-person perspective. But unlike Wittgenstein, Austin, Ryle, Searle, and other philosophers of language, Sacks used tape recordings of singular conversations rather than recollected examples of characteristic expressions and typical situations. For him, recorded materials provided a strong point of leverage for an explicative investigation, since their intuitively transparent details greatly surpassed the kinds of typical expressions, conversational exchanges, proverbs, and the like that can be recollected when "language is on holiday."

Sacks's preference for tape recording can also be understood – far too easily, it seems – in terms of a Baconian rejection of "speculation" in favor

[46] Sacks, "Notes on methodology," p. 26.

of naturalistic investigations: "The subtlety of nature is greater many times over than the subtlety of the senses and understanding; so that all those specious meditations, speculations, and glosses in which men indulge are quite from the purpose, only there is no one by to observe it."[47] After conversation analysis developed into a research program, such an empiricist and naturalistic understanding of the value of naturalistic "data" (in the form of machine-recorded audio- and videotapes used as a strong simulacrum of "natural conversation") began to supplant a more elusive "reflexive" rationale for consulting the electronic text.

The competencies for performing and understanding the most mundane and impersonal of discursive acts furnished a set of ordinary objects and analytic warrants for Sacks's and his students' investigations. In their terms, ordinary methods for opening and closing conversations, negotiating the transfer of turns, and correcting and avoiding various errors and misunderstandings all employ verbal and gestural components that "anyone" is competent to analyze, and such ordinary analyses are part and parcel of the "naturally occurring" production of the activities themselves.[48] Sacks's celebration of the simple, trivial, and surface understandings through which members conduct their everyday actions was more than an expression of his interest in particular interactional structures because it provided a starting point for a profoundly antitheoretical challenge to prevailing genealogies of social order.

In his own dispute with the inheritors of the Hobbesian conception of the problem of order, Sacks took Boyle's side in the Boyle–Hobbes controversy.[49] Since at least Parsons's *Structure of Social Action,* discussions of the problem of social order had been dominated by a metaphysical picture of the "scientist" observing the world through a conceptual framework or, in more recent terms, a "paradigm."[50] Parsons argued on good neo-Kantian grounds

[47] Francis Bacon, *The New Organon and Related Writings,* Aphorisms, bk. 1, X, ed. Fulton H. Anderson (Indianapolis: Bobbs Merrill, 1971).

[48] "Anyone" does not mean every single person, but any competent member. What "anyone" knows is not established by a statistical survey, since it is reflexive to the situated demonstration of competence.

[49] Shapin and Schaffer, *Leviathan and the Air Pump,* identify a key point of historical rupture in Hobbes's legacy. In the aftermath of his dispute with Boyle, Hobbes's conception of social order became established as the cornerstone for subsequent developments in social and political theory, and his views of natural philosophy were, for the most part, dismissed. Boyle, of course, succeeded in promoting the experimental way of life to the point of displacing the natural–philosophical mode of investigation. Like Boyle, Sacks began his investigations with an interest in the natural–philosophical phenomenon of witnessing, and like Boyle he insisted on a mode of procedure that began with heterogeneous facts rather than a pervasive theoretical scheme. See Steven Shapin, "Robert Boyle and mathematics: reality, representation, and experimental practice," *Science in Context* 2 (1988): 23–58

[50] See especially the first two chapters of Talcott Parsons's *The Structure of Social Action,* vol. 1 (New York: McGraw Hill, 1937). Parsons's account of science is far more resilient than

that there was no getting around the theory-ladenness of perception, so that the first order of business was to take charge of the implicit conceptual framework that guides and governs the categorical structure of scientific observation. Although he could provide no guarantee that an explicitly constructed theory would avoid contamination from residual commonsense assumptions, Parsons tried to reconstruct the observer's implicit knowledge into a logically ordered set of conceptual elements and empirical propositions.

Parsons transposed a theory-centric view of science into a general conception of social action in which the ordinary actor became the bearer of a moral order.[51] The actor's orientation incorporated a complex normative framework, including cultural norms and values, anticipations of sanctions, and learned dispositions for performing appropriate role behaviors. The actor's internal model of social structure was not scientific, since it was predominantly normative and its conceptual elements were protected from rigorous critical scrutiny, but the role of the actor's model in ordinary action was analogous to that of a theory in a deductive system of empirical explanation.

In both cases, a systematic conceptual framework is granted a guiding role for directing the agent's attention to relevant facts, and it is responsible for the alignments among members of an epistemic–moral community. In such a conception of action, "unreflective" understandings can never be taken at face value, because what may appear simply to be "out there" to the naive observer can be traced back to a scheme of interpretation that orients the observer's attention, selectively organizes available information, and imposes categorical and normative judgments on the perceptible manifold. The task for research and education is to bring to light and to reexamine critically the assumptions in such schemes.

many sociologists realize. Although Parsons's "theory" of social structure and social action is often said to be but one of the several theoretical "paradigms" in contemporary sociological theory, the very notion of "paradigm" that sociologists commonly use is more indebted to Parsons than to Kuhn. Kuhn's affiliation with a theory-centric sociology is articulated in Jeffrey Alexander, *Positivism, Presuppositions, and Current Controversies*, vol. 1: *Theoretical Logic in Sociology* (Berkeley and Los Angeles: University of California Press, 1982).

[51] The inverse also applies: Parsons's conception of science was an instance of his functionalist theory of society. Parsons (*The Structure of Social Action*, vol. 1, pp. 6ff.) portrayed scientific fields as functional systems of empirical propositions. A modification of any proposition in such a system implicated changes to a greater or lesser degree in the other propositions in the system. The propositions in such a system were related to and contingent on empirically observable facts, but the system was also, in Parsons's terms, an "independent variable" in the development of a science. Although Parsons distinguished scientific knowledge and rationality from commonsense knowledge and substantive rationality, in its general outlines his theory of the social system also emphasizes a system of interrelated statements (in this case norms rather than empirically verifiable statements), which orients the actor to relevant aspects of the everyday world.

Although Sacks expressed no interest in valorizing common sense, he did challenge the tendency to discount the "mere appearance" of an intelligible world while searching for an abstract, reflectively examinable foundation for such intelligibility. For instance, before one of his lectures he gave his students an assignment that required them to observe and describe people exchanging glances in public places. After reading the students' reports, he made the following remarks to his class:

> Let me make a couple of remarks about the problem of "feigning ignorance." I found in these papers that people will occasionally say things like, "I didn't really know what was going on, but I made the inference that he was looking at her because she's an attractive girl." So one claims to not really know. And here's a first thought I have. I can fully well understand how you come to say that. It's part of the way in which what's called your education here gets in the way of your doing what you in fact know how to do. And you begin to call *things* "concepts" and *acts* "inferences," when nothing of the sort is involved. And that nothing of the sort is involved is perfectly clear in that if it were the case that you didn't know what was going on – if you were the usual made up observer, the man from Mars – then the question of what you would see would be a far more obscure matter than that she was an attractive girl, perhaps. How would you go about seeing in the first place that one was looking at the other, seeing what they were looking at, and locating those features which are perhaps relevant?[52]

Sacks was not proposing an epistemological grounding for the validity of observation; rather, he was pointing to the utterly "groundless" and naive intelligibility of social objects and social acts.[53] When he admonished his students for calling "*things* 'concepts' and *acts* 'inferences'," he called into question the educated precautions by which they subverted the categorical intelligibility of actions *seen at a glance* and those *seen in the glancing*.[54] He suggested that in their efforts to formulate a methodologically "reasoned"

[52] Harvey Sacks, "On exchanging glances," lecture 11, pp. 335–36, in Jefferson, ed., *Harvey Sacks – Lectures 1964–1965.*

[53] See Dusan Bjelic, "On the social origin of logic" (Ph.D. diss., Boston University, 1989). Also see Eric Livingston, *Making Sense of Ethnomethodology* (London: Routledge & Kegan Paul, 1987), chaps. 12 and 13.

[54] For an account of "seeing at a glance," see David Sudnow, "Temporal parameters of interpersonal observation," pp. 259–79, in D. Sudnow, ed., *Studies in Social Interaction* (New York: Free Press, 1972). The reference to "inferences" in the preceding passage from Sacks's lecture can be read as a critical reference to Erving Goffman's interactionist studies. Read in this way, Sacks's rebuke to the student in his class applies no less forcefully to Goffman, since it questions the assumption that orderly interactional practices can be analytically explicated by speaking of a complex relationship between "impressions" given (and given off) by a person and the "inferences" made by that person's witnesses. See Erving Goffman, *The Presentation of Self in Everyday Life* (Garden City, NY: Doubleday, 1959), pp. 2–3.

version of what they otherwise saw without having to think about it, the students were pursuing a kind of educated agnosia in which they forgot the commonplace phenomena they set out to analyze. A vivid, albeit tragic, example of such "exact" description is provided by "Dr. P.," the brain-damaged "man who mistook his wife for a hat," described in neurologist Oliver Sacks's clinical tales:

> "What is this?" I asked, holding up a glove.
> "May I examine it?" he asked, and, taking it from me, he proceeded to examine it as he had examined the geometrical shapes.
> "A continuous surface," he announced at last, "infolded on itself. It appears to have" – he hesitated – "five outpouchings, if this is the word."
> "Yes," I said cautiously. "You have given me a description. Now tell me what it is."
> "A container of some sort?"
> "Yes," I said, "and what would it contain?"
> "It would contain its contents!" said Dr. P., with a laugh. "There are many possibilities. It could be a change purse, for example, for coins of five sizes. It could . . ."[55]

Dr. P.'s utterly correct description expresses a deep aberration that estranges him from the known-in-common and taken-for-granted naiveté through which persons typically see things without having to decompose them into constituent elements. It is as though Dr. P's *Lebenswelt* is reduced to the elementary sense data conjured up by Frege and Russell. Harvey Sacks, to the contrary, enjoined his students to include their "preconceived ideas" in their observational accounts. In his view of observation, the highly prejudicial category "attractive girl" was seen before any analysis. Moreover, it was instantaneously seen on behalf of others and before any Cartesian separation of the elements in the "actual" scene from those inherent in the subject's perspective.

For Sacks, the naive facility through which one person sees an object and sees that it is an object for others is a matter of membership more than of perception and cognition; persons are in the social world not as sentient bodies absorbing information but as members (in the sense of being the surface "organs" of a pervasive and unremitting molecular production) whose accountable acts contribute to the "assembly" of naturally organized ordinary activities.[56]

[55] Oliver Sacks, *The Man Who Mistook His Wife for a Hat and Other Clinical Tales* (New York: Harper & Row, 1987), p. 14.

[56] The phrase "naturally organized ordinary activities" is one of Garfinkel's characteristic usages. Sacks and his colleagues were more inclined to use phrases like "naturally occurring activities."

The molecular techniques through which such "assembly" is accomplished are conceptually distinct from conscious or unconscious beliefs because there is no implication of a single reasoning (or unreasoning) agent directing the action. Instead, constituent acts are produced in actual assemblages in which they "latch" together in rapid sequence like molecules in an organic chain (see the Appendix to this chapter). "Inference" and "cognition" are implicated only as secondary products or analytic reconstructions of how particular assemblages of acts must have been produced.[57] The speed of the assembly outstrips any effort to reason abstractly about it. In a particularly revealing passage, Sacks suggests that it makes no more or less sense to speak analytically of the "brains" of a molecule than of those of a human agent:

> There is no necessary fit between the complexity or simplicity of the apparatus that you need in order to construct some object, and the face-value complexity or simplicity of the object. These are things which you have to come to terms with, given the fact that this has indeed occurred. And in so far as people are doing lay affairs, they walk around with the notion that if somebody does something pretty simply, pretty quickly, or pretty routinely, then it must not be much of a problem to explain what they've done. There is no reason to suppose that is so. I'll give an analogical observation. In a review of a book attempting to describe the production of sentences in the English language – a grammar, in short – the reviewer observes that the grammar, though it's not bad, is not terribly successful, and it remains a fact that those sentences that any 6-year-old is able to produce routinely, have not yet been adequately described by some persons who are obviously enormously brilliant scientists. Of course the activities that molecules are able to engage in quickly, routinely, have not been described by enormously brilliant scientists. So don't worry about the brains that these persons couldn't have but which the objects seem to require. Our task is, in this sense, to build their brains.[58]

For Parsons, a complex social structure is microcosmically represented in the conceptual framework through which actors reproduce that structure, but

[57] Stanley Fish (*Doing What Comes Naturally: Change, Rhetoric, and the Practice of Theory in Literary and Legal Studies* [Durham, NC: Duke University Press, 1989], p. 386) uses the analogy of a "chain" when speaking of reasoning in the practice of law: "The agent embedded in a chain enterprise is the natural heir of the constraints that make up the chain's history. As a link in the chain he is a repository of the purposes, values, understood goals, forms of reasoning, modes of justification, etc., that the chain at once displays and enacts." Such an agent has no need to consult a fully articulated model or theory of the practice when acting, since the action is made relevant by its historical place in the chain. In conversation analysis, the molecular links in the chain are not agents but constituent actions performed by multiple agents.

[58] Harvey Sacks, "The inference making machine," pp. 199–200, in Jefferson, ed., *Harvey Sacks – Lectures 1964–1965.*

for Sacks, simple molecular acts performed routinely and in locally organized combinations generate a complex product. This antitheoretical picture can be illuminated by referring to a study by David Turnbull on how the great gothic cathedrals were built.[59] Turnbull observes that historians of architecture have been bedeviled by the assumption that the enduring, highly complex, and geometrically exact structures of the great cathedrals must have been generated from elaborate plans incorporating sophisticated engineering principles. The absence of evidence for such plans from the historical record has led some historians to conclude either that such plans existed and were destroyed or that the cathedrals were a mysterious result of nonscientific trial-and-error methods. Turnbull refers to recent laboratory studies in order to attack the opposition between engineering knowledge and *bricolage,* and he argues that the cathedral builders had no need for elaborate plans and that their atheoretical methods were by no means unscientific. Instead of assuming that the cathedral builders started with elaborate plans and complex mathematical principles, he observes that the cathedrals resulted from the builders' and masons' localized uses of stencils or "templates" from which they developed standardized shapes for stones and a simple set of tools and reckoning devices:

> In the absence of rules for construction derived from structural laws, problems could be resolved by practical geometry using, compasses, a straightedge, ruler, and string. The kind of structural knowledge that is passed on from master to apprentice relates sizes to spaces and heights by ratios, such as half the number of feet in a span expressed in inches plus one inch will give the depth of a hardwood joist. These rules of thumb are stated as, and learnt as, ratios; for, as the span gets larger, the depth of the joist will too. This sort of geometry is extremely powerful. It enables the transportation and transmission of structural experience. It makes possible the successful replication of a specific arrangement in different places and different circumstances.[60]

This picture of activity is congruent with two of the more vivid images in Wittgenstein's *Philosophical Investigations (PI).* First, the templates – inscriptions from which the shapes of the stones were cut – had a role in cathedral building much like that of the utterances, "slab," "block," "pillar," and "beam" in Wittgenstein's imaginary primitive language game (sec. 2 ff.) in which a builder calls out a succession of these names to his assistant to signal him to bring an appropriately shaped stone. Wittgenstein designed this

[59] David Turnbull, "The ad hoc collective work of building gothic cathedrals with templates, string, and geometry,"*Science, Technology, and Human Values.* 18 (1993):315–40.
[60] Ibid., p. 323.

language game as something of a parody of a traditional version of language as a collection of names for objects. However, as Norman Malcolm points out, even the "slab" game is not as restrictive as it might seem at first. The limited repertoire of terms can conceivably take on all sorts of pragmatic functions when employed as part of the builders' routines.[61] In this severely restricted language game, the utterances have a conventional role as names for objects, but at the same time we can imagine a builder using them as verbal tokens in a sequential activity in which he requests, corrects, or affirms a reciprocal act by an assistant. By extending Wittgenstein's example, one could imagine a dialogue like the following interchange between the builder and his assistant:

Builder: "Slab."
Assistant: (Hoists block and offers it to builder.)
Builder: "Slab!" (Shakes head and points back to pile.)
Assistant: "Slab?" (Puts block away and picks up a differently shaped stone.)
Builder: "Slab." (Nodding and smiling while taking slab from assistant.)

A more complicated instance of such an exchange was recorded while two moving company employees lowered a refrigerator down a narrow and winding staircase. As in the idealized "slab" example, the utterances were sensibly bound to the presumptively "evident" properties of the object being lifted, along with the stairs, the confining walls, and the developing "methods" for contending with their contingent presence. The dialogue was produced as a performance in which instructions on how to do that very performance were passed between an "old hand" (A) and a novice assistant (B) who performed the "heavy lifting":

A: Okay, now,
(1.4)
A: I'm gonna <u>lift</u>.
(0.8)
(B): (uh huh)
A: <u>Okay</u>?
((Thump – as refrigerator is audibly moved))
A: And <u>you</u> lift <u>too</u>. ((More thumping noises))
B: (tell me where the ...)
A: (<u>Up</u>?)
B: Yeh
A: Okay,
((Thump))

[61] See Norman Malcolm, "Language without conversation," *Philosophical Investigations* 15 (1992): 207–14.

A: (let's do it)
(3.8) ((intermittent thumping))
A: We got it, we're doin' good.
(0.4)
A: We're doin' a good job
(B): (em hmm)
(4.0)
A: Okay, now,
(0.2)
A: I'm gonna set it down again,
((Loud Thump))
A: I'm gonna tryan' do the same thing again, okay?
(0.5)
B: Pick this up?
(0.4)
A: Yup. Now it's down.
(0.8)
A: (now)
(0.4)
A: Up teh me,
(0.2)
A: That's it, okay.
(0.8)
A: Now,
(0.4)
A: let's try an' do the same thing.

The "utterances" in this case act as moments punctuating and coordinating the serial raising and lowering of the object, and they take on a pace and rhythm that is as much a function of the stairs as it is of any "conversational" mechanism.

In Turnbull's example, the cathedral builders' templates were textual devices featured in the communicational and disciplinary routines through which a master mason and a staff of builders coordinated their activities. The templates acted as plans only in a limited and "indexical" sense because they were subsumed within the traditional skills and tools of the trade.

The other relevant analogy from Wittgenstein is that of a toolbox including a hammer, pliers, saw, screwdriver, rule, glue pot and glue, nails, and screws.[62] Wittgenstein invites his readers to consider the "functions of words" to be like the heterogeneous functions of tools in constructive activity. In the concrete case that Turnbull discusses, he emphasizes that the cathedral builders' *bricolage* practice made flexible use of compasses,

[62] Wittgenstein, *PI,* sec. 11.

straightedges, rulers, and string in an open-ended practice. These tools were not "dedicated" to particular tasks; rather, they were adapted to an unforeseen range of tasks and contingencies. The simple designs of the templates and the tools in the toolbox did not "represent" a practical objective in the way that an elaborate plan or theory is said to govern, explain, define, or represent the goal of a relevant activity, but their competent use required no plan and no explanation in order to produce and reproduce effectively an elaborate and emergent architecture. Nor does the absence of such a plan or explanation imply that one was missing.

With his view of social order as an immense grammatical "cathedral" built up from a heterogeneous collection of simple devices, Sacks began to describe some of the most primordial of these devices. In one of his most remarkable demonstrations, he explained the intelligibility of a two-year-old child's utterance, "The baby cried. The mommy picked it up."

> When I hear "The baby cried. The mommy picked it up," one thing I hear is that the "mommy" who picks the "baby" up is the mommy of that baby. . . . Now it is not only that I hear that the mommy is the mommy of that baby, but I feel rather confident that at least many of the natives among you hear that also.[63]

Sacks observed that the story includes two sentences and that the "occurrences" in the narrative follow one another in the same order that the sentences follow one another. He added that the first occurrence "explains" the second occurrence (the mommy picked up the baby because the baby cried). These observations were not offered as "social science findings" but as explications of intelligible features of the story that anyone should be able to recognize:

> All of the foregoing can be done by many or perhaps any of us without knowing what baby or what mommy it is that might be talked of. . . . They "sound like a description," and some form of words can, apparently, sound like a description. To recognize that some form of words is a possible description does not require that one must first inspect the circumstance it may be characterizing.[64]

In other words, Sacks raised what has become a familiar claim in contemporary literary theory, that a text can be "iterated" without referring to such contextual matters as the time at which it was uttered, the identity of the speaker,

[63] Harvey Sacks, "On the analysability of stories by children," p. 216, in Roy Turner, ed., *Ethnomethodology* (Harmondsworth: Penguin Books, 1974); originally in John J. Gumperz and Dell Hymes, eds., *Directions in Sociolinguistics: The Ethnography of Communication* (New York: Holt, Rinehart and Winston, 1972), pp. 329–45.

[64] Ibid., p. 217. It may seem as though Sacks were basing his analysis on a Schutzian account of the "reciprocity of perspectives." See Alfred Schutz, "The dimensions of the social world," in his *Collected Papers,* vol. 2 (The Hague: Nijhoff, 1964), pp. 20–63. There is a key difference, however, between Schutz's discussion of the interpretive understanding of a social scene on the basis of a fund of knowledge and Sacks's account of a preinterpretive seeing at a glance. For a nonpositivistic account of such intelligibility, see Wittgenstein's discussion of "seeing-as" in *Philosophical Investigations,* pp. 193–208.

or the speaker's intentions. Unlike many enthusiasts of literary theory, however, Sacks was intrigued by the possibility of building a social science grounded in descriptions. Although this aim might seem naively realistic, it should be clear from the following passage that Sacks was not a "realist," at least in the conventional sense of that word. Instead, he was more inclined to investigate the conventional ways in which "realistic" descriptions are organized.

> If . . . members have a phenomenon, "possible descriptions" which are recognizable *per se,* then one need not in the instance know how it is that babies and mommies do behave to examine the composition of such possible descriptions as members produce and recognize. Sociology and anthropology need not await developments in botany or genetics or analyses of the light spectra to gain a secure position from which members' knowledge, and the activities for which it is relevant, might be investigated. What one ought to seek to build is an apparatus which will provide for how it is that any activities, which members do in such a way as to be recognizable as such to members, are done, and done recognizably. . . . The sentences we are considering are after all rather minor, and all of you, or many of you, hear just what I said you heard, and many of us are quite unacquainted with each other. I am, then, dealing with something real and something finely powerful.[65]

Again, Sacks was alluding to the possibility of an analysis here, and at the end of this passage he went only so far as to indicate the existence of a "finely powerful" machinery. He proposed that it was now possible to "build . . . an apparatus" that recovers the recognizable features of descriptions. For Sacks, such an apparatus is implied by the prereflective way in which members hear utterances and act in accord with the "heard" order of events in a conversation.[66] This order is a demonstrably impersonal and iterable order, a machinery that organizes ordinary interactional events.

[65] Sacks, "On the analyzability of stories by children," p. 218. Livingston (*Making Sense of Ethnomethodology,* p. 76) points out that "analyzability" in this instance implied a practical objectivity that is distinct from any claim that one or another academic analysis of the utterance is objectively correct: "Nothing critical depended on his [Sacks's] analysis being absolutely correct in this one instance. The phenomenon that he had begun to elucidate is that the analyze-ability, or story-ability, or hear-ability, or objectivity of the sequence is part of the sequence itself. The 'mommy' is the mommy of the 'baby,' and she picked her baby up. That analyzability is part of the way the story was told and heard." Livingston (p. 76) goes on to say that some of Sacks's collaborators confused the demonstrable analyzability of tape-recorded utterances with a methodological ground for particular analytic overhearings of those utterances: "They used the notion of what a 'member,' i.e. a co-conversationalist in a conversation's local production cohort, definitely hears as a means of justifying their work practices. Their notion of a 'member' became a straightforward analytic device that they enforced as the grounds for collaborative discussion and research." For a different view of Sacks's and his colleagues' shift in "analytic stance and procedure," see E. A. Schegloff, "Introduction" to Sacks, *Lectures on Communication,* pp. xliii–xliv.

[66] The preinterpretive intelligibility of conversation in action is comparable to what Wittgenstein says about following a rule "blindly" and acting intelligibly but without "reasons" (*Philosophical Investigations,* sec. 211, 219). Unlike Sacks, however, Wittgenstein is studiously reticent about making any suggestion about the existence of an "apparatus" governing such "blind" actions.

The mechanistic imagery in this passage is typical of the references to machinery, mechanism, device, apparatus, and system that pervade conversation-analytic writings. Given these mechanistic vocabularies, one might figure that Sacks's program is vulnerable to the following criticism, expressed by Peter Winch:

> It is quite mistaken in principle to compare the activity of a student of a form of social behaviour with that of, say, an engineer studying the workings of a machine. . . . His understanding of social phenomena is more like the engineer's understanding of his colleagues' activities than it is like the engineer's understanding of the mechanical system which he studies.[67]

A mechanistic account of an engineer's understanding of her colleagues' activities may reflect what Gilbert Ryle called a *category mistake,*[68] but by Sacks's reckoning a grammatical "apparatus" is implied by the way that engineers manage to reproduce mechanical structures from one another's descriptions. Such an apparatus would describe what Ryle himself proposed to systematically investigate: "the logical regulations" governing the practical uses of concepts.[69] Sacks argued that it should be possible to bring to light the systematic organization of descriptions through which a competent member informs and instructs relevant colleagues about how to produce their collective activities. Of course, Sacks did not attempt to develop a study of engineers' communicational activities. Rather, he hoped to produce systematic descriptions that "anybody" could use as instructions for performing commonplace activities.

If Sacks's argument about methodological description in science also applies to engineering, then engineers' accounts of mechanical systems should enable their colleagues, students, and hired technicians to understand and reproduce what those accounts describe and instruct. By analogy, possible descriptions of the machineries of conversation should recover how competent speakers of a language manage to collaborate with relevant "colleagues" to understand and reproduce the relevant conversational actions. Professional engineers often do not write their own methods texts; they hire technical staff to draw blueprints for other engineers, compose instructions for students, and write manuals for machine users. Similarly, participants in ordinary conversation rarely take the trouble to codify the systematic features of their methods. This leaves the door open for professional

[67] Peter Winch, *The Idea of a Social Science* (London: Routledge & Kegan Paul, 1958), p. 88.
[68] Gilbert Ryle, *The Concept of Mind* (Chicago: University of Chicago Press, 1949), pp. 16 ff.
[69] Ibid., p. 7. It should be mentioned, however, that Sacks did not limit his investigation to "concepts" in Ryle's sense. Instead, he subsumed conceptual analysis into an investigation of the sequential organization of talk. See Jeff Coulter, *Rethinking Cognitive Theory* (New York: St. Martin's Press, 1983).

conversation analysts to take up the task of writing the methods texts and users' manuals for such activities.

Sacks's argument about the role of description in science raises a number of interesting ambiguities. Although he preferred to draw an analogy between "scientists'" and conversation analysts' descriptions of methods, different implications arise when we compare the formal descriptions provided by conversation analysts with such cases as the following: a blueprint instructing competent engineers on the design of a machine, an industrial engineering plan for coordinating and pacing the actions and machineries on an assembly line, a users' manual instructing novices on how to operate a personal computer, or a set of instructions for a staff of technicians on how to perform and monitor selected events in an experiment. Each of these instances implicates a different division of labor and social distribution of knowledge between those who write the methods texts and their "colleagues" who perform the activities described. The analogy with conversation analysis is further complicated by the fact that the actions described are actions that the "technicians in residence" are already competent to do; indeed, their competent activities supply the "data" analyzed in the first place. Unlike the industrial engineer who uses formal description with the explicit aim of "extracting" the craft from a group of practitioners so as to reorganize and "rationalize" those skills, with the exception of some confused attempts at social criticism, conversation analysts express no aim to build a technology separate from the locally produced order of activities described.

Sacks tried to describe a sort of "machinery" different from the mechanical systems represented in engineering blueprints; he tried to construct accounts of how the "technicians in residence" at the conversational worksite assemble their ordinary communicational activities.[70] "Ideally, of course, we would have a formally describable method, as the assembling of a sentence is formally describable. The description not only would handle sentences in general, but particular sentences. What we would be doing, then, is developing another grammar. And grammar, of course, is the model of routinely observable, closely ordered social activities."[71]

According to the engineering analogy, participants in conversation assemble activities, and while doing so they analyze one another's utterances in order to determine who should speak next, when they should start talking,

[70] I owe the expression "technicians in residence" to David Bogen (personal communication). Gail Jefferson uses the term *template* to describe a particular analytic operation in conversation, but in the above passage I am using it in a more general way, to describe the patterns from which complex conversational orders are built up moment by moment. See Gail Jefferson, "On the sequential organization of troubles talk in ordinary conversation," *Social Problems* 35 (1988): 418–42.

[71] Sacks, "Notes on methodology," p. 25.

and what they should say. Again according to the analogy of cathedral builders, this "analysis" is implicated in the orderly shaping and conjoint placement of standard building blocks. It is not, for the most part, thoughtful work, or even unconsciously or prereflectively thoughtful work, but neither is it thoughtless or mindless.

By proposing to "build the brains" that organize the molecular acts through which conversations become organized and accountable, Sacks laid the groundwork for a descriptive program. This would not, however, be the sort of behaviorist or materialist program that reduces the world that a naive observer sees at a glance to a preconceived field of "really real" objects, identities, and relationships. Instead, it would be a descriptive program in which a fixation on observable detail takes into account a full range of intuitively evident and naturally accountable "existential" predicates.[72] In principle, everything would be included among the observable and describable objects for such a program, such as "observation," "description," "analysis," "evidence," and "raw data accounts."[73] There would simply be no time out, and no privileged space, for constructing rectified versions of these vernacular activities for use as exclusively scientific resources.[74]

Sacks did not let the often-mentioned fact that observation and description are reflexive features of the activities that sociologists describe prevent him from treating those activities as ordinary organizational things. That persons are able to see at a glance what others are doing and that these "others" are able to account for that observability in the very way they act were simple facts of life in accordance with his analytic policies. By transforming the heterogeneous phenomena of practically observable and practically analyzable actions into "data" for laboratory-based descriptions and analyses, Sacks

[72] Gian-Carlo Rota distinguishes what he calls "existential observation" from the more familiar types of realistic observation in science. In contrast with many readers of Heidegger, Rota treats the existential critique of materialism not to be an extension of the idealist tradition but, instead, to be an unprecedented insistence on a kind of hyperrealism that is indifferent to reductionist concepts of a "reality" behind appearances. See Gian-Carlo Rota, "The end of objectivity," a series of lectures for The Technology and Culture Seminar at MIT, Cambridge, MA, October 1973. Although Sacks was silent about existentialism and seemed far more attuned to the analytic traditions in philosophy, his descriptivism was similarly nonreductionistic and hyperrealistic.

[73] Some of the more ethnomethodologically attuned studies in CA exhibit this hyperobjectivist, as opposed to positivist stance. Anita Pomerantz, for instance, discusses the occasions on which speakers present descriptions of "just the facts." See Anita Pomerantz, "Pursuing a response," pp. 152–63, in Atkinson and Heritage, eds., *Structures of Social Action*, p. 163, n. 1. Also see A. Pomerantz, "Telling my side: 'limited access' as a 'fishing' device," *Sociological Inquiry* 50 (1980): 186–98. Pomerantz describes "just the facts" and "raw data" as members' usages in particular situations while remaining indifferent to any invidious distinction between such members' usages and scientific facts and data.

[74] See Don Zimmerman and Melvin Pollner, "The everyday world as a phenomenon," pp. 80–103, in Jack Douglas, ed., *Understanding Everyday Life: Toward the Reconstruction of Sociological Knowledge* (Chicago: Aldine, 1970).

offered a descriptive program in which the themes and entitlements under which positive sciences are conducted became pragmatic data in the world described.

Phase 2: An analytical discipline

Sacks said little about how primitive sciences become "professionalized" into technical disciplines, but it is clear that he and his colleagues rapidly moved beyond the "short-term possibility" of a primitive science by developing a mode of study that aimed to surpass vernacular intuitions with more refined observational and analytical technologies. Initially, Sacks's descriptivism coexisted, and comfortably so, with a sophisticated understanding of the literatures and themes that later came to be associated with "poststructuralism" in the humanities and social sciences. While "doing mere description," he and his colleagues displayed the blasé detachment of persons who knew the game they were playing all too well. "Doing description" was a necessary condition for making a reflexive exhibit of the constituent structures of positive science: observation, description, verification, raw data, and the rest. These structures of accountability were embodied in the conversation-analytic laboratory that Sacks set up in the late 1960s and early 1970s at the University of California at Irvine. This laboratory provided a setting in which the molecular constituents of "positive science" were themselves made observable.[75]

The conversation-analytic laboratory became an installation for exhibiting and examining the "technical" production of ordinary actions. The laboratory was outfitted with equipment for recording, playing back, and editing audiotape and videotape recordings, and it housed an archive of tapes and transcripts that had been amassed by Sacks and his students. The transcripts were written in accordance with a unique notation system developed by Gail Jefferson. These data were indexed and filed, and they were circulated in the small community of copractitioners. Sacks devised a program for training his students to be "technicians" who labored intensively on the tapes and cultivated unique abilities to "hear" and transcribe subtle features of the sequence, pace, timing, and voicing of the utterances recorded on the tapes. Although the laboratory contained various items of equipment, its centerpiece was a more abstract "machinery" for (re)producing matters of fact. Like Boyle's pictures and descriptions of the air pump presented to a

[75] In his more recent writings and lectures, Garfinkel uses the term *perspicuous setting* to point to the existence of organized settings in which "classical" themes from the literatures in philosophy and social science are featured as ordinary practical accomplishments. When I speak of Sacks's laboratory as a perspicuous setting, I am suggesting that it was designed to examine the elementary themes of a positive philosophy of science: observation, description, and replication, among others.

community of virtual witnesses, Sacks's machine organized a domain of facts that for all appearances were palpable and subject to material inspection and manipulation while at the same time these facts were essentially bound to a literary mode of presentation.[76] To say this does not detract from the innovative and powerful way in which the machinery organized the practices of a community of analysts.

By the early 1970s, Sacks and his colleagues had assembled a self-consciously "technical" discipline that included an established set of investigative procedures, an analytic discourse, and a communal organization. As their findings began to accumulate, they consolidated those findings into a set of formal "systems" for the sequential organization of talk. These systems included rules and "machineries" governing how speakers in conversation construct turns at talk, open and close conversations, repair conversational errors and trouble, initiate and sustain topical talk, and organize "adjacency pair" utterances such as greetings and return greetings, questions and answers, requests and responses, and other reciprocally organized discursive structures. A relatively well defined community of conversation analysts emerged, and its more active members attended specialized conferences, produced collections of CA studies, trained their students to be specialists in the discipline, and formed an intensive and relatively exclusive co-citation network.

Whereas Sacks's initial discussions of conversational "things" and "machineries" once provided a kind of Wittgensteinian counterpoint to the metaphysics of subjective agency, the "things" and "machineries" described in CA gradually were treated as an objective grounding for corporately enforced technical practices in a new social science discipline.[77] To make a long story short, CA's practitioners turned the thematic foci of their reflexive investigations into the programmatic grounding for a science of interactional behavior. The objectified findings produced in CA were valued in their own right, and this drew attention away from a more "alchemical" interest in the crafting of observation, description, and replication.[78] Observation, descrip-

[76] Literary and rhetorical features of conversation analytic work are discussed in R. J. Anderson and W. W. Sharrock, "Analytic work: aspects of the organization of conversational data," *Journal for the Theory of Social Behaviour* 14 (1984): 103–24; Erving Goffman, "Replies and responses," pp. 5–77, in E. Goffman, *Forms of Talk* (Philadelphia: University of Pennsylvania Press, 1981); Elliot G. Mishler, "Representing discourse: the rhetoric of transcription," *Journal of Narrative and Life History* 1 (1991): 255–80; and David Bogen, "The organization of talk," *Qualitative Sociology,* 15 (1992): 273–96.

[77] Cf. Livingston, *Making Sense of Ethnomethodology,* p. 85. As Livingston acknowledges, his critical remarks on conversation analysis are drawn from Garfinkel's lectures and unpublished writings on the topic of formal analysis.

[78] The reference to alchemy derives from Trent Eglin's insight that an alchemical order – a reflexive program for exhibiting and analyzing the constituents of material "craft" – remains tacitly embedded in the local production of natural science. This sense of alchemy differs profoundly from the popular image of alchemy as a misguided prescientific program for transmuting lead into gold. See Trent Eglin, "Introduction to a hermeneutics of the occult: alchemy," pp. 123–59, in H. Garfinkel, ed., *Ethnomethodological Studies of Work* (London: Routledge & Kegan Paul, 1986).

tion, and replication gradually became instrumental features of a conversation-analytic expertise. Consequently, the ethnomethodological initiatives that had once been prominent in CA became buried in a positivist packaging of findings on the structures of naturally occurring conversation.[79]

The turn-taking machine

The research in conversation analysis continues to be very diverse, and it would be inaccurate to suppose that it is governed by a coherent set of methodological prescriptions. Nevertheless, a substantial body of work in the field has coalesced around a formal model of the conversational structures, which was presented in Harvey Sacks, Emanuel Schegloff, and Gail Jefferson's 1974 paper, "A Simplest Systematics for the Organization of Turn-taking in Conversation."[80] The paper reviewed a large body of research on audiotaped phone calls, group therapy sessions, service encounters, and other routine modes of interaction. Sacks and his colleagues' model articulated a basic set of rules[81] governing the orderly administration of speaking turns in conversation, and just as important, the paper itself became a model (or exemplar) for CA's corporate community.[82]

Aside from its substantive claims about the organization of "talk in interaction," the paper's prolix technical style and method for presenting transcribed data set a standard for subsequent research in the field. The stylistic and analytic organization of the paper demonstrated that CA had moved beyond a natural-philosophical mode of investigation that was once

[79] Perhaps a similar fate befell Andy Warhol's factory (cf. Carolyn Jones, "Andy Warhol's factory," *Science in Context* 4 [1991]: 101–31). The factory initially was designed as a productive installation, where "factory" was the artistic theme and the standardized artistic products were part of the parody represented by the entire scene. But when the artistic products of the factory became valued as commodities in their own right, the factory was no longer the primary installation of Warholian art. Instead, it became a place where publicly valued artifacts were produced.

[80] Harvey Sacks, Emanuel Schegloff, and Gail Jefferson, "A simplest systematics for the organization of turn-taking in conversation," *Language* 50 (1974): 696–735.

[81] The concept of "basic rules" is discussed in Felix Kaufmann, *Methodology of the Social Sciences* (New York: Humanities Press, 1944). Kaufmann contrasts the basic rules of a game with what he calls *preference rules.* The basic rules of chess define the game in a context-free way, whereas preference rules cover options that open up in the course of play. Schegloff ("An introduction/memoir") alludes to Sacks's familiarity with Kaufmann's work, and Sacks was evidently influenced by Garfinkel's application of Kaufmann's distinction between "basic" and "preference" rules. See Harold Garfinkel, "A conception of, and experiments with, 'trust' as a condition of stable concerted actions," pp. 187–238, in O. J. Harvey, ed., *Motivation in Social Interaction* (New York: Ronald Press, 1963).

[82] The suggestion of a "corporate community" is provided in the following sentence from Garfinkel and his colleagues, "Respecifying the natural sciences as discovering sciences of practical action," app. I: "Postscript and preface," p. 65: "Latter day CA which, since Harvey Sacks's death, insists upon coded turns' sequentially organized ways of speaking of talk and structure, makes talk out as structure's mandarins: ruling it, insiders to everything that counts, dreaming science, all dignity, pedantic, and corporately correct. These ways make talk out as really the just what all concerns with structure could have been about, and, to the point of these remarks, the just what ethnomethodological concerns with structure could have been about."

closely affiliated with Garfinkel's ethnomethodology and post-Wittgensteinian ordinary language philosophy and now aspired to be a social science discipline that investigated the "systematics" of talk.

Sacks and his colleagues (pp. 699 ff.) define conversation as a "speech exchange system" for allocating participants' rights and obligations to take turns at speaking. As they describe it, this system is organized as "an economy" whose orderly administration accounts for a set of "grossly apparent facts" that are evident to "unmotivated observations" of tape-recorded data. These facts include, for instance, that speaker change occurs in conversation, that predominantly one party speaks at a time, that transitions between speakers occur with no gap or overlap, and that turn order, turn size, length of conversation, what parties say, the relative distribution of turns, and the number of parties in a conversation are not fixed but vary. Such facts testify to the methodic way in which participants manage to conduct their exchanges with minimal gap and overlap. Sacks and his colleagues then describe a "context-free" machinery that accounts for the methodic production of these facts while allowing a "context-sensitive" use of that machinery by participants in singular conversations. This machinery consists of a set of "components" and "rules" that together describe a hierarchically ordered set of options through which participants in any conversation construct turns at talk and establish an orderly succession of speakers.[83]

The use of the term *fact* in this context is somewhat confusing. Sacks and his colleagues say that the facts constitute "critical tests" of the model. Although such language might suggest a superficial analogy with a "crucial test" of a scientific theory, as Jeff Coulter noted, some of the "facts" described by the turn-taking model are definitional for conversation.[84] An observable fact, such as that speaker change regularly occurs, can be cited as a criterion for identifying an occasion of talk as a conversation (or at least as something other than a lecture or monologue). The empirical occurrence or nonoccurrence of such "facts" does not provide a contingent test of the correctness of the model for conversation; instead, it defines whether or not the event under investigation should be counted as a "conversation" in the first place.

Several of the "facts" in Sacks, Schegloff, and Jefferson's inventory are stated negatively as parameters that are "not fixed" for conversation but are "allowed" to vary. For example, a substantive feature of formal debates is that the size and order of the speaking turns typically are specified in

[83] The list of facts includes specifications of the techniques for allocating turns, the linguistic "units" for constructing turns, and the "repair mechanisms" for resolving turn-taking errors and violations.

[84] Jeff Coulter, "Contingent and *a priori* structures in sequential analysis," *Human Studies* 6 (1983): 361–74.

advance, and a substantive feature of an interview is that question-and-answer turns are "preallocated" respectively to the interviewer and interviewee. In Sacks, Schegloff, and Jefferson's list, such general features of other "speech exchange systems" become positive "facts" for conversation. By using a language of "fact" and transforming the criteria for identifying alternative speech systems into "facts" about conversation (when these features are not prespecified), the paper lays the foundation for an analytic model that "accounts for" the facts. And because Sacks, Schegloff, and Jefferson define conversation as the most general type of speech exchange system, their model for conversation presumably occupies a higher level of abstraction than does a model that would account for the more restricted forms of talk.[85]

Sacks and his colleagues' model is designed to account for the methodic production of the grossly apparent facts of conversation by specifying a context-free and context-sensitive machinery. This apparatus consists of a set of "components" and "rules" describing a hierarchically ordered set of options for constructing turns at talk and selecting next speakers in the course of singular conversations. This machinery consists of two components and a set of rules for allocating turns to speakers. The components are separated into "turn-constructional units" and "turn-allocation techniques." Turn-constructional units are defined syntactically, but they are not limited to any single syntactic unit. They can include sentences, clauses, phrases, single words, and even nonlexical expressions.

An important feature of any of these "unit types" is what Sacks and his colleagues call their "projectable" completion, which means that "whatever the units employed for the construction, and whatever the theoretical language employed to describe them, that they have points of possible unit completion, points which are projectable before their occurrence" (p. 720). For instance, when said in reply to certain types of question, one-word utterances like "yes" and nonlexical items like "uh huh" can act as turn-constructional components. This is because the completion of a turn can be projected to occur at the completion of the expression. At the other extreme,

[85] Sacks and his colleagues define the "nonprespecified" parameters of conversation as though they were positive "facts" about *that* system, rather than criteria for distinguishing it from other systems. This way of conceptualizing conversation as a kind of master speech-exchange system sets up an entire research program through which one after another form of talk is shown to be a derivative of the primary form of "ordinary" conversation. John Heritage, for instance, points to the ontological and methodological significance of the "primacy" of what he calls "mundane conversation" when he argues that "institutionalized" forms of talk in educational settings, court hearings, government tribunals, political speeches, and news interviews: (1) employ selective reductions of the "full range of conversational practices available in mundane interaction" and (2) specialize on particular procedures "which have their 'home' or base environment in ordinary talk." See J. Heritage, *Garfinkel and Ethnomethodology* (Oxford: Polity Press, 1984), pp. 239–40.

turn-construction units can extend well beyond the limits of a sentence, such as when a speaker elaborates a story or joke consisting of a series of sentences.

Turn-allocation techniques are divided into two types, (1) those in which a current speaker selects the next speaker and (2) those in which the next speaker self-selects. Questions, greetings, summonses, and invitations can, although they do not always, select a particular recipient who then speaks next, whereas parties in a conversation may self-select at the projectable end of a prior speaker's story, joke, answer, or any other type of utterance that does not select a particular recipient. The components provide conditions for the operations of the basic rules for turn taking. In other words, the rules defining the turn-taking machine's cycle of operations are engaged whenever conversationalists approach a transition-relevance place, and the options for turn transition at any such juncture are defined by the turn-constructional techniques:

1. For any turn, at the initial transition relevance place of an initial turn-constructional unit:
   a. If the turn-so-far is so constructed as to involve the use of a "current speaker selects next" technique, then the party so selected has the right and is obliged to take next turn to speak; no others have such rights or obligations, and transfer occurs at that place.
   b. If the turn-so-far is so constructed as not to involve the use of a "current speaker selects next" technique, then self-selection for next speakership may, but need not, be instituted; first starter acquires rights to a turn, and transfer occurs at that place.
   c. If the turn-so-far is so constructed as not to involve the use of a "current speaker selects next" technique, then current speaker may, but need not continue, unless another self-selects.
2. If, at the initial transition relevance place of an initial turn-constructional unit, neither 1a nor 1b has operated, and, following the provision of 1c, current speaker has continued, then the rule-set a–c reapplies at the next transition-relevance place, and recursively at each next transition-relevance place, until transfer is effected. (p. 704)

The model is a hierarchical and closed system, since the ordering of the rules serves to constrain each of the options that the rules provide. It also is a *normative* machinery, since the options in the rule set define the participants' "rights" and "obligations" to speak and to listen. In a distinctive way, this system is a "social system" in Parsons's sense: It displays the "double contingency" of social interaction, by which is meant that an act performed by speaker A is oriented in its course to the possibility of a normatively guided reaction (or sanction) by recipient B, and B in turn comprehends A's act in reference to a common normative grounding.[86] Misunderstandings, and normative misalignments can, of course, occur, but they too are subject to adjudication and "repair" in reference to what John Heritage calls "the seamless web of accountability" provided by conversational structures.[87] The various rule sets that are said to make up the "interactional order" of conversation are distinguished from the normative order of "pattern variables" that Parsons constructed for the "overall" social system, but in their own domain they operate as context-free norms that are integrated with the detailed actions and reactions performed *in situ.*

The turn-taking model specifies two mechanisms for determining rights of succession in conversation: the designation of the next speaker by the current speaker (Rule 1a) and self-selection on a first-come, first-served basis (Rules 1b and 1c).[88] In both cases, exclusive rights to a turn apply only for the duration of the immediate turn-constructional unit. Rights are designated, secured, or renewed whenever a transition-relevance place is reached (or approached). Although the rule set is implemented only at points of projectable transition between speakers, there is no time out from its governance. In brief, the turn-taking machine operates in a closed system of normative possibilities, working incessantly, recursively, and compulsively to ensure "clean" transfers of the floor from one speaker to the next.

Having listed the facts and then offered a model to account for them, Sacks and his colleagues devote the bulk of their paper to showing "how the system accounts for the facts." For instance, in their treatment of how the turn-taking system accounts for the fact that one speaker speaks at a time, they say the following:

> Overwhelmingly one party talks at a time. This fact is provided for by two features of the system: First, the system allocates single turns to single speakers; any speaker gets, with the turn, exclusive rights to talk to the first possible completion of an initial instance of a unit-type – rights which are renewable for single next instances of a unit-type under the operation of rule 1c. Second, all turn-transfer is coordinated around transition-relevance places, which are themselves determined by possible completion points for instances of the unit-types. (p. 706)

Each of the other grossly apparent facts is similarly addressed, often through the analysis of transcribed examples. In the course of this exposition,

[86] Talcott Parsons, *The Social System* (Glencoe, IL: Free Press, 1951).
[87] Heritage, *Garfinkel and Ethnomethodology,* p. 239.
[88] Rule 1c can be treated as a special case of Rule 1b. According to Rule 1c, if current speaker does not select the next speaker, then he or she may resume speaking at a transition relevance place "unless another self-selects." In other words, current speaker is bound by the same first-come, first-served rule as are other participants.

Sacks and his colleagues describe two subsidiary systems, the organizational rubric of "adjacency pairs" and the mechanisms for the "repair" of errors and disruptions in conversation.

Adjacency pairs include a variety of units composed of paired acts produced by different speakers. A simple case is the mere exchange of greetings:

A: Hello.
B: Hello.

Other adjacency pairs include summons–answer, question–answer, and invitation–acceptance/decline. An important feature of adjacency pairs is the "conditional relevance" of the two "pair-parts" that make up the adjacency pair. In conversation-analytic terms, an initial greeting is a "first-pair part," and a responsive greeting is a "second-pair part." The use of a first-pair part is an important "device" for selecting a next speaker (Rule la), but more than that, it establishes the general kind of action to be produced in a relevant response. For instance, a greeting sets up the relevance for a response in kind; a question provides conditions for an answer; and so forth. A recipient's response to a first pair-part does not necessarily impose reciprocal constraints on the first speaker. When more than two speakers are present, an answer does not typically "select" the questioner to speak next, nor does it necessarily constrain what the next speaker will say.

Repair is a name for a variety of "mechanisms" for dealing with turn-taking errors and violations, as well as errors in word selection, terms of address, and so forth. Sacks and his colleagues argue that certain possibilities for repair are built into the basic economy of turn taking. So, for instance, if a current speaker selects a next speaker (Rule la) who fails to respond, the current speaker can employ Rule 1c, continuing to speak after the lapse, perhaps to prompt the initial recipient. Or if one speaker interrupts an ongoing turn by another speaker, the overlap can be resolved quickly by either speaker's cutting off his or her own utterance. Moreover, because repairs prompted by speakers other than the current speaker generally are not initiated until after the completion of a current turn, the initial speaker has the "right" to perform a "self-repair" before the end of his or her current turn. Accordingly, speakers can correct incipient errors without being sanctioned by their recipients. Sacks, Schegloff, and Jefferson treat this as an inherently rational organization for behavior "which accommodates real worldly interests, and is not susceptible of external enforcement [and which] . . . incorporates resources and procedures for repair of its troubles into its fundamental organization" (p. 51).[89]

[89] A more elaborate treatment of "self" and "other" repair is given in E. A. Schegloff, G Jefferson, and H. Sacks, "The preference for self-correction in the organization of repair in conversation," *Language* 53 (1977): 361–82.

A liberal economy

The turn-taking paper's formal approach was so influential that it attracted many ethnomethodologists away from the more ad hoc variety of ethnomethodological study that focused on the local achievement of observation, description, replication, and other epistemic themes. Gradually, the turn-taking machine became an established foundation for subsequent studies. Turn-taking and other systemic structures were identified as "mechanisms" and placed in the grammatical role of agents. Consider, for example, the following passage from Sacks, Schegloff, and Jefferson's paper:

> In characterizing the turn-taking system we have been dealing with as a "local management system," we take note of the following clear features of the rule set and its components:
>
> 1. The system deals with single transitions at a time, and thereby with only the two turns which a single transition links; i.e., it allocates but a single turn at a time.
> 2. The single turn it allocates on each occasion of its operation is "next turn."
> 3. While the system deals with but a single transition at a time, it deals with transitions:
>   a. comprehensively – i.e., it deals with any of the transition possibilities whose use it organizes;
>   b. exclusively – i.e., no other system can organize transitions independent of the turn-taking system; and
>   c. serially, in the order in which they come up – via its dealing with "next turn."
>
> These features by themselves invite a characterization of the system of which they are a part as a local management system, in that all the operations are "local," i.e., directed to "next turn" and "next transition" on a turn-by-turn basis. (p. 725)

In this passage, deterministic, bureaucratic, and mechanistic idioms describe the local "operations" of an impersonal formalism, an "it" that "allocates," "deals with," and "methodizes" the sequential assembly of discourse. This inversion of conventional notions of human agency and intentionality may at first seem to "dehumanize" conversational practice – turning conversational participants into "methodized" dopes – but before concluding this we should recall that the machinery involved is nothing other than a moral economy, a domain of "natural rights." Don Zimmerman and Deidre Boden, for instance, are alert to this point when they cite their colleague Thomas Wilson to support a version of conversational "agency" based on turn-taking considerations, which is very much in line with classical humanism:

> From the viewpoint of Wilson (1989), what we have just described is *human agency,* understood to be intrinsic to the machinery that organizes social interaction. He suggests that there can be no social interaction without participants *acting* on the assumption that they as well as their co-participants are autonomous, morally responsible agents whose actions are neither determined nor random. The organization of interaction builds in such an assumption as a fundamental principle.[90]

Once formulated in terms of such a "fundamental principle," the turn-taking machine's economy becomes the basis for a liberal ethos. The rule set combines the free-market mechanism of first-come, first-served (Rule 1a) with a more traditional property right, according to which the current "owner" of a turn can pass the right of ownership directly to others. Although the system presupposes a competitive interest in securing "turns at talk," it works in a cooperative and collaborative way. The detailed transference of a turn from one speaker to another requires a precisely coordinated and reciprocally displayed "orientation" by all parties to the exchange. Compared with more heavily regulated systems of speech exchange, in which turns and options are preallocated, the turn-taking system for conversation allows a relatively "free" or entrepreneurial mode of management that is regulated by an autonomous system.

In numerous studies, the contractual language in which the central rule set is expressed has been treated literally as a basis for criticizing systematic restrictions of conversational "rights" in particular speech economies. Zimmerman and Candace West contend that women's rights to take turns without interruption are curtailed in cross-sex conversations[91]; West and Angela Garcia argue a similar case about women's rights to initiate topical development[92]; West[93] and Kathy Davis[94] detail how patients' speaking rights are restricted in clinical discourse; and Harvey Molotch and Dierdre Boden[95] and Alec McHoul[96] discuss the exploitation and transgression of discursive

[90] Don Zimmerman and Deirdre Boden, "Structure-in action: an introduction," pp. 3–21 in D. Boden and D. Zimmerman, eds., *Talk and Social Structure* (Oxford: Polity Press, 1991), quotation p. 11. Rererence in the quote is to Thomas P. Wilson, "Agency, structure and the explanation of miracles," paper presented at the Midwest Sociological Society meetings, St. Louis, MO, 1989.

[91] Don Zimmerman and Candace West, "Sex roles, interruptions and silences in conversation," pp. 225–74, in Barrie Thorne and Nancy Henley, eds., *Language and Sex: Difference and Dominance* (Rowley, MA: Newbury House, 1975).

[92] Candace West and Angela Garcia, "Conversational shift work: a study of topical transition between women and men," *Social Problems* 35 (1988): 551–75.

[93] C. West, *Routine Complications: Troubles with Talk Between Doctors and Patients* (Bloomington: Indiana University Press, 1984).

[94] Kathy Davis, *Power Under the Microscope* (Dordrecht: Foris, 1988).

[95] Harvey Molotch and Deirdre Boden, "Talking social structure: discourse, domination and the Watergate hearings," *American Sociological Review* 39 (1985): 101–12.

[96] Alec McHoul, "Why there are no guarantees for interrogators," *Journal of Pragmatics* 11 (1987): 455–71. McHoul treats the turn-taking system for conversation as a principled basis for designing "transgressive" alternatives to centrally administered discursive systems, but

power during interrogation sequences. As Sacks, Schegloff, and Jefferson define it, conversation is an especially flexible system in which turn size, turn order, and speakership are "allowed to vary." Consequently, when speech exchange systems such as interviews, interrogations, and clinical examinations are analyzed against the normative backdrop of "ordinary conversation," they appear to be artificially imposed and asymmetrically restricted systems of communication.

Schegloff effectively disputes Zimmerman's and West's analytic procedures by demonstrating how they fail to take account of ordinary conversational structures that confound the relevance of "gender" in the organization of the conversational data,[97] and in an analysis of a highly publicized "news interview" he discounts a popular impression that the interview was a unique political spectacle.[98] In both cases, Schegloff demonstrates great facility in showing how apparently "interesting" or "spectacular" instances of talk involving notable individuals and momentous occasions derive their analytic organization from ubiquitous structures of "mundane" conversation.[99]

Schegloff's demonstrations provide cogent warnings against particularized interpretations of conversational materials, but like the normative analyses he criticizes, his arguments remain firmly committed to the idea that analytic structures of "talk in interaction" supply a determinate structural foundation for diverse activities. In a supportive commentary on Lucy Suchman and Brigitte Jordan's conversation-analytic treatment of social science interviews,[100] Schegloff recites a passage from the turn-taking paper: "Since all sorts of scientific and applied research use conversation now, they all employ an instrument whose effects are not known. This is perhaps unnecessary" (Sacks, Schegloff, and Jefferson, pp. 701–02).

Members' intuitions and professional analyses

Contrary to Schegloff, one might figure that for speakers of a natural language, conversation is an "instrument" whose "effects" are quite well known, since those effects arise and are controlled through the competent use of that instrument. Schegloff has argued repeatedly, however, that a

because he allows "conversation" itself to be a contingent achievement of an order, relative to other discursive orders, his treatment differs from a mere "application" of the turn-taking model to a specific institutionalized system.

[97] E. A. Schegloff, "Between micro and macro: contexts and other connections," pp. 207–34, in J. Alexander, B. Giesen, R. Munch, and N. Smelser, eds., *The Micro–Macro Link* (Berkeley and Los Angeles: University of California Press, 1987).

[98] Schegloff, "From interview to confrontation."

[99] See Bogen, "The organization of talk."

[100] Lucy Suchman and Brigitte Jordan, "Interactional troubles in face-to-face survey interviews," *Journal of American Statistical Association* 85 (1990): 232–41.

vernacular mastery of language does not entitle a speaker to the kind of analytic knowledge attained through the study of tape-recorded data.[101] Although ordinary conversationalists presumably "know" how to take turns and conversation analysts treat evidences of "participants' orientations" as a "proof criterion" for specific analytic characterizations,[102] Schegloff draws a fundamental distinction between an *analytical* understanding of the abstract components, rules, and recursive operations that describe systems of talk in interaction, and a *vernacular* understanding of talk.[103] According to this distinction, the societal member exhibits a naive mastery of the techniques that the scientist describes formally. The member may be competent to instantiate the describable techniques of conversation, but the scientist builds a formal apparatus that subsumes the member's local practices.

This distinction between vernacular intuition and scientific analysis indicates the extent to which conversation analysis has become a professionalized analytic endeavor. Conversation analysts no longer regard their studies to be explicating features of linguistic intelligibility that any competent member should be able to recognize. They do not aim to be doing a primitive science, which, as Sacks initially described it, is grounded in the intelligibility of descriptions for any competent language user. Instead, two separate orders of

[101] Schegloff, "From interview to confrontation," and "Goffman and the analysis of conversation," pp. 28–52, in P. Drew and A. Wooton, eds., *Erving Goffman: Perspectives on the Interaction Order* (Oxford: Polity Press, 1988); "On some questions and ambiguities in conversation," pp. 28–52, in Atkinson and Heritage, eds., *Structures of Social Action;* and "Introduction" to Sacks, *Lectures on Conersation,* vol. 1, pp. xl–xliv.

[102] As Sacks, Schegloff, and Jefferson ("A simplest systematics," pp. 728–29) show, the subsequent treatment of an utterance by participants in a conversation provides a "proof criterion" for analytic characterizations of that utterance in the sequence of utterances in which it is placed. So, for instance, a characterization of an utterance as a "question" or "invitation" is not determined by formal syntactic or semantic criteria alone but by how the utterance is treated by its recipient(s). If the utterance is apparently "answered," this acts as a criterion for characterizing it as a "question." See Schegloff and Sacks, "Opening up closings," *Semiotica* 7 (1973): 289–327, for further elaboration on this point, and Coulter, "Contingent and *a priori* structures in sequential analysis," for a critique of the analytic "proof" procedure.

[103] It can be argued that a distinction between members' intuitions and professional insight is a necessary part of any analytic enterprise. Ryle's (*The Concept of Mind,* pp. 25 ff.) distinction between "knowing how" and "knowing that" provides a resource for an effort to investigate systematically the logical grammar of ordinary concepts, and in the social sciences, *docta ignorantia* (Kaufmann, *Methodology of the Social Sciences*) sets up analytic specifications of what members "know without knowing." Although the distinction is well established and perhaps unavoidable in the social sciences, too often it becomes a device for claiming special validity for a professional enterprise whose subject matter and observation language nevertheless immerses practitioners in a horizon of "ordinary" concepts, judgments, historical understandings, and modes of practical reasoning. Both Schutz and Garfinkel identified this as a fundamental phenomenon to be investigated, rather than a methodological problem to be superseded by the technical development of a social science.

technical competence are now implicated in a hierarchical relationship: the vernacular competence to produce and recognize particular techniques in conversation, and an analytic competence to subsume those techniques into collections of similar cases. As Zimmerman summarizes it, such analytic competencies are grounded in the specialized practices of a professional social scientific community:

> To be sure, initial purchase on some phenomena may be gained on intuitive grounds, but this is merely the beginning. From this point on, the phenomenon is "worked up" by searching across many conversations, resulting in increasing empirical control and a more general understanding of the process that generates it. When applied to new cases (which, of course, could undercut the formulation and force its revision) such empirically grounded formulations furnish a warrant for the identification of particular conversational events. Indeed, the cumulative results of conversation analytic research should permit a detailed understanding of particular, singular conversations.[104]

For Zimmerman, what counts as adequate analysis is warranted by referring to an emerging set of professional conventions for identifying and recording relevant data, composing transcripts, building collections of equivalent cases, and contributing to the literature.[105] The naive mastery by which ordinary conversations are produced becomes relegated to the far side of a strict demarcation between professional analysis and commonsense understandings of social structure. Only professionals are entitled to criticize the published reports: "Any critique of analytical results in the tradition must itself be empirically grounded, that is, based on alternative analysis of appropriate materials."[106]

An even stronger emphasis on the advantages of technical analysis over commonsense intuition is given in the editors' introduction to an influential anthology of conversation-analytic studies:

[104] Don Zimmerman, "On conversation: the conversation analytic perspective," pp. 406–32, in J. Anderson, ed., *Communication Yearbook II* (London: Sage, 1988).

[105] For an instructive critical commentary on Zimmerman's (ibid.) argument, see D. L. Wieder, "From resource to topic: some aims of conversation analysis," pp. 444–54, in Anderson, ed., *Communication Yearbook II.* Different views of the role of transcription in the production of conversation-analytic findings are discussed by Christopher Pack, "Features of signs encountered in designing a notational system for transcribing lectures," pp. 92–122, in H. Garfinkel, ed., *Ethnomethodological Studies of Work* (London: Routledge & Kegan Paul, 1986); George Psathas and Tim Anderson, "The 'practices' of transcription in conversation analysis," *Semiotica* 78 (1990): 75–99; and Mishler, "Representing discourse." Gail Jefferson has also turned her analytical attention to the practices of transcribing. See her "An exercise in the transcription and analysis of laughter," pp. 25–34, in T. Van Dijk, ed., *Handbook of Discourse Analysis,* vol. 3: *Discourse and Dialogue* (London: Academic Press, 1985).

[106] Zimmerman, "On conversation: The conversation analytic perspective." See also Wieder, "From resource to topic," pp. 447 ff.

> In sum, the use of recorded data serves as a control on the limitations and fallibilities of intuition and recollection; it exposes the observer to a wide range of interactional materials and circumstances and also provides some guarantee that analytic conclusions will not arise as artifacts of intuitive idiosyncrasy, selective attention or recollection, or experimental design. The availability of a taped record enables *repeated* and *detailed* examination of particular events in interaction and hence greatly enhances the range and precision of the observations that can be made. The use of such materials has the additional advantage of providing hearers and, to a lesser extent, readers of research reports with *direct* access to the data about which analytic claims are being made, thereby making them available for public scrutiny in a way that further minimizes the influence of individual preconceptions. Finally, because the data are available in raw form, they are cumulatively reusable in a variety of investigations and can be reexamined in the light of new observations or findings.[107]

The emphasis on direct access to "raw data" and the corresponding distrust of theoretically or intuitively mediated modes of observation, representation, and inference is quite striking in this passage. Note the significance placed on the fact that in conversation analysis, naturally occurring "data" are directly gathered by recording machines, without being contaminated by idealization, intuition, intervention, or interpretation. The distrust of vernacular intuition extends to a suspicion about some of the typical modes of abstraction, recollection, and reconstruction used in experimental research.

In accordance with Sacks's program for a primitive natural science, this passage expresses a disinterested orientation to a mundane world that can be inspected and described in a verifiable way. However, we can now see that the replication and verification of observations and observational results are no longer open to just "anybody." Instead, an entitlement to participate in the epistemic community is restricted to certified practitioners of conversation analysis. The circulation of descriptions is now technically and professionally mediated: Tape recordings make data "available in raw form," and these data are circulated along with the findings described in research reports to other members of the scientific community. Not only do the reports instruct other observers on how to "go out and look" for the described findings; the raw data from which the findings are drawn also are included in the reports (and the accompanying tapes) themselves. Accordingly, the circulation of

[107] John Heritage and J. Maxwell Atkinson, "Introduction," pp. 2–3, in Atkinson and Heritage, eds., *Structures of Social Action.*

texts and tape recordings integrates the conversation-analytic research community and grounds its findings: "The analytic intuitions of research workers are developed, elaborated, and supported by reference to bodies of data and collections of instances of phenomena. In this process, an analytic culture has gradually developed that is firmly based in naturally occurring empirical materials."[108] As this passage makes clear, "intuitions" are still relevant to the research process, but they are now cultivated in a specialized "analytic culture."

Assuming that the conversation-analytic community actually is a coherent "analytic culture," the preceding quotations portray what might loosely be called a "positivistic" (or perhaps more accurately a "logical empiricist") discipline.[109] To say this is not to suggest that conversation analysts produce erroneous empirical findings, since such an indictment would presuppose a technical commitment to amassing correct empirical findings. What is more troubling about the logical empiricist direction that conversation analysis has taken is its problematic relation to the solution that Sacks initially offered to the foundational question "How is a social science possible?" Recall that for Sacks the very existence of primitive natural science demonstrates that methods can be described in ordinary language in such a way that others can reproduce those methods.

An important feature of such methods accounts is that they are internal to the community of practitioners who compose and use them. That is, descriptions of members' competencies are presented as intelligible instructions for other members. By distinguishing the analytic competency of members of the conversation-analytic community from the vernacular competency of the ordinary conversationalists described, conversation analysts have segregated their technical reports from the communal practices they describe.[110]

[108] Ibid., p. 3.

[109] Conversation-analytic writings exhibit some of the tendencies that Ian Hacking (*Representing and Intervening: Introductory Topics in the Philosophy of Science* [Cambridge University Press, 1983], pp. 41–42) lists under the heading of the "six instincts" associated with positivism. These include the emphasis given to direct observation, the tendency to avoid postulating theoretical entities, and the preference for description over explanation. This is not the entire picture, however, since the positivistic themes in CA coexist with a legacy from ethnomethodology's constitutive treatment of "social facts." Schegloff ("From interview to confrontation," p. 203) distinguishes conversation analysis from positivistic social science by noting that conversation analysts endeavor to control their descriptions of conversational data by reference to (evident) "participants' orientations," but he also distances his position from Garfinkel's "antipositivistic" ethnomethodology.

[110] One might argue that the very point of developing an analytic culture is to achieve just such a segregation. For instance, the Analytical Society formed by early-nineteenth-century British mathematicians tried to divorce the symbolic operations of algebra from intuitive concepts of number. This separation of professional analysis from lay intuition, according to

The adequacy of such accounts no longer depends on their effective use as instructions for reproducing the practices described; instead, judgments about empirical adequacy are reserved for other members of the analytic culture, thereby entitling them, and them alone, to decide on nonintuitive (or specialized intuitive) grounds how well any technical report represents the collection of data it describes.

Whereas in the "natural historical" phase of investigation, singular instances of recorded conversation were used "therapeutically" to examine received scholastic wisdom about common sense and ordinary language, in "latter day CA"[111] the analytic value of any instance is established over and against "intuition," and the analysis is organized in reference to a professionally assembled collection of similar instances.[112] Consequently, by distinguishing their expertise from the "intuitive" competencies that make up the

David Bloor ("Hamilton and Peacock on the essence of algebra," pp. 202–32, in H. Mehrtens, H. Bos, and I. Schneider, eds., *Social History of Nineteenth Century Mathematics* [Boston: Birkhauser, 1981]), reflected an interest in enhancing the self-sufficiency of professional mathematicians. Regardless of whether such an interest can be imputed to conversation analysts, the point of my criticism is that severing professional analysis from lay intuitions – however felicitous for the disciplinary prospects of CA – virtually guaranteed that CA would no longer sustain the kind of "epistemic" investigation that Sacks's early research opened up. There is also a key difference, however. Unless mathematicians were to adhere to the Husserlian idea that their practices are essentially founded in the ordinary skills of counting, measuring, practical geometry, and related modes of mathematical "application," there is little to restrain their analytic culture from becoming an esoteric (and even a bizarre) form of life. Conversation analysts, on the other hand, are committed to the idea that their "object" domain is itself produced through situated lay analyses, so that an entirely autonomous conversational "algebra" would be absurd. A more relevant analogy would be to an algebra that attempted to subsume the production of ordinary mathematics under a more abstract rubric, but this was not what the Analytical Society intended to do. I am indebted to a graduate term paper by Aditi Gowri, Science Studies Program, University of California at San Diego, for alerting me to these aspects of early-nineteenth-century algebra.

[111] This expression is taken from Garfinkel et al., "Respecifying the natural sciences," p. 65.

[112] Conversation analysts do analyze single cases, but when they do so, the observable properties of the case are made meaningful not through an "intuitive" explication of the transcribed details but through comparisons of the details of the case and others in the CA corpus. An apparent exception to this rule is E. A. Schegloff's "On an actual virtual servomechanism for guessing bad news: a single case conjecture," *Social Problems* 35 (1988): 442–57, in which a case is presented as a "transparent" instance in which a "discovery of something new" becomes possible without initial recourse to its convergence with "analytic resources developed elsewhere" (p. 442). Schegloff goes on to say, however, that his analysis draws on an exemplary "analytic tool" from CA that enabled him to discern the operations of a particular mechanism in the transcribed fragment, which then enabled him to search for comparable instances. It is clear that Schegloff's analysis operates within the circuit of previous CA findings (how could it not?) and is packaged in CA's analytic rubrics. From reading Schegloff's paper and finding his reading of the transcript to be less than compelling, I contend that the fragment's "transparency" to his analysis is not equivalent to the transparency described in Sacks's program for primitive science, in which "anybody" can intuit what an observer describes when presented with the evidence. For me to say this is not to discredit Schegloff's analysis; rather, it is to credit him with an intuitive understanding that is accountable strictly within a small community of professional analysts.

activities studied, members of the analytic culture have attempted to create a disciplined approach to ordinary practical action.

By using specialized equipment, observational techniques, and analytic language the conversation analysts are able to construct formal descriptions of the vernacular "analyses" implicit in the reproduction of ordinary conversational actions.[113] The adequacy of such descriptions is often said to depend on the extent to which they recover the member's practical orientation. In CA, *analysis* is the pivotal term for linking the study of conversation to its subject matter. Analysis is also, of course, what scientists and logicians do when they break things down into their essential components, but for conversation analysts it is also a ubiquitous feature of the practices they investigate. Whereas Sacks once focused on a range of epistemic topics – description, measurement, categorization, observation, reproduction – the term *analysis* eventually came to stand as a master category, an omnirelevant bridge between the intelligibility of talk and the rigor of professional investigations. For CA, participants in conversation analyze one another's talk, and their ordinary, or vernacular, analyses are displayed in the orderly way in which different speakers collaborate in the production of conversations. This analytic competence is somewhat similar to the competent production and recognition of grammatical sentences described by linguists, except that the syntax described in CA covers the mechanisms by which two or more speakers coordinate their contributions to talk in interaction.[114] Although analysis is said to be *in* the conversational data as well as in the scientific description *of* those data, a division of labor emerges between the social scientist's and the member's analytic work: The member exhibits a naive mastery of the techniques that the scientist describes formally. In Sacks's terms, the scientist "builds the brains" that the member could not possibly have possessed, as brains.

Vernacular and analytic categories of speech acts

Although conversational analysts recognize that conversation is produced locally as a vernacular accomplishment and that the "materials" of conversation are ordinary forms of expression, they describe with abstract technical terms the components and rules that make up the turn-taking machine. For example, *question* is a vernacular term for a familiar linguistic phenomenon. In conversational analysis, questions are subsumed in the type of turn-

[113] See Erving Goffman's remarks in *Frame Analysis: An Essay on the Organization of Experience* (New York: Harper & Row, 1974), p. 5, in which he (wrongly) attributes such an aim to Garfinkel's program of ethnomethodology. Goffman apparently is referring to Garfinkel's early work on "trust" when he states that Garfinkel aimed at one time to "look for rules which, when followed, allow us to generate a 'world' of a given kind."

[114] See E. A. Schegloff, "The relevance of repair to syntax-for-conversation," *Syntax and Semantics* 12 (1979): 261–86.

allocational technique called "current speaker selects next speaker." The obvious fact that in conversation questions oblige a recipient not only to speak but also to give an answer is further elaborated under the technical specifications for adjacency pairs.

Schegloff expounds on the relationship between technical and vernacular accounts of activities in a brilliant criticism of linguistically based programs like Searle's speech-act theory.[115] He makes two related points based on conversational-analytic investigations of tape-recorded dialogue: (1) Utterances that do not take the syntactic form of "question" often act as questions in conversation, and (2) utterances that do take the syntactic form of "question" do not necessarily act as questions. Schegloff demonstrates both points in reference to the following example:[116]

B_1: Why don't you *come* and see me some times
[
A_1: I would like to
B_2: I would like you to. Lemme just
[
A_2: I don't know just where the- us- address *is.*

> Where are the questions here? Is there a question here? For a participant whose next utterance or action may be contingent on finding whether a current utterance is a "question" – because, if it is, an "answer" may be a relevant next thing for him to do, does syntax, or linguistic form, solve his problem? *Not only does our intuition suggest* that, although no syntactic question (nor question intonation, for that matter) occurs in A's second utterance of the excerpt, a question–answer (Q–A) sequence pair has been initiated, a request for directions if you like; more important, it is so heard by B, who proceeds to give directions. And although B's first utterance in the excerpt looks syntactically like a question, it is not a "question" that A "answers" but an "invitation" (in question form) that she "accepts." (Emphasis added.)

Schegloff further argues that the expression "Why don't you *come* and see me some times" is not inherently ambiguous.[117] The recipient who hears the utterance *in situ* responds without hesitation, apparently hearing it as an invitation and thus providing evidence of the relevant categorical identity of that utterance. Ambiguity arises only when the particular utterance is isolated from the sequence and inspected for its syntactic, semantic, intonational, or pragmatic *form.*

Up to this point, Schegloff's argument exemplifies the critical mode of

[115] Schegloff, "On some questions and ambiguities in conversation," pp. 29 ff.; J. R. Searle, *Speech Acts* (Cambridge University Press, 1969).
[116] Schegloff, "On some questions and ambiguities in conversation," p. 31.
[117] The apparent pronunciation of "times" rather than "time" in the transcript is odd, though perhaps due to a noisy tape. In any case, it is not germane to Schegloff's analysis.

investigation that I earlier identified with the "primitive phase" of CA. He uses tape-recorded data to conduct an original variant of the kind of grammatical investigation that Wittgenstein recommended: "Such an investigation sheds light on our problem by clearing misunderstandings away. Misunderstandings concerning the use of words, caused, among other things, by certain analogies between the forms of expression in different regions of language."[118] In this case, the misunderstandings that Schegloff identifies are based on formal analogies between isolated sentences. Like Wittgenstein, Schegloff uses ordinary examples,[119] and he appeals to his readers' intuitions to demonstrate that a more differentiated understanding of language use is necessary if we are to avoid the mistake of assuming that utterances with a similar syntactic form always play the same role when used *in situ*.

Schegloff's argument subsequently veers off the "antiformalist road."[120] Having demonstrated that abstract linguistic form is insufficient to explain the pragmatic role of an expression in a sequence of utterances, he reinstates formal determinacy at another level of abstraction: "For a substantial part of what we might expect to be available to us as understanding of questions as a category of action is best and most parsimoniously subsumed under the category of 'adjacency pairs'; much of what is so about questions is so by virtue of the adjacency-pair format."[121] Although focused on the "adjacency-pair format," rather than the more encompassing operations of the turn-taking machine, his argument is similar to one made by Sacks and his colleagues: "Thus, while an addressed question requires an answer from the addressed party, it is the turn-taking system, and no syntactic or semantic features of the 'question,' that requires the answer to come 'next'."[122] In both cases the vernacular category of question is subsumed under a more abstract technical description: "adjacency pair first-pair-part" or "current-speaker-selects-next-technique."

Note, however, that a technical respecification of a vernacular category of speech act (e.g., "invitation" respecified as an "adjacency pair, first-pair-part") does not discount the local relevancy and intelligibility of the particu-

[118] Wittgenstein, *PI,* sec. 90.

[119] Schegloff's examples differ from Wittgenstein's in one very significant respect: Whereas Wittgenstein compares familiar expressions and recalls some of their various uses in everyday situations, Schegloff draws his materials from tape recordings of singular conversations and thereby gains the advantages of recognition over recollection.

[120] I am borrowing Stanley Fish's expression (*Doing What Comes Naturally,* pp. 1 ff.). Fish places several programs in philosophy, literary theory, and critical legal studies under the "antiformalist" heading, and he mentions ethnomethodology among them. In his critical essays, he shows again and again that an expressed commitment to the "antiformalist road" offers no guarantee of staying on it.

[121] Schegloff, "On some questions and ambiguities," p. 34.

[122] Sacks, Schegloff, and Jefferson, "A simplest systematics," p. 86, n. 46.

lar vernacular categories.[123] Indeed, vernacular understandings are not the object of Schegloff's criticism; instead, he admonishes speech-act grammarians for ignoring vernacularly produced and intuitively transparent features of actual conversations. As he describes it, the expression "Why don't you *come* and see me some times" was treated as an invitation in the original setting in which it occurred. Misunderstandings arise when a *technical* definition of the isolated expression as, for example, a syntactic question, is irrelevant to the evident use of the expression as an invitation or, rather, when the technical definition is relevant to that use only in a counterfactual sense. A recipient who "answered" the invitation as though it were a question would be making an inappropriate response.[124] Schegloff relies on his readers to recognize that "Why don't you *come* and see me some times" was originally and unambiguously not (or not only) a question. Further, he relies on us to recognize that his understanding of the expression's use is congruent with the treatment that the original recipient accorded to it. This evident concordance among Schegloff, his readers, and the participants in the tape-recorded conversation is not grounded in any technical expertise, but in the vulgar intuitions through which we are able to read the transcript and "hear" it as a fragment of an ordinary conversation.

Schegloff does not demand that we replace a vernacular understanding of the expression with a technical one; rather, he appeals to our vernacular intuitions in order to convince us to reject a grammatical definition of the expression based on sentence grammar alone. It is only at that point that he subsumes the intuitively correct vernacular category ("invitation" and not "question") in an alternative conception of syntax ("adjacency pair").[125] We should not forget that our vernacular intuitions have already proved adequate to the characterization of the expression that Schegloff's formal analysis gives back to us under a more general rubric. Our intuitions fail only when we are asked to infer the sequential use from the form of an isolated expression or when we are asked to define "question" abstractly.

Intuitions about the syntactic form of questions are not always irrelevant to their use, however. By rule and by definition, interrogators are supposed to

[123] Of course, conversational analytic work on "invitations" does not end when that vernacular category is subsumed under a more abstract rubric. See, for instance, Judy Davidson, "Subsequent versions of invitations, offers, requests, and proposals dealing with potential or actual rejection," pp. 102–28, in Atkinson and Heritage, ed., *Structures of Social Action.*

[124] This is not entirely clear. One could answer "Why don't you come and see me some time?" with an excuse: "I've just been so busy these days." But the identification of the answer as an excuse (and not simply "giving information") presupposes an understanding of the question as, if not an invitation, a complaint.

[125] Schegloff ("Goffman and the analysis of conversation") explicitly characterizes CA's domain to be one of the "syntax " of "talk in interaction." Note that the "proof procedure" is not unlike that used in linguistic studies of sentence grammar, except that rather than resting on intuitive judgments on the grammaticality of isolated sentences, it uses transcribed tape recordings and intuitive recognitions of the actions thereby documented.

ask questions. Nevertheless, interrogators commonly present witnesses with assertions as a way of, for example, going over "old ground," eliciting confirmation, and inviting admissions.[126]

In the following sequence, an interrogator recites a documented "fact" for a witness's confirmation and is then challenged by the witness's counsel to ask a question. After some hesitation, he reformulates his assertion as a question, simultaneously demonstrating an intuition about the abstract syntactic requirements of "questioning" and claiming that what he was doing all along was, for all practical purposes, "questioning" the witness:

Nields: And it's dated the Seventh of April, Nineteen Eighty-Six.
(0.6)
North: Righ:t.
(1.4)
Nields: And that's::, three days after the date of thee, (0.5) term- terms of <u>ref</u>erence (.) on Exhibit O:ne.
(2.5)
Nields: You can check if you wish or you can take my word for it, it's dated April Four.
(0.4)
North: Will you take my word.
((Slight background din; pages turning))
(11.0)
(North): °(Okay, (1.0) good.)°
(7.0)
(North): °((whispering)) that's wha:::t?
(0.6)
( ): ( )
(4.5)
Sullivan: °(whu-)° What is your question, uh
Nields: I haven't asked a question yet, I'm simply: uh:: (0.8) uh::: (0.4) Well, the question is, isn't this three days after (.) the date on the term of reference on Exhibit One?
North: Apparently it is::.[127]

The problem is not that our vernacular understandings are inadequate but that our intuitions are given truncated examples and peculiar tasks in a

[126] See J. Maxwell Atkinson and P. Drew, *Order in Court: The Organisation of Verbal Interaction in Judicial Settings* (London: Macmillan, 1979).

[127] "Testimony at Joint Hearings Before the House Select Committee to Investigate Covert Arms Transactions with Iran and the Senate Select Committee on Secret Military Assistance to Iran and the Nicaraguan Opposition," July 7, 1987, morning session. Transcribed by M. Lynch and D. Bogen. See M. Lynch and D. Bogen, *The Ceremonial of Truth at the Iran Contra Hearings* (Durham, NC: Duke University Press, forthcoming).

program of general linguistics. Furthermore, such characterizations as "adjacency pair first-pair-part" or "current-speaker-selects-next-technique" are no more precise than vernacular categories like "invitation" and "joke."

As Schegloff demonstrates, questions and invitations are differentiated at the "surface level" of the conversations he presents in transcripted form, and he relies on his readers' intuitive recognitions of the subtle differences that are available in an inspection of those examples. This is not to deny the value of the technical rubric of "adjacency pair" for bringing to light a common aspect of a diverse set of phenomena. Categories like "adjacency pair" provide reasonable accounts of social activities. Their reasonableness is endogenous to the practices of what Garfinkel sometimes calls *constructive analysis.* An explanation of this is given in an incisive, although neglected, paper by M. D. Baccus:

> That "reasonableness" relies on the indexicality of the account as an essentially vague referencing device which allows the imagined availability of properties and features of the phenomenon. The accomplishment of an analytic account is that it is removed from the phenomenon in such a way that "cases," or analytic instances of the events of a phenomenon, are now things to be measured *against the account* and not against each other as in the production of the account, or, as in deciding the equivalency of events or actions. Nor are cases matched to an account to find *its* adequacy; accounts are read to find the *adequacy of the case* and as *an instance of the account.* Thus cases are read in or out of relevance with respect to the adequacy of the account to delineate their cogent features; and, each case is relevant only as one of a *collection* of instances which are *adequately equivalent,* that is, made so by collecting naturally available properties into some accountable unit which stands as "a case."[128]

The passage is not explicitly about CA's analytic procedures, but it can be read to apply to the way that the analytic rubrics become adequate descriptions of a collection of cases of, for example, "adjacency pairs" of a particular type, while becoming progressively removed from the endogenous production of any singular "case" that is thus subsumed under the category. This is not to devalue conversational-analytic studies of the distinctive features of particular sequential phenomena like greetings and return greetings, questions and answers, assessments and second assessments, announcement sequences, and the like. Far from revealing the inadequacy of vernacular intuitions, such studies provide detailed explications of familiar social phenomena.

As we noted earlier, Schegloff's arguments lead us a considerable distance down the antiformalist road, but if we stay on that road a bit longer than he

[128] M. D. Baccus, "Sociological indication and the visibility criterion of real world social theorizing," pp. 1–19, in Garfinkel ed., *Ethnomethodological Studies of Work;* quotation from p. 5.

does, we can direct antiformalist arguments at his own brand of analysis. Schegloff's demonstration enables us to recognize that the vernacular identification of a "question" is reflexively bound to the sequential context in which it is used. He then subsumes "questioning" under technical rubrics like "turn at talk," "turn-constructional unit," "turn-transition relevance place," and "adjacency pair." These terms differ from some of the technical names for logical, syntactic, or pragmatic units like "proposition," "question," "request," and so on because (with the exception of "turn") they are not as readily confused with vernacular categories of ordinary action. Nevertheless, when conversation-analytic terms are examined in reference to tape recordings and transcripts, it becomes clear that they refer to locally accomplished and vernacularly accountable activities. Indeed, the tightly contingent and recognizably contiguous organization of "adjacent" utterances, silences, and gestures provide "proof criteria" for technical analyses as well as evidence for the contextually sensitive operations of context-free machinery.

Following the critical discussion of formulating in conversation by Garfinkel and Sacks (see Chapter 5), however, questions can be raised about what members[129] are doing when they do [taking a turn at talk]. This is not the same as asking what the turn-taking machine is doing when members talk to one another, although the question is similar in form to Schegloff's inquiry about what members are doing when they do [asking a question or answering it] or [answering a question].[130] Schegloff's critique of Searle enables us to see that the intuitive identity of a "speech act" cannot be derived from the intentional structure of individual consciousness, but far from revealing the

[129] Garfinkel and Sacks, "On formal structures of practical actions." *Members* is Garfinkel and Sacks's term for "masters of natural language." A member is distinguished from a person or individual because its identity as such is always relative to an organized setting. Superficially, "members" is a synonym for "parties," a term preferred in conversational analysis. It differs substantially, however, in the way it connotes not a party to a contract but a status taken on trust that comes with the discursive territory; membership implies the phenomenological horizons of an organizationally specific "precontractual solidarity."

[130] Although his argument reveals profound misunderstandings of CA, Goffman ("Replies and responses") effectively demonstrates the extent to which a formal analysis of "adjacency pairs" fails to come to terms with the locally produced relevance of "questions" and "answers" in conversation. Goffman (p. 34) applies the sort of argument that Schegloff uses against speech-act theory against Schegloff's interactional analysis: "Unlike the self-sufficient sample sentences referred to by traditional grammarians, excerpts from natural conversations are very often unintelligible; but when they *are* intelligible, this is likely to be due to the help we quietly get from someone who has already read the situation for us." Schegloff ("Goffman and the analysis of conversation," p. 110) objects to Goffman's argument by saying, "The point of introducing the notion of 'adjacency pairs' is, in part, to circumvent the problem of treating some particular type of sequence unit as a serious prototype." However, as I understand it, the point in contention is that the analytic identification of a question–answer sequence unavoidably makes use of a "typology" to discern just how participants in a conversation are "achieving" an adjacency pair. By definition, there are no members of the general class of adjacency pairs other than those whose first-pair parts are characterized (and presumably recognized *in situ*) by type (greeting, complaint, question, etc.). The names for these types are vernacular names, and necessarily so.

fallibility of vernacular intuition, Schegloff more effectively attacks a rival analytic system.[131]

Displacing analysis

On the face of it, there is nothing particularly wrong or unusual about CA's having followed the path that I have described. It could easily be argued that CA's tacit adherence to a logical–empiricist conception of science has little bearing on the quality of Sacks's and his colleagues' scientific achievements. As Sacks put it, the actual achievements of the program would depend "on research, and on the findings of that research."[132] Even if CA's research program embodied a "mythology" of scientific praxis, this would not mark its failure as a science. Indeed, it would place it in very good company.

To borrow Shapin and Schaffer's terms, CA has succeeded in devising a particular variant of "the mechanics of fact making"[133] that currently includes a "material technology" of magnetic tape-recording and playback machinery that preserves singular conversations for detailed inspection; a "literary technology" consisting of the detailed transcription system developed by Gail Jefferson for codifying the lexical and nonlexical features of the talk described in conversation-analytic studies; and a "social technology" through which the members of the analytic culture circulate tapes and transcripts, cultivate common sensibilities to subtle features of their data, read and cite one another's publications, address a common set of technical concerns, and compose their research papers in a distinctively technical vocabulary and style. CA became a conventional social science in the sense that what came to count as relevant topics, adequate data, adequate transcription, and adequate analysis was established conventionally in and through the exemplary shoptalk, literary strategies, and representational practices developed by the active participants in the community.

But why should we be bothered by the conversation analysts' failure to maintain a "reflexive" orientation to the analytic culture they have established, as though they were anthropologists studying their own scientific culture?[134] The natural sciences do not do this, and perhaps no science should be expected to do so. The problem in this case, however, is not that

[131] For an extended critique of Searle from an ethnomethodological/Wittgensteinian point of view, see David Bogen, "Linguistic forms and social obligations: a critique of the doctrine of literal expression in Searle," *Journal for the Theory of Social Behaviour* 21 (1991): 31–62.

[132] Sacks, "Introduction," p. 212.

[133] Shapin and Schaffer, *Leviathan and the Air Pump*, pp. 25 ff.

[134] For a critical review of some recent contributions to "anthropology of science," see Bruno Latour, "Postmodern? No, simply amodern! Steps toward an anthropology of science," *Studies in the History and Philosophy of Science* 21 (1990): 145–71.

proponents of CA fail to account for their "actual" research methods but that their substantive findings are infused with the principled operations that Sacks assumed were built into the practice of science. The principled, and privileged, relation to a domain of "analysis" that currently makes up CA's unexamined social situation leaves an indelible impression on the way that research findings are construed and presented.[135]

The problem as I see it is that Sacks's initial conception of the intersubjectivity of scientific method continues to influence the way that conversation analysts conceive of ordinary methods: Method has come to stand as the model of conversational grammar. By this I do not mean that they employ a "man the scientist" model but that in their studies abstract rules of method have come to stand as adequate accounts of the local "administration" of ordinary actions. As I noted earlier, Sacks treats it as "obvious" that scientists' "reports of their own activities are adequate, i.e., they provide for the reproducibility of their actions on the part of themselves or others, by the use of methods."[136]

Schegloff takes this further when he summarizes Sacks's argument and emphasizes "accounts" of methods (as opposed to "the use of" methods): "A way (perhaps *the* way) to have such stable accounts of human behavior is by producing accounts of the methods and procedures for producing it."[137] There is now far more interest in local, pragmatic, and rhetorical features of scientific methods than there was during Sacks's lifetime, and with the benefit of hindsight, it is now possible to say that the picture of science that Sacks presented was a "mythological" one. The important point is not that his was an erroneous view of scientific activity but that it became a warrant for setting up the scientific status of conversation-analytic "accounts of the methods and procedures for producing" vernacular activities. In published accounts, formal descriptions of methods, such as Sacks and his colleagues' description of the rules for turn taking, are said to provide the grounding for the regularity and reproducibility of the activities described. This contrasts with the Wittgensteinian/ethnomethodological policy (see Chapter 5) to the effect that such formal statements are internal to the practices they describe, instruct, or regulate and that they do not "account for" the methodic practices

[135] Although Bourdieu's understanding of ethnomethodology leaves much to be desired, he clearly articulates the problem I have just raised (in this case he is criticizing structural linguistics and anthropology): "The practical privilege in which all scientific activity arises never more subtly governs that activity (insofar as science presupposes not only an epistemological break but also a *social* separation) than when, unrecognised as privilege, it leads to an implicit theory of practice which is the corollary of neglect of the social conditions in which science is possible." See Pierre Bourdieu, *Outline of a Theory of Practice,* trans. Richard Nice (Cambridge University Press, 1977), p. 1.

[136] Sacks, "Introduction," p. 214.

[137] Schegloff, "An introduction/memoir," p. 203.

in which they are used, except in and through the competent performance of those very practices.

Whereas Sacks treated the very fact of the existence of science as a grounding for a possible natural observational science of sociology, a different picture of method now emerges: Accounts of findings become adequate in light of the work of reproducing those findings, but the adequacy of such accounts and of such work is reflexive to a disciplinary matrix.[138] The import of this is that structures of accountability are intertwined with the local practices and phenomena of a discipline, and there is no reason to suppose that scientific practices can be adequately described in terms of formal accounts of "human behavior." On the other hand, for members who are competent and entitled to read them, accounts of findings can serve as adequate accounts of how to repeat those findings.[139]

It might seem unduly tendentious, and even unfair, for me to critique CA's scientism at such length when there are so many other worthy targets in the social sciences. Why contest CA's claims to analytic expertise, and why go on at such length about how members of the analytic community have created a rather small social science discipline? It seems that I am demanding that CA should somehow "rise above" the professional trappings of a social science, and my criticisms suggest what to many readers must seem to be a ridiculous alternative – a mode of investigation that requires no principled distinction between professional analysis and ordinary practical reasoning. I shall reserve further discussion of this ridiculous alternative for the next chapter, but for now it is worth mentioning that my tendentiousness has to do with both a reverence for a past (the genealogical connection between CA and ethnomethodology) and a hope for a contingent future (the prospect for a kind of natural historical investigation of epistemology's topics).

The professional success of CA has colored contemporary views of what ethnomethodology always has been and what it might become.[140] This has not been such a terrible eventuality, since CA deserves recognition as an innovative approach to the study of social order, and ethnomethodology might not have survived without the success of its "positivistic" offspring. It is nevertheless a source of some consternation that CA's exemplary studies have largely superseded the "classic" agenda in ethnomethodology.[141]

[138] See Thomas Kuhn's "Postscript" to the 1970 edition of his *The Structure of Scientific Revolutions*, in which he speaks of paradigms as a "disciplinary matrix." The most relevant sense of the term *matrix* for our purposes is not the mathematical one but the organic sense of an environment in which a living community is embedded.

[139] See Eric Livingston, *The Ethnomethodological Foundations of Mathematics*, for a demonstration of the instructable reproducibility of a "proof statement" in and as the locally organized, first-time-through, lived work of proving.

[140] Heritage's *Garfinkel and Ethnomethodology* has become the definitive version of CA's ethnomethodological roots.

[141] See Pollner, "Left of ethnomethodology."

Although I share this concern, as suggested in Chapter 4, the classic studies in the ethnomethodological canon are subject to their own impasses. What is needed is not a return to a purer or more orthodox implementation of the program announced in *Studies in Ethnomethodology.* As Garfinkel himself acknowledges, that program is radically incomplete, and as I understand it, the task of "completing" it will require something more than an accumulation of empirical studies to fill out the specifications of a theory of social structure. An ethnomethodological research program that is not bound by principled interpretations of a canon requires a continual deepening and critical respecification of the "epistemic" themes that make up the heart of the program: method, analysis, accountability, and the like. Such a program requires that we call a moratorium on treating "science" as a source of grounded investigation. Instead, a more appropriate (and indeed an inescapable) starting point for investigations would be the transparently intelligible and intuitively obvious – and yet defeasible – workings of language and practical action that compose an uninvestigated and unjustified "situation of inquiry."

The overall aim of ethnomethodological studies of work in the sciences has been to respecify many of the classic themes from intellectual history.[142] Briefly characterized, this is a matter of demonstrating how an epistemic theme is practically situated in a characteristic set of practices, for example, how a mathematical "proof" or an experimental "observation" is constituted as a temporally elaborate assemblage of activities, equipment, and literary residues. Of course, these themes need not be investigated only in the domains of science and mathematics. Sacks's conception of a primitive natural science points the way to a researchable domain of practices through which members come to see what others describe and to describe what others see. Although I have contested the idea that these structures of accountability provide a generic grounding for a possible natural observational science of sociology, this does not make them any less interesting as investigative topics. And given its historical intimacy with ethnomethodology, there is no reason to disregard CA's accumulation of studies. The question is, what more can be done with them?

Appendix: molecular biology and ethnomethodology

In his early lectures and writings Sacks used two dominant scientific metaphors: the metaphor of primitive natural science and that of molecular

[142] Harold Garfinkel, "Evidence for locally produced, naturally accountable phenomena of order, logic, reason, meaning, method, etc., in and as of the essential quiddity of immortal ordinary society, (I of IV): an announcement of studies," *Sociological Theory* 6 (1988): 103–6.

biology. He also used the metaphor of a "machine" or "machinery," sometimes combining it with the biological metaphor. To speak of metaphors here is somewhat misleading, since Sacks was making a stronger claim than that a possible natural observational science of human behavior could be designed to be like the existing natural sciences. He was arguing instead that this "behavioral" science was nothing other than the very doing of science.[143] Sacks did not go much further than to mention that molecular biology provided a model for conversation analysis (or what later came to be called conversation analysis), but his insight was impeccable. The further development of molecular biology has vindicated Sacks's arguments by developing along lines that oddly resonate with ethnomethodology's programmatic interest in the relations between instructions and order.

Following Sacks's early ethnomethodological initiatives, conversation analysts rejected the idea that the micro–macro problem in sociology should be solved by theoretical schemes for "connecting" the individual to the overall society.[144] Instead of viewing the individual as an entity whose cognitions and emotions are shaped into a microcosmic representation of the normative order of the society as a whole, conversation analysts came to regard models of *the* individual and *the* society as constructs that gloss over a substratum of *molecular* techniques that pervade the body social. In CA's conception of social order, persons become armatures of context-free and context-sensitive machineries, and descriptions of "the society as a whole" are dismissed as remote ways of speaking about an assemblage of concerted techniques whose molecular organization has yet to be deciphered. Whereas the functionalist program placed an indelible stamp on contemporary sociology's conception of the problem of order by drawing its principal metaphors from holistic biology and the problem of connecting the microorgan to the macroorganism (or, on a different level, the organism to its ecological niche), conversation analysis drew much of its scientific imagery from molecular biology.[145]

A key difference between a microsociology, in which individual actors are the most elementary constituents, and a molecular sociology, in which

[143] Consider the term *science* here to be describing an ordinary phenomenon, for example, the ordinariness of the "science" we can witness by visiting our colleagues in the biology department and watching what they do in their labs. This is separate from the definitional question, "What is science?"

[144] See Emanuel Schegloff, "Between macro and micro: contexts and other connections," and Richard Hilbert, "Ethnomethodology and the micro–macro order," *American Sociological Review* 55 (1990): 794–808.

[145] The "organic" metaphor applies more often to functionalist conceptions of "structure." In his theory of action, Parsons (*The Structure of Social Action,* vol. 1) uses the analogy of classical mechanics when discussing the role of theory in empirical research and formulating the basic conceptual elements of the "unit act."

embodied techniques are foundational, is that the latter units are essentially plural and heterogeneous. There is no idealized concept of the fundamental sociotechnique parallel to that of the social actor.[146] Instead, CA's molecular sociology begins with a conception of social order in which different combinations of heterogeneous techniques produce an endless variety of complex structures. This conception is distinctive for the way it is social structural all the way down. The basic unit of analysis is not an ideal–typical "actor" or "self" but a plurality of socially structured techniques through which orderly social activities are assembled. The research agenda is to unpack these molecular sequences.

Conversation analysts attempt to characterize a simple order of structural elements and rules for combining them, and thus they undertake a reductionist program not unlike that of molecular biology. Ever since Watson and Crick characterized the molecular structure of DNA, many fields of biology have taken a reductionist turn. Instead of viewing the helical strands and base pairs that make up the DNA molecule as microorgans whose forms and functions are defined by their relations to a whole organism, molecular biologists now treat DNA sequences as detachable structures that cut across and sometimes dissolve established conceptual divisions between organ systems and entire organisms.[147]

Although molecular structures and rules of combination help explain significant holistic problems related to reproduction, inheritance, and disease, these structures are not "micro" reflections of "overall" organismic functions. There is no homunculus inscribed in the molecule that resembles the macroscopic order it helps reproduce. Instead, a sequential "code" is said to provide a "set of instructions" that are translated and transcribed by an organic "machinery." Difficult questions linger about whether organismic unities can ultimately be reduced to molecular structures and rules of combination, but for all practical purposes molecular biologists put such questions aside when they pursue their intensive investigations. For the most part, molecular biologists presume a pervasively ordered universe, such that an intensive effort to decipher the molecular constituents of a particular strain of bacterium implicates a vast array of organic life forms.

The analogy I have drawn between molecular biology and CA's molecular sociology is based less on the doctrines of molecular biology than on the

[146] For the time being, I will not consider Latour and Callon's use of the term *actor* to describe heterogeneous agents, agencies, and stable assemblages. Instead, I refer to the more familiar use of the term in American sociology to describe a theoretical model of an agent that acts in a given situation.

[147] Lily Kay, "Life as technology: representing, intervening, and molecularizing," *Rivista di Storia della Scienza*, October 1992.

production of routine molecular biological techniques.[148] CA adopts an ethnomethodological conception of social activity in which the most elementary acts are intelligible by reference to the coherent structures of accountability that those acts help produce. Praxiologically, the field of molecular biology is defined in terms of both the DNA sequences that molecular biologists describe and the repeatable instructions for routine laboratory techniques that they perform. In order to describe the sequential "instructions" by which an egg produces an organism, the technicians at the lab bench must produce standardized sequences of action in accordance with the instructions in their laboratory manuals. The "social molecules" in this case are not small "things" to be scrutinized scientifically but are sequences of observable and reportable technique that compose a scientific investigation.

Like molecular biology, molecular sociology is not atheoretical; its unity is praxiological. In contrast with a "grand" synthetic theory of the "system," the molecular picture is held together by a heterogeneous array of techniques:

> Molecular biology is unified not by a central theory, but by an approach to explaining and altering organismic function by reference to, and use of, an *omnium gatherum* of detailed molecular mechanisms. In this respect, molecular biology is less like Newtonian mechanics than it is like auto mechanics: what it studies are *mechanisms* and it *uses* those mechanisms to intervene in nature. Indeed, the subject matter of molecular biology is detailed mechanisms – and what it studies is mechanisms all the way down (but also . . . all the way up).[149]

An important point here is that the conception of unity is not based on a theoretical representation of a totality that is impressed in the molecular unit (like a homunculus inscribed in the "germ cell"); instead, an assemblage, or set of syntagms, provides a "set of instructions" for making the organism, and the genetic engineer's task is to decipher those sequences in order to elucidate, and eventually take over, that "making." This agenda is expressed concisely by the following question raised by a prominent molecular biologist at a public colloquium: "How do you make a large-scale organism from some set of instructions that is contained in a single egg?"[150] The pronoun

[148] See K. Jordan and M. Lynch, "The sociology of a genetic engineering technique," and K. Jordan and M. Lynch, "The mainstreaming of a molecular biological tool: a case study of a new technique," pp. 160–80 in G. Button, ed., *Technology in Working Order* (London: Routledge, 1992).

[149] Richard M. Burian, "Underappreciated pathways toward molecular genetics," paper presented at the Boston University Colloquium for the Philosophy of Science, Boston, April 15, 1991.

[150] Walter Gilbert, "The scientific origins of the human genome initiative," paper presented at the Boston University Colloquium for the Philosophy of Science, Boston, April 16, 1991.

"you" identifies the molecular biologist's task with that of the "egg," and the "egg" in turn becomes a manual of instructions to be read by the molecular biologist. In the case of the human genome project, it is a manual for making "Man."

According to this conception of the unity of molecular biology, praxiological structures – the instructable reproducibility of routine sequences – identify the biologist's actions with those of the "natural order" investigated: "Sequencing" describes both the endogenous arrangements of DNA base pairs and the laboratory techniques by which such arrangements are deciphered and reproduced. When viewed in this way, molecular biology and molecular sociology are related by more than an analogy; the practice of molecular biology includes a locally organized set of sociotechniques that use and reflexively "orient to" the instructable reproducibility of social actions. "Sequencing," "transcription," and "translation" are inscribed on a double register: A material register on which instructions are written for generating natural structures, and a methodological register on which humans write instructions enabling others "artificially" to replicate those structures.

To appreciate how these registers are intertwined, we can examine a brief "personal perspective" article in which Stanley Cohen, an eminent member of the tribe of molecular biologists, reviews progress in the field since the late 1960s and early 1970s. Cohen describes practices that by now are well established and that enable particular sequences of genetic DNA to "express" themselves by endowing strains of *E. coli* bacteria with resistance to selected antibiotics. Cohen's article outlines a historic series of experiments that utilize "a heterogeneous population" of molecular agencies to "increase the ease and flexibility of gene manipulation, so that segments of DNA molecules can now be taken apart and put together in a variety of ways" (p. 4). The article gives a "whiggish" account, but this is not relevant to our purposes:

> In 1972, my collaborators and I, using a modification of the procedure worked out by Mandel and Higa, found that *E. coli* could take up circular plasmid DNA molecules and that transformants in the bacterial population could be identified and selected utilizing antibiotic resistance genes carried by plasmids. . . . Cells transformed with plasmid DNA reproduced themselves normally and produced a clone of antibiotic resistant bacteria. Since each cell in the clone contained a DNA species having the same genetic and molecular properties as the plasmid DNA molecule that was taken up by the initial transformant, the procedure made possible the cloning (and thus, the biological purification) of individual plasmid molecules present in a heterogeneous population.[151]

[151] Stanley N. Cohen, "DNA cloning: a personal perspective," *Focus* 10 (1988): 1–4.

Throughout the article, Cohen describes how naturally occurring entities, their constituents, and their normal life-processes are renormalized and reorganized in the service of a production process. They become "manipulable" with greater or lesser "ease"; they act as extensions and articulations of an engineering process, of taking apart and reassembling. Such manipulation is conjoined with "analysis," which in turn is facilitated by the literary features of the materials (e.g., their markers, duplicative mechanisms, and transcriptions).

Cohen describes techniques for *joining* or *linking* strands of DNA, *converting* hydrogen-bonded DNA "circles" to covalently closed "circles," and *introducing* DNA segments into bacterial cells by means of *transformation.* He speaks of various discoveries made along the way, discoveries of *useful* agents or techniques that enable researchers to *manipulate* genes in order to *accomplish* covalent linkage or *achieve* ligation. Cohen portrays DNA strands as segments or blocks of material with properties (complementarity, blunt endedness, duplex regions) that afford various practical accomplishments. (The terms I have emphasized are Cohen's terms, and it takes no special sociological insight to assign a praxiological sense to them.)

This language suggests a kind of hands-on engineering with miniscule building blocks. These blocks are constructed so that they can be linked or spliced together end to end, to form circles or strands. They can also be taken apart and put together again in different combinations. It is clear that physically joining and taking these things apart is not done simply by picking up segments and latching them together like pop beads, since the media for doing so are various viral and bacterial "vectors," chemical reagents, and catalysts.

Cohen also uses terms associated with agriculture: "Adventitious DNA segments" are introduced into "living" cells and are "propagated" there. Many of the practical difficulties and remedies he mentions have to do with disciplining and cultivating microbes.[152] Naturally occurring constituents of cells (genetic DNA, plasmids) and normal cellular processes are organized in the service of a production process. Some of Cohen's language also suggests a textual or literary field: Bacteria contain copies of plasmids; DNA sequences can be transcribed; and antibiotic-resistant genes are used as markers for keeping track of cloned DNA.

There is no apparent incongruity in Cohen's account of cultivating "living" microbes, reading and transcribing texts, and manipulating things. He describes microbes as complex organizations of "machinery" consisting of arrangements of DNA and material reactions to chemical agents that can be used to cleave, link together, and mark strands of DNA. It is a soft machinery

[152] See Latour, *The Pasteurization of France.*

whose moving parts are composed of living bacteria and bacteriophages, whose life-processes (like reproduction and death) are put in the service of manipulation and analysis.

The overall narrative is a history of invention, in which the successive steps are bridged by discoveries. It seems that there is little or no gap between discovery and invention. The discoveries are set up and immediately appropriated as necessary steps on the way to building a "tool." In Cohen's terms, a machinery is discovered, although not as an off-the-shelf microbial mechanism to be put into the immediate service of science, medicine, and industry. Rather, what is discovered is a series of *in vivo* affordances for *in vitro* actions (cleavage, marking, transcription, transformation, etc.). The discovered machinery is gradually shaped into a production line, a series of steps in which molecular constituents are combined, cultivated, and systematically reorganized. The bioengineered production line organizes, cultivates, restricts, and controls legions of bacteria and their constituent activities. Bacteria and plasmids are domesticated, redesigned, and put to work. Constituents are denatured, but far from dehumanizing traditional praxis, the mechanization process humanizes bacteria.

Rather than distinguishing a domain of human behavior from a domain of objects particular to molecular biology, Cohen's description points to a concerted "cultivation" in a novel domain of social praxis. A "molecular sociology" of a rather literal kind takes place, in which familiar structures of accountability – assemblages of discursive and embodied actions, architectures, spaces-and-classes – colonize a previously unspeakable and uninhabited domain. But this molecular sociology is inseparable from the molecular orders it makes accessible: the intelligible, describable, reproducibly observable, and practically manageable contents of investigation. Although a science of human behavior is implied in the story, it is not a general science that can be described in isolation from the thick environment of cultured things, bioengineered tools, and cultivated practices that give rise to it. It is a "homegrown" science of practical actions that is autochthonous to molecular biology.

Sacks apparently recognized not only that science is an interesting institution to be studied by sociologists but also that each natural science includes a sociology of science as a necessary feature of its production. His was not a "naive social theory" of scientific practice.[153] The problem with his conception of science and his effort to "ground" a possible natural science of human behavior on "the very existence of science" is that he assumed that descrip-

[153] Sacks, "Sociological Description" (p. 15, n. 13) attributes "the 'end of naive social theory'" to Marx *(Theses on Feuerbach),* quoting Marx's dictum: "The question whether human thinking can pretend to objective truth is not theoretical but a practical question. Man must prove the truth, i.e., the reality and power, the 'this-sidedness' of his thinking in practice."

tions of procedural rules could stand as self sufficient accounts of ordinary and scientific practice. Indeed, such accounts can be adequate whenever members are able to use them in accordance with an established practice that they have mastered.

In the case of molecular biology, though there are intriguing parallels between molecular biology's and CA's conceptions of order, neither molecular biologists' accounts of genetic sequences nor the manuals in which they describe sequences of technical action gives an exhaustive account of what a practitioner needs to do in order to "sequence" a fragment of DNA and/or perform the sequential operations that enable such a task to be accomplished. Or rather, the adequacy of the objective and procedural accounts is "discovered" and demonstrated in the course of a practice that remains, and can only remain, obscure to a general description of structures of social action. The pairing of instructions with product (whether these are written on a "genetic" or a "technical" register) offers no "ground" for a general science of practical action. Instead, it provides an inexhaustible topic of investigation both in and about a science.

Molecular biology thus becomes a case of, rather than a model for, a molecular sociology. The primitive social objects in this case are instructively reproduced techniques, for example, the techniques for cloning molecules, sequencing strings of DNA, or duplicating such sequences. A technique like the polymerase chain reaction (PCR) is thus an elementary social object in the classic sense of being formal, institutionalized, and produced on different occasions despite changes in personnel, materials, and immediate practical context. The standardized unity of the technique – that is, that any given performance can be recognized as another instance of the "same" technique – is not God-given or established by definition but is instead accomplished as part of the production of the technique. An attempt to use the technique can fail to become an instance of "it." Consequently, the ethnomethodological study of molecular sociology attempts to recover the formal technique together with its contingent performance as an elementary constituent of the routine practice of molecular biology.

CHAPTER 7

From quiddity to haecceity: ethnomethodological studies of work

Ethnomethodologists emphasize that the stable, constraining, recognizable, rational, and orderly properties of "social facts" are local accomplishments, whereas sociologists of scientific knowledge claim that "natural facts" are social constructions. Investigators in both fields try to show how taken-for-granted facts arise from concerted human activities, and they explicitly turn away from the idea that facts are manifestations of a transcendent natural order demonstrated through a rational method of inquiry. In a way, both ethnomethodology and the sociology of scientific knowledge are part of what Stanley Fish calls the "intellectual left" opposition to philosophical essentialism. Both insist that

> the present arrangement of things – including, in addition to the lines of power and influence, the categories of knowledge with their attendant specification or factuality or truth – is not natural or given, but is conventional and has been instituted by the operation of historical and political (in the sense of interested) forces, even though it now wears the face of "common sense."[1]

Among the members of the intellectual left, Fish lists "Marx, Vico, Foucault, Derrida, Barthes, Althusser, Gramsci, Jameson, Weber, Durkheim, Schutz, Kuhn, Hanson, Goffman, Rorty, Putnam, and Wittgenstein, and their common rallying cry would be 'back (or forward) to history.'"[2] He goes on to say, however, that programmatic opposition to essentialism, foundationalism, formalism, and positivism does not necessarily put those "isms" out of play in avowedly constructivist and deconstructionist studies:

> Now this [opposition to essentialism, etc.] is a traditional enough project – it is the whole of the sociology of knowledge; it is what the Russian Formalists meant by defamiliarization, and what the ethnomethodologists intend

[1] Stanley Fish, *Doing What Comes Naturally* (Durham, NC: Duke University Press, 1989), p. 225. Note, however, that ethnomethodologists do allow for a "natural and given" order of ordinary activities, but they treat the "givenness" of such order as a commonplace achievement.

[2] Ibid. Clearly, this list of heroes can fragment along numerous lines of dispute and difference.

> by the term "overbuilding"; and it is the program, if anything is, of deconstruction – but . . . it [often] takes a turn that finally violates the insight on which it is based. That turn turns itself, in part, on an equivocation in the use of the word "constructed."[3]

This "equivocation" has to do with how the word *construction* can sometimes imply a deliberate manufacture or manipulation of an object in accordance with a plan of action. Commonly, constructivist theories depict socially organized actions as though they actually or potentially pursued tangible objectives, were based on clear-cut interests, and involved deliberate choices of means to facilitate those interests and objectives. This is suggested when everyday terms like *invention, inscription, manufacture, machination, manipulation,* and *intervention* are theoretically preferred over equally familiar idioms like *discovery, description, observation, testing, proving,* and the like. As a consequence, whenever interests and choices are nonobvious and practitioners give no explicit account of the existential (as well as objective) circumstances of their practices, analytic terms like *tacit, taken for granted,* and *unconscious* can be used to fill the void with a model of action in which choices are (or could have been) made and tangible interests and motivations govern those choices.[4]

In contrast, for a phenomenologically informed view of social action, the term *construction* (or *constitution*) describes an achievement from which there is no time out. Constructive (or constitutive) actions thus are not limited to "political" programs that are (or can be) implemented in plans of action and deliberate decisions. In this sense, the construction of facts implies no antonym, since the possibility of unconstructed facts drops out of relevance.

When a social analyst asserts that a particular fact was constructed, this can, but does not necessarily, imply that the fact is arbitrary, suspect, politically motivated, or otherwise doubtful, and therefore not really a fact. Oftentimes, such as when sociologists give constructivist explanations of phenomena like "mental illness," and "drug abuse" – social categories that are evidently and controversially bound to specific historical eras and institutional arrangements – the terms of analysis appeal to public skepti-

[3] Ibid., p. 226. The reference to "overbuilding" is obscure to me, though perhaps it refers to the practical *over*determination of social facts in ordinary scenes (the "order at all points") that Garfinkel and Sacks emphasize.

[4] One way to understand Wittgenstein's (*Philosophical Investigations,* ed. G. E. M. Anscombe [Oxford: Blackwell Publisher, 1958], sec. 211–19) famous discussion of acting without reasons or justification ("I obey the rule *blindly*") is to read it as a critical comment about the analytic policy in much of philosophy and social science of treating recognizable features of actions as evidence for underlying choices, decisions, and intentions. Although Wittgenstein does not dispute the occasional relevance of choices, decisions, and so forth to linguistic actions, he does question the analytic extension of the imagery of deliberation and ratiocination to describe "tacit" or "unconscious" sources of action.

cism about these categories.[5] But when constructivism is advanced as a theory that applies across the board, it becomes a metaphysical challenge to "objectivism" roughly akin to idealism, which is adopted before the sociologist or social-historian sets out to demonstrate how a given fact was "constructed."

"Construction" is only one of several terms that seem to imply the possibility of an opposite, of "unconstructed" reality. *Bricolage* contrastively implies the possibility of engineering; theory-ladenness is set off against a passive apprehension of sense data; representation contrasts with a direct "contact" with nature, and self-reflection with unreflective action. But as soon as social construction, *bricolage,* theory-ladenness, representation, and reflexivity become all-inclusive terms for describing actual scientific practices, the oppositions implied by those terms collapse. As Derrida says, if "the engineer and the scientist are also species of *bricoleurs* then the very idea of *bricolage is* menaced and the difference in which it took on its meaning decomposes."[6] At that point a "constructive" analysis loses its critical leverage because it can always be asked: "So, now that we agree that everything is 'constructed,' does this tell us anything in particular about the 'constructions' that scientists make?" No particular epistemic or political criticism would seem to follow from the announcement that "science is a social construction," nor would it imply that scientists could possibly choose to act differently.

One way to counteract the Derridean "menace" is to preserve the various antinomies between construction and reality, *bricolage* and engineering, and belief and knowledge, by "bracketing" them. Although associated with the mythologies of positivism – mythologies that are presumably out of play for an up-to-date sociological analysis of scientific knowledge production – the classic antinomies can be held to be substantively and rhetorically in play for the members of the scientific fields studied. The themes associated with

[5] See Melvin Pollner, "Sociological and commonsense models of the labeling process," pp. 27–40, in Roy Turner, ed., *Ethnomethodology* (Harmondsworth: Penguin Books, 1974); and Steve Woolgar and Dorothy Pawluch, "Ontological gerrymandering: the anatomy of social problems explanations," *Social Problems* 32 (1985): 214–27. Likewise, critical analyses of scientists' "stories," such as Donna Haraway's *Primate Visions: Gender, Race, and Nature in the World of Modern Science* (New York: Routledge, Chapman & Hall, 1989), are convincing because they reveal that the details of particular stories – such as some of those told in well-regarded primatology texts – are closely tied to currently controversial themes. The surface structure of the writings can be read with the aid of feminist literary criticism as transparent ancestor tales that could have been written differently. It remains to be seen, however, that such effective criticisms justify a sweeping conclusion about the "gendered" or "constructed" nature of all science.

[6] Jacques Derrida, "Structure, sign, and play in the discourse of the human sciences," p. 256, in R. Macksey and E. Donato, eds., *The Structuralist Controversy: The Language of Criticism and the Sciences of Man* (Baltimore: Johns Hopkins University Press, 1970). See my discussion on this point in Chapter 4.

positivism become part of the "positivistic common sense" of mundane reason,[7] of the "objectivistic self-understanding in the sciences,"[8] of the "ideology of representation" espoused in scientists' discourse,[9] or of "the scientist's account" of the recent history of physics.[10] It is tempting to suppose that "the scientist" is a species of lay philosopher or a tribal cosmologist whose beliefs are cast into relief against the backdrop of a more disinterested and sophisticated sociological understanding. And insofar as "the scientist's" understanding of the history, philosophy, and sociology of science can be shown to be limited and partisan, the door is now open for historically, philosophically, and sociologically informed criticisms of the limitations of technoscientific reason and the hegemony of that mode of reason over all the major "spheres" of modern life.[11]

There is a sense in which the history, philosophy, and sociology of science should not be entrusted to practicing scientists. Just as a brilliant painter can turn out to give an inarticulate, egotistical, and unrevealing verbal account of how she "thinks with the end of her brush," so a Nobel laureate can give a self-serving and question-begging account of experimentation: "There's nothing to it, really; all you do is construct a hypothesis, set up an experiment to test it, and if it works, you've got something."[12]

Although famous scientists often devote their retirement from the laboratory to writing "reflections" and "reminiscences" for popular consumption, as often as not such reminiscences testify to the fact that prowess with mathematical models and experimental equipment does not readily transmute into the kind of insight required for composing critical and compelling ethnographic and historical descriptions. There is another sense, however, in which the history, philosophy, and sociology of science can be entrusted only to practicing scientists. The general terms and themes that make up the

[7] Melvin Pollner, "'The very coinage of your brain': the anatomy of reality disjunctures," *Philosophy of the Social Sciences* 4 (1975): 411–30, esp. p. 424.

[8] Jürgen Habermas, *Philosophical–Political Profiles,* trans. F. Lawrence (Cambridge, MA: MIT Press, 1983), p. 16. For a critical discussion, see David Bogen, "A reappraisal of Habermas's *Theory of Communicative Action* in light of detailed investigations of social praxis," *Journal for the Theory of Social Behaviour* 19 (1989): 47–77.

[9] Steve Woolgar, *Science: The Very Idea* (Chichester: Ellis Horwood; London: Tavistock, 1988), pp. 99 ff.

[10] Andrew Pickering, *Constructing Quarks* (Chicago: University of Chicago Press, 1985), pp. 3 ff.

[11] Steve Fuller, for instance, moves very quickly from accepting a generalized "social" critique of scientific autonomy and authority and reinstating a normative (and administrative) program for placing science under "public" control. See his "Social epistemology and the research agenda of science studies," pp. 390–428, in A. Pickering, ed., *Science as Practice and Culture* (Chicago: University of Chicago Press, 1992).

[12] This is a paraphrase of comments made by a Nobel laureate about several presentations by historians and sociologists of science at the RPI Laboratory Life Symposium, Rensselaer Polytechnic Institute and General Electric Corporation, Troy, NY, 1981.

lingua franca of science studies – discovery, invention, theory-laden observation, experimental design, controversy and its resolution, and the "pathways" of innovation – are also featured in practitioners' research writings and shoptalk.

Moreover, practitioners have "first crack at" the claims and counterclaims associated with discoveries, controversies, and the rest, and their local–historical understandings of these matters are reflexively embedded in their practices and deposited in the archives left behind by those practices. This does not mean that practitioners are necessarily "able to reflect" in a coherent and revealing way about how in general discoveries are made and controversies are resolved, nor does it imply that their practices are performed "unreflexively"; instead, it means that their collective practices reflexively use local–historical understandings. Such reflective understandings settle for all practical purposes how particular results stand with respect to prior results, how "our" lab's findings contribute to the discipline, how "this" phenomenon is or is not "the same" as a phenomenon described in a written report, or how "this" sample may or may not have been "contaminated" during a novice technician's possibly "incompetent" performance of an experimental run. The local historical and social associations implied and presupposed by such indexical expressions differ markedly from the storyable events and honorific personages that typically inhabit practitioners' reminiscences. The relevant mode of practical and local–historical "reflection" is not a matter of an individual's insight into his or her own achievements and relationships; rather, it has to do with how any single account, utterance, claim, or material product acquires its historical significance by being placed in a collective and potentially contentious order of accounts, claims and products.

It is perhaps misleading to draw pointed contrasts between versions of science generated in the course of practices versus those generated retrospectively by scientists "on holiday."[13] As many sociologists of science have argued, there is no a priori basis for insisting that laboratory shoptalk is a more authentic part of science than Nobel speeches, proposals for funding to congressional committees, and popularized accounts and pedagogies. However, if we follow Wittgenstein by insisting that words issue originally and unavoidably from particular deeds,[14] we should be able to comprehend that an academic discipline that aims to recover "the content" of other disciplines

[13] For example, see Bruno Latour's didactic contrast between "science in action" and "ready-made science." *Science in Action* (Cambridge, MA: Harvard University Press, 1987), chap. 1.

[14] This differs from saying that an author's or speaker's "intention" is necessary for the comprehension of a text or utterance. "Intention" is an overly generalized, and inappropriately mentalistic, construction for understanding the way that indexicals are embedded in the course of readings, writings, speakings, hearings, seeings, and the rest.

is faced with an intractable set of problems. As soon as the "statements" associated with the various natural sciences (the utterances, speeches, equations, writings, citations, and other documents collected in the archives) are (re)collected and placed in semiotic schemes, stable representations of historical lineages and semantic networks, social historical and ethnographic narratives, cognitive maps, and so forth, they become dissociated from the varieties of deeds that give them their life.

What eminent scientists have written about their practices and social relationships does not necessarily stand proxy for the language they use in the heat of arguments with colleagues. Nor does it recover the pragmatic language that comes into play when practitioners inspect results in hand and instruct their associates how to perform routine procedures. Consequently, though there may be plenty of documentary evidence for saying that practicing scientists have a "positivistic self-understanding," it is not at all clear whether such a coherent metaphysical commitment is ubiquitously relevant to the details of scientific work. Indeed, the most common refrain from "laboratory studies" is that scientists act differently than their reports, biographies, and methodological writings say they do. This is often taken as support for an ironic contrast between an official version of logically defensible and consensually validated science and an actual science that is "in fact" messy and contentious. But such a conclusion concedes far too much to "Western scientific rationality" by presuming that in its absence a disorderly mess would prevail. A less presumptuous, although far less exciting, conclusion would say merely that philosophers, historians, and sociologists who encounter science exclusively through its literature will not "know their way about" the laboratory.[15] In other words, their understanding of science would be divorced from the language and practical skills that generate experiments, demonstrations, proofs, and the like.

Considerations like these were at issue when Garfinkel suggested that a "gap" existed in the sociology of occupations and professions literature, including the sociology of science.[16] As he explained it, this gap was created by the methods and interests of studies "about" occupations, as compared with the methods and interests that make up the "what" (or *quiddity)* of the

[15] By saying this, I do not mean to imply that philosophers, historians, and sociologists of science do not "know how to do" science by virtue of doing philosophy or whatever. Instead, I mean that the debates and dichotomies that arise in erudite discussions of science are divorced from the sites, occasions, and techniques of laboratory projects, blackboard demonstrations, computer programming, and the like.

[16] To my knowledge, Garfinkel began speaking of this "gap" in lectures, public presentations, and informally circulated drafts in the early 1970s. I believe that he raised the issue during a plenary address to the 1974 American Sociological Association Meetings in Montreal. An early published mention of these issues is H. Garfinkel, "When is phenomenology sociological?" a panel discussion with J. O'Neill, G. Psathas, E. Rose, E. Tiryakian, H. Wagner, and D. L. Wieder, in *Annals of Phenomenological Sociology* 2 (1977): 1–40.

practices themselves. To a limited degree, Garfinkel's proposals for investigating this "missing what" resembled Bloor's better-known injunction to investigate "the very content and nature of scientific knowledge." But just as some of the other parallels between the two programs I discussed in previous chapters turned out to gloss over deep-seated differences, so is it the case for the content of science.

The missing what

Garfinkel introduced his proposal to study the "missing what" of organized complexes of activity by crediting Harvey Sacks with an insight to the effect that virtually all the studies in the social and administrative sciences literatures "miss" the interactional "what" of the occupations studied: Studies of bureaucratic case workers "miss" how such officials constitute the specifications of a "case" over the course of a series of interactions with a stream of clients; studies in medical sociology "miss" how diagnostic categories are constituted during clinical encounters; and studies of the military "miss" just how stable ranks and lines of communication are articulated in and as interactional work.

For example, according to Garfinkel, a curious feature of Howard Becker's study of dance band musicians and their audiences is that Becker, himself an accomplished jazz musician, described many of the linguistic and customary practices by which dance band musicians attempt to distance themselves from the "squares" who make up the typical audience.[17] Becker informs his readers about numerous interesting aspects of the culture of jazz musicians, but he never discusses how they manage to play music together. The interactional and improvisational "work" of playing together – a social phenomenon in its own right – was somehow "missed" by Becker and other sociologists of music.[18] Moreover, as Garfinkel pointed out, the fact that this "missing interactional what" is missed is not an acknowledged problem in the social sciences, since among other things, the proposal that the "work" of

[17] Garfinkel attributed this critique of Howard Becker to David Sudnow. To appreciate the point, compare Becker's chapters on dance band musicians (*Outsiders* [New York: Free Press, 1963]) to Sudnow's *Ways of the Hand* (Cambridge, MA: Harvard University Press, 1978). Note, however, that Sudnow's phenomenological account of learning to play jazz on the piano does not describe the "interactional" playing of jazz – a more difficult project he reserved for later studies that he has thus far not completed.

[18] Alfred Schutz ("Making music together: a study in social relationship," pp. 159–78 of *Collected Papers*, vol. 2, ed. M. Natanson [The Hague: Nijhoff, 1964]) provided a sketchy basis for such a project. It was perhaps unfair to pick on Becker, since among Garfinkel's contemporaries, Becker has been among the most tolerant of ethnomethodology. For instance, in the Purdue Symposium on Ethnomethodology, Becker acted largely as a translator and mediator between Garfinkel and other ethnomethodologists and a group of baffled sociologists.

playing music together is a proper topic for sociology challenges the idea that sociology is a unique discipline with a distinctive corpus of topics, methods, and findings. The whole point of using conventional sociological methods and findings is to relate various extant social practices back to a context-free "core" of rules, norms, and other social structures.[19]

What Garfinkel seemed to be suggesting was nothing less than an abandonment of a sociological "core" in favor of an endless array of "wild sociologies"[20] existing beyond the pale of sociological empiricism. This challenge to the epistemic "empire" of a coherent and unitary sociology did not go unheeded; it was denounced in vociferous terms by some of the more prominent spokespersons for established social science disciplines. Many of the complaints leveled against ethnomethodology included charges of antiprofessionalism that went well beyond "theoretical" or "methodological" matters. The crudest diatribes – such as those by Lewis Coser and Ernst Gellner – were especially indicative because they abandoned any pretense of reasoned argument in favor of indignant objections and sarcastic caricatures of how the ethnomethodologists conducted their affairs.[21]

Aside from the gratuitous name calling, gross misunderstandings, and egregious conceits that accompanied their charges against ethnomethodology's "unprofessionalism," Coser, Gellner, and other prominent critics were far from off base in recognizing that what Garfinkel and his cronies were proposing would, if it ever caught on, be the death of sociology as they knew it. In more recent years, the rancor and mutual distancing have subsided, owing to incessant efforts by professional sociologists to "theorize" and "methodize" ethnomethodology, as well as to the empiricist tendencies that have become prominent in conversation analysis.[22]

[19] There was thematic continuity between this proposal and the story of "Shils's complaint" that Garfinkel mentions when recounting the "invention" ethnomethodology (see Chapter 1).

[20] See John O'Neill, *Making Sense Together: An Introduction to Wild Sociology* (New York: Harper & Row, 1980).

[21] Lewis Coser, "ASA presidential address: two methods in search of a substance," *American Sociological Review* 40 (1975): 691–700; Ernst Gellner, "Ethnomethodology: the re-enchantment industry or the California way of subjectivity," *Philosophy of the Social Sciences* 5 (1975): 431–50. These criticisms are discussed in Chapter 1.

[22] Jeffrey Alexander and Bernhard Giesen, for instance, separate "late" ethnomethodology into two lines of development, one of which remains closely associated with Garfinkel's early attack on rule-based social science models, with the other being the "offshoot" of conversation analysis. Because the latter line of research describes interactional practices in terms of "constraining rules," Alexander and Giesen argue, it can more easily be subsumed within normative models of the social system. See Alexander and Giesen, "From reduction to linkage: the long view of the micro–macro link," pp. 1–42, in J. C. Alexander, B. Giesen, R. Munch, and N. J. Smelser, eds., *The Micro–Macro Link* (Berkeley and Los Angeles: University of California Press, 1987), esp. p. 28. For further discussion on this point, see David Bogen, "The organization of talk," *Qualitative Sociology* 15 (1992): 273–96.

A strong antiprofessional stance still characterizes Garfinkel's proposals for studying diverse activities as vulgar but relatively autonomous sociologies that have not been, and ultimately cannot be, colonized by a core sociological discipline. Much in the way that Wittgenstein – acting out of a deep commitment to philosophical investigations – dismissed professional philosophy as a peculiar way to use the vernacular, Garfinkel – acting out of his commitment to investigations of the production of social order – dismissed professional sociology as a literary enterprise whose thin and meager voice is drowned out by the noisy din of undocumented and unexplored "wild sociologies."

The proposal that ethnomethodology's task was to "discover" the myriad "missing whats" inhabiting specialized practices dispersed throughout the ordinary society was also cause for confusion even among sociologists who expressed sympathy for the project. For instance, in his otherwise helpful and well-researched review of ethnomethodological studies, Heritage misconstrues the "studies of work" program by giving a scientistic spin to Garfinkel's proposals.[23] To a large extent he treats the available studies of work in the sciences as specialized applications of the program for a natural observational science of human behavior exemplified by conversation analysis (see Chapter 6). As Heritage sees it, the "missing what" has to do with sociology's failure "to depict the core practices of occupational worlds," and he adds that in occupational sociology, "the lived realities of occupational life are transmuted into objects suitable for treatment in the accounting practices of professional social science."[24] Presumably, an observational science would be needed to provide a "natural observational base" to fill the gap in the sociological literature. Heritage recognized that Garfinkel's students would have to school themselves in the specialized language and practices of the particular disciplines they were studying, rather than relying on an ordinary linguistic access to structures of conversation as a starting point for making analytic observations and reports, and using such reports as a base for further observations.

As should be obvious to anyone who has attempted to read specialized scientific journals, a mastery of disciplinary techniques is required for making adequate sense of the prose, graphics, and mathematical expressions. To comprehend the unique "what" at the core of each coherent discipline requires a reciprocally unique method for coming to terms with it. Such a method is inseparable from the immanent pedagogies by which

[23] John Heritage, *Garfinkel and Ethnomethodology* (Oxford: Polity Press, 1984), pp. 292–311. Heritage does qualify his account by saying that at the time he wrote it there were few publications available from which to reconstruct what Garfinkel and his "second generation" of students were up to.

[24] Ibid., p. 300.

members master their practices. In plain ethnographic terms, Garfinkel seemed to be insisting on a strong participant-observation requirement, through which his students would gain "adequate" mastery of other disciplines as a precondition for making ethnomethodological observations and descriptions. Garfinkel suggested as much by insisting that students who decided to study the natural sciences, legal professions, and mathematics disciplines should take the relevant courses of training, and some of his students did pursue such a program of study for their dissertation and postdoctoral research.[25]

Considered as I have just outlined it, Garfinkel's "unique adequacy requirement of methods" differed from more familiar ethnographic policies mainly by the stringency of its injunction to master the practices studied (rather than simply learning to talk "about" them) and by its complete disavowal of all established methods for mapping, coding, translating, or otherwise representing members' practical reasoning in terms of established social science schemata. Sociology's failure "to depict the core practices of occupational worlds," was therefore not something that ethnomethodologists would attempt to set right by more accurately "depicting" such "core practices," since any attempt to do so would construct yet another social science representation abstracted from the embodied practices described. Instead, Garfinkel seemed to have devised a program for "going native," disappearing into the field without delivering social scientific accounts of the "field experience."

In his later writings, Garfinkel made suggestions for hybridizing ethnomethodology with other disciplines (mathematics, natural sciences, legal studies, etc.), so that the "product" of the research would not take the form of reports about exotic practices; instead, it would consist of efforts to develop hybrid disciplines in which ethnomethodological studies of, for example, lawyers' work would contribute to legal research.[26] Had this program taken seed throughout sociology, its effect would have been to disperse the "home" discipline into innumerable hybrids initially

[25] Eric Livingston was trained in mathematics; Stacy Burns enrolled in law school; Melinda Baccus worked as a paralegal secretary; David Weinstein enrolled in (and flunked out of) a truck driver's training program in South Dakota; Albert B. Robillard and Chris Pack took jobs in pediatrics departments; and George Girton pursued training in the martial arts. As one can readily imagine, there were other incentives for undertaking such training and employment, and the students did not always take up the activities for the sole purpose of pursuing ethnomethodologial investigations. Incidentally, my own studies of laboratory work were not "grounded" in such mastery, and this became a source of criticism. See H. Garfinkel, E. Livingston, M. Lynch, D. Macbeth, and A. B. Robillard, "Respecifying the natural sciences as discovering sciences of practical action, I & II: doing so ethnographically by administering a schedule of contingencies in discussions with laboratory scientists and by hanging around their laboratories," unpublished paper, UCLA, 1989, pp. 10–15.

[26] Ibid., p. 14.

held together by the familiar themes that mark ethnomethodology's recognizable discourse. Because Garfinkel's variant of ethnomethodology eschews the specification of a core of methods and theoretical concepts,[27] the effect would have been to dissolve any semblance of a foundation in the academic social sciences.[28] Contrary to Heritage's summary, such a program could not establish a "natural observational base" for a science of occupations because that "base" would dissolve into a veritable ecology of local conspiracies that organizes and distributes the work of the various disciplines studied. Unless ethnomethodology were to establish itself as a master discipline – a discipline of all disciplines – Garfinkel's unique adequacy requirement would become a pretext for a one-way journey out of sociology.[29]

From essential contents to iterable epistopics

Before we conclude that the unique adequacy requirement is simply a way to accelerate the demise of sociology, we should keep in mind that it is not a methodological criterion for a general science of practical action. Garfinkel makes this clear when he says, "Each natural science is to be recovered in the entirety of its identifying, technical material contents as a distinctive science of practical action . . . which is not interchangeable with any other discovering science."[30] What is to be recovered should not be likened to a transportable "content" to be unearthed through archaeological investigations. In-

[27] As noted in Chapter 1, indexicality and reflexivity are core themes in discussions of ethnomethodology, and generally ethnomethodologists try to observe closely and describe social actions, often using videotape and audiotape recordings as a documentary basis. However, the thematic and methodological initiative in these studies leads away from a foundational theory or rule-based method in order to show in the circumstantial details of each case how social order is endogenously produced.

[28] It is difficult to tell whether Garfinkel's project failed or succeeded. For the most part, the cohort of "second-generation" students who pursued doctorates under Garfinkel in the 1970s and 1980s vanished from sociology departments and took up other occupations in and out of academia. It is impossible for me to say whether these former students hybridized ethnomethodology with the occupations into which they settled or whether they simply abandoned it. In some cases, they left sociology (and the academic study and teaching of ethnomethodology) not by design but out of desperation for employment. Garfinkel became infamous for the way he inspired and encouraged his students to pursue their studies while at the same time forcibly weaning them from any conventional career ambitions in sociology. This did not have happy consequences for some of Garfinkel's most dedicated and accomplished students. In my own case, I can testify that my failure to enact fully Garfinkel's program contributed to my continuous (although somewhat tenuous) employment in sociology departments.

[29] Note that Arthur Frank's "Out of ethnomethodology" (in H. J. Helle and S.N. Eisenstadt, eds., *Micro-Sociological Theory: Perspectives on Sociological Theory,* vol. 2 [London: Sage, 1985]) can be read as a recommendation to return "into" the sociological fold.

[30] Garfinkel et al., "Respecifying the natural sciences," p. 2.

stead, each disciplinary "content" might better be understood as itself an immanent archeology of knowledge.

Accordingly, to be able to recover a "distinctive science of practical action" in "the entirety of its identifying, technical material contents," ethnomethodologists would have to situate their inquiries in the identifying details of each science studied. Their descriptions and analytic formulations would thus need to rely on those details as both topic and resource, and there would be no gap, boundary, or discontinuity between adequate analysis and the member's language and practical mastery. The ethnomethodologist would not "go in" to the discipline studied in order to "come back" with a cognitive map or other representation of the culture, since no map would be sufficiently complete to recover the scenic details implicated by a competent reading of the map's semiotic features. Short of delivering an entire constellation of details that make up the practical worksite, the only uniquely adequate "news" that could be delivered to professional sociologists would be an apology to the effect, "You would have had to have been there."

In order to avoid confusion about "unique adequacy," it is important to avoid the assumption that ethnomethodology is, or might turn out to be, a "natural observational science of human behavior" that seeks to penetrate to the "core of practices" in other disciplines. As Garfinkel has himself pointed out, the term *quiddity* can encourage misunderstandings of what ethnomethodological studies might be about.[31] References to a "missing what" and "unique adequacy" encourage a conception of each disciplinary speciality as a unique species of practice defined by a singular essence that can be comprehended only by "getting inside" the relevant epistemic circle. If we were to suppose that ethnomethodology could become an epistemic center from which inquiries into all other disciplines could be conducted, we might conclude that Garfinkel's ambition was to build a science capable of grasping the genetic essence of each praxiological species. Something like this is suggested when it is supposed that ethnomethodology's aim is to "look for rules which, when followed, allow us to generate a 'world' of a given kind."[32] When viewed in this way, the unique adequacy requirement would be part of a grandiose "social genome project" whose data base would be sets of instructions for generating different social worlds. As I see it, however, such a project would forget one of the principal lessons from

[31] Garfinkel's reasons for abandoning the term *quiddity* are elaborated in H. Garfinkel and D. L. Wieder, "Evidence for locally produced, naturally accountable phenomena of order*, logic, reason, meaning, method, etc., in and as of the essentially unavoidable and irremediable haecceity of immortal ordinary society: IV two incommensurable, asymmetrically alternate technologies of social analysis," pp. 175–206, in G. Watson and R. Seiler, eds., *Text in Context: Contributions to Ethnomethodology* (London: Sage, 1992).

[32] Erving Goffman, *Frame Analysis: An Essay on the Organization of Experience* (New York: Harper & Row, 1974), p. 5.

ethnomethodology, that a competent reading and writing of any set of instructions itself presupposes the "scenic" details of the occasions in which members consult the instructions.

If the unique adequacy requirement of methods were to imply that each scientific discipline "contains" a "unique content" that makes it what it is, the policy would contradict the nonessentialist and antifoundationalist picture of scientific disciplines suggested (however inconsistently) in science studies research. Since ethnomethodology has also contributed to that alternative picture, the "contradiction" could perhaps be attributed to a lingering residue of ethnomethodology's Schutzian heritage (see Chapter 4).

Alfred Schutz's conception of science, which he developed in part from Felix Kaufmann's methodological writings, included the idea that a "scientific corpus" – a body of propositions accumulated over the history of a discipline that provides the working scientist with a science-specific "stock of knowledge at hand" – distinguishes disciplinary-specific theorizing from ordinary practical action. This conception assumes that each science develops a teleological unity, which in turn enables a series of stable methodological and cognitive demarcations to be erected between its corpus and those in the other established disciplines. This contrasts with the emphasis in more recent science studies on the openness of the specialized sciences to practical techniques, ideological influences, and discursive formations that are not so neatly bounded or contained. In Bruno Latour's and Michel Callon's influential accounts of "science in action," for instance, the practices and technologies of the laboratory are said to be inexorably linked to chains of literary inscription and translation, which transact and transgress the "boundaries" between science, technology, popular movements, government, industry, medicine, agriculture, and daily life. Viewed in this way, the "autonomy" and "boundedness" of any scientific discipline is a secondary product of a multifaceted movement through which the practices and products of that discipline are historicized into "black boxes" whose integrity and established contents are taken for granted.[33] If Garfinkel is understood as proposing a program for recovering the contents of such black boxes, he can fairly be accused of searching for the Holy Grail, the (non)existence of which rests entirely on a mythology created in and through the successful institutionalization of the sciences. Understood differently, and in my view more correctly, the unique adequacy requirement should imply that the basis for the coherence of a discipline is not to be discovered *in* the black box; instead, it is established through the very existence *of* the black box.

[33] See Bruno Latour, "Give me a laboratory and I will raise the world," pp. 141–70, in K. Knorr-Cetina and M. Mulkay, eds., *Science Observed: Perspectives on the Social Study of Science* (London: Sage, 1984).

Just as Wittgenstein is sometimes falsely accused of proposing that "language games" are impenetrable and incommensurable monads and of denying the possibility of "translating" from one language community to another, so can Garfinkel's conception of the *quiddity* of the "discovering sciences" be faulted for presupposing a picture of each science as a bounded epistemic container with well-defined zones of "inside" and "outside." Such accusations can be turned aside, however, by reading Wittgenstein and Garfinkel in the way recommended in Chapter 5.

Wittgenstein's concept of language games (if, indeed, it is a concept) is subject to innumerable interpretations, but it should be clear that he does not depict language games as coherent stocks of knowledge "shared" by well-defined communities of practitioners. Instead, as he makes explicit, "the term 'language-game' is meant to bring into prominence the fact that the *speaking* of language is part of an activity, or of a form of life." Wittgenstein makes it difficult to infer that language games are enclosed by disciplinary or cultural boundaries when he lists the following examples:

> Giving orders, and obeying them –
> Describing the appearance of an object, or giving its measurements –
> Constructing an object from a description (a drawing) –
> Reporting an event –
> Speculating about an event –
> Forming and testing a hypothesis –
> Presenting the results of an experiment in tables and diagrams –
> Making up a story and reading it –
> Play-acting –
> Singing catches –
> Guessing riddles –
> Making a joke; telling it –
> Solving a problem in practical arithmetic –
> Translating from one language into another –
> Asking, thanking, cursing, greeting, praying.[34]

Some of these games are more or less closely associated with science (e.g., forming and testing a hypothesis, presenting the results of an experiment in tables and diagrams), but virtually all of them can be said to occur both "in" and "beyond" science. As studies of daily life in laboratories show, while going about the work of "testing hypotheses," scientists joke around, trade

[34] Wittgenstein, *PI,* sec. 23. Even this list only suggests the complexity of the field. As conversation-analytic research has demonstrated, phenomena like stories, jokes, and giving and receiving orders are gross orders of activity that can be further differentiated in terms of the routines that make them up and the circumstances in which they occur.

insults, curse at their equipment, play act, and sing jingles.[35] Moreover, some distinctively "scientific" activities are performed through the aid of artists and draftsmen, computer programmers, animal trainers and tenders, instrument makers and repairers, accountants, secretaries, and other specialists and technicians who are not entitled to call themselves scientists.[36] When laboratory researchers describe the appearance of an object or give its measurements, they often employ specialized instruments and disciplinary-specific metrical units and standards, but the sensibility of their activity is not "contained" exclusively in a single discipline. Description and measurement are part of daily life, although what counts as a sensible description or adequate measurement can vary considerably from one circumstance to another.[37]

Wittgenstein's examples of language games partly converge with the "molecular" activities – sequential structures for giving orders and obeying them; asking, thanking, cursing, greeting, joking, telling stories, and so forth – described by conversation analysts (see Chapter 6). In CA, such activities are said to be organized by context-free rules, which permeate intuitively recognized "boundaries" between different activities and situations while at the same time these rule-governed structures are articulated in a context-sensitive way; that is, constructed, used, and understood in accordance with the immediate circumstances. However, another well-known analogy from Wittgenstein – that of "family resemblances" – should warn us away from any idea that language games are distinct "packets" of activity that retain an essential (context-free) form whenever they are "implemented."[38]

[35] See M. Lynch (*Art and Artifact in Laboratory Science: A Study of Shop Work and Shop Talk in a Research Laboratory* [London: Routledge & Kegan Paul, 1985], pp. 169–70) for an example. Note that such activities as telling jokes and "horsing around" do not necessarily occur outside the frame of an experiment, since they often are performed by sabotaging the equipment, playfully modifying data displays, teasing laboratory animals, and the like. Although some practical jokes and other spontaneous productions of humor are recognizably "transgressive" of "serious" laboratory work, that recognizability is itself a product of laboratory work. Also see G. Nigel Gilbert and Michael Mulkay, *Opening Pandora's Box: A Sociological Analysis of Scientists' Discourse* (Cambridge University Press, 1984).

[36] Garfinkel summarizes these skills under the heading of "the dependence on bricolage expertise." See Garfinkel et al., "Respecifying the work of the natural sciences," p. 24.

[37] See M. Lynch, "Method: measurement – ordinary and scientific measurement as ethnomethodological phenomena," pp. 77–108, in G. Button, ed., *Ethnomethodology and the Human Sciences* (Cambridge University Press, 1991).

[38] Sacks sometimes used the analogy of a "warehouse" of techniques from which conversationalists drew when assembling their activities. Although this analogy has been gainfully employed in conversation-analytic studies, it too easily suggests that the relevant techniques are available as fixed packets of standard activity rather than as identities assigned (often retrospectively) to conjoint activities constructed locally, with every detail standing as a detail within a here-and-now assemblage that surpasses any general definition of a conversational technique.

When Wittgenstein likens the relationships among different "games" to "family resemblances" – the occasional and cross-cutting similarities among the different members of a family – he explicitly discourages his readers from thinking that a "set" of similar games is defined under a single criterion. By extension, the various practices that can be described under the heading "giving orders and obeying them" do not necessarily share a structure defined by a general set of rules or mechanisms. Instead, the description "giving orders and obeying them" provides a way of speaking about an open-ended array of practices that can take innumerable forms. What counts as a sensible and accountable "order" and what counts as a relevant and acceptable mode of "obedience" do not necessarily get resolved by a general norm, disposition, or system; rather, "orders" and "obedience" can do distinctive jobs on different occasions.

Although the sciences may not be defined by unique sets of language games, particular epistemic themes are recurrently associated with the discourses about and in the various disciplines. Some of these "epistopics" were summarily treated in Sacks's account of primitive natural science (see Chapter 6), and they include familiar themes from epistemology and general methodology: observation, description, replication, testing, measurement, explanation, proof, and so on.[39] They are the common terms of a lingua franca that seemingly "transcends" the various specialized disciplines; they provide headings, points of contact, and contentious topics in an interdisciplinary discourse among historians, philosophers, and sociologists of science, and they make up "metatheoretical" terms of trade for theory building in the social sciences. But when I speak of these themes as *epistopics,* I mean to divorce them from a "metatheoretical" aura and to attend to the manifest fact that they are *words.*[40] This is not to say that they are "merely" words, nor would I want to dissociate them from material phenomena and embodied praxis. Instead, I am opposing the tendency, for example, to treat *observation* as though the word guaranteed a unity of activity whenever it could be said that "observing" takes place or to assume that *representation* names a coherent category of activity or that a concept of "measurement" defines the

[39] My use of an "etcetera clause" is itself an important characteristic of this list of epistopics: They are not a finite set; they cannot be exhaustively listed; and their use is not limited to "scientific" discourse. At the same time, I assume that the epistopics I have listed should be recognizable as names for recurrent themes, topics, and concepts in discussions of methodology and epistemology. Note, however, that Wittgenstein tends to use these terms as transitive verbs rather than nouns – for example, measuring rather than measurement – in order to dissociate them from general methodology and epistemology and to place them more tangibly within commonplace activities.

[40] A similarly blunt proposal about "motives" was made by C. Wright Mills in "Situated actions and vocabularies of motive," *American Journal of Sociology* 5 (1940): 904–13.

core meaning of diverse measuring activities in carpentry, engineering, laboratory and field research, and other less specialized domains.[41]

As suggested in Chapter 5, the project of extending Wittgenstein's occasional proposals into an empirical study of language games can involve an effort to take up some of the rubrics he identifies (e.g., "presenting the results of an experiment in tables and diagrams") by examining those occasions on which that language game can fairly be said to be practiced. The aim of such an investigation would not be to develop general models of scientific and ordinary activity but to explain how a general epistemic theme is a part of a local complex of activities.

Just as Wittgenstein's discussion of language games problematizes established philosophical assumptions about linguistic meaning, judgmental criteria, and structures of rationality, so ethnomethodological investigations of language games provide an antidote to extant proposals regarding epistemology, scientific communities, scientific "discourse," and the like, all of which treat the sciences as a unified ideological or methodological field. Accordingly, ethnomethodological studies can contribute to a more highly differentiated sense of what we are saying when we speak of description, observation, discovery, measurement, explanation, and representation. Instead of assuming that science, or for that matter any other coherent activity, possesses an epistemological or cognitive unity, the project becomes one of respecifying the "epistopical" coherence of language games like "observing," "measuring," or "constructing an object from a drawing." I use the neologism *epistopic* to suggest that the topical headings provided by vernacular terms like *observation* and *representation* reveal little about the various epistemic activities that can be associated with those names. The epistopics are classic epistemological themes, in name only. Once named as – or locally identified as a competent case of – observing, measuring, or representing, an activity and its material traces can be shown to be governed by a set of rules, a body of knowledge, a method, or a set of normative standards associated with the particular theme. But once we assume that nominal coherence guarantees nothing about localized praxis, we can begin to examine how an activity comes to identify itself as an observation, a measurement, or whatever without assuming

[41] My list of epistopics includes familiar themes from epistemologically relevant science studies. In his recent writings, Garfinkel provides another open-ended list of what he calls "order topics," which critically recalls and respecifies the foundational issues of grand social theory. His list of themes includes logic, meaning, method, practical action, the problem of social order, practical reasoning, detail, and structure. See Garfinkel and Wieder, "Evidence for locally produced, naturally accountable phenomena of order*." Garfinkel and Wieder warn that ethnomethodology's interest in these terms is not as "topics" for scholarly exposition but as "phenomena" for detailed investigation and demonstration.

from the outset that the local achievement of such activities can be described under a rule or definition.

It is crucial to understand that a focus on epistopics has nothing to do with a nominalist program. As Wittgenstein remarks, "Nominalists make the mistake of interpreting all words as *names,* and so of not really describing their use, but only, so to speak, giving a paper draft on such a description."[42] A nominalist treatment of the epistopics might proceed according to the following steps: (1) Take note of the fact that the activities of natural scientists, philosophers, sociologists, industrialists and others can be described under common headings like making representations, writing and interpreting texts, and so on; (2) observe that what counts as a "representation" or "text" is defined by local "constructive" activities; (3) describe how the practical and conceptual linkages among a network of locally organized activities are negotiated through the translation of a "same" representation from one domain of activity to another; and (4) explain the coherence of the network by referring to how a mode of representation becomes a stable symbolic commodity. The transactions among disciplinary activities thus become products of "definitional" and "interpretive work," thereby preserving a theoretical unity at the level of the *signifier* (an ideology of representation, a semiotic network, a discourse with determinate "effects," or a grand narrative), despite variations in the way that the *signified* is produced and understood. An ephemeral unity thus covers over a practical diversity. But once we treat epistopics not as metaphysical unities but as recurrent themes that gloss over the "work" of their local production, there is no longer any assurance that an ephemeral "sameness" will secure the analytic transition from the iterable terms of a lingua franca to the substantive fields of practice named by those terms.

By saying that the epistopics "are words," I certainly do not mean to imply that they are of no interest in themselves. Words like *observation, representation, measurement,* and *discovery* are used as recurrent topical headings and classic themes in the history, philosophy, and the sociology of science, and they also identify an unavoidable situation of inquiry for ethnomethodology and social studies of science. Even when these epistopics are dismissed as "philosophical" preoccupations of no interest to empirical researchers, they remain solidly entrenched in the programmatic claims, methodological justifications, and analytic language of empirical social science investigations. There is no avoiding them in academic discussions, even when it is acknowledged that the epistopics do not provide a metaphysical foundation for scientific activities.

Although the often-mentioned differences between actual scientific practices and idealized accounts of observation and testing seems to allow us to

[42] Wittgenstein, *PI,* sec. 383.

say that "actual scientists do not do as they say," it makes just as much sense to conclude that what scientists actually do when they accountably "make observations" and "test hypotheses" has an unknown relation to general epistemological treatments (including those of a critical epistemology). An investigation of the matter would not stop at the point of showing that the "notion" of scientific observation is actually a rhetorical construct or ideological tool. Instead, such an investigation would seek to demonstrate how a vernacular use of the term *observation* is uniquely adequate to some practice. To insist that observation is nothing other than what counts locally as observation does not entail an ironic view of actual practice. What else could observation be? Rather, it is to *respecify* what can be meant by that term.

What I have just said is by no means an original insight; it summarizes what I take to be the gist of Garfinkel's proposals for "respecifying the natural sciences as discovering sciences of practical action." Rather than saying that the discovering sciences are culturally unified disciplines whose "unique contents" can be recovered though ethnographic study, Garfinkel begins by displacing the sociology of science into a deep integration with natural scientific practices. As the title of one of Garfinkel's writings suggests – "Respecifying the Natural Sciences as Discovering Sciences of Practical Action"[43] – the discovering sciences can be said not only to discover the laws and objective phenomena associated with their achievements but also to discover and rediscover reflexively how the classic themes in studies in the history, sociology, and philosophy of science can be made to apply to the current situation of inquiry.[44] The discovering sciences unavoidably forge their histories and articulate their institutional linkages; they specify what counts as observation, adequate measurement, and reproducible findings; and they investigate what it takes to make experiments work. No single scientist or collectivity of scientists works in a void, and so the "reflexive achievement" of history, findings, experimental demonstrations, and the like always involves singular actions produced *in relation to* the "historical" specifications articulated and enacted by predecessors, colleagues and rivals. Nobody has a free hand, including the historian who comes on the scene some years later.

Perhaps because of some of the confusing implications I have mentioned, Garfinkel eventually dropped the term *quiddity* in favor of its more obscure Latin cousin *haecceity,* meaning the "just thisness" of an object.[45] Although the two terms can be used as synonyms meaning "what makes an object what it uniquely is," Garfinkel used the term *haecceity* to point more clearly to a

[43] Garfinkel et al., "Respecifying the natural sciences."

[44] This applies to the theme of "discovery" itself, as Augustine Brannigan demonstrated in *The Social Basis of Scientific Discoveries* (Cambridge University Press, 1981).

[45] Garfinkel (personal communication, 1989) joked that he disowned the term *quiddity* after realizing that Quine had used it in the title of one of his books.

pronomial or indexical "making of meaning" that no longer owes a debt to essentialism. This sense of *haecceity* differs from that of a stable "core meaning" inhering in the experience of an object. As Sacks argued in reference to cases of indicator terms like *this place* and *here,* these terms can be used without determinately "standing for" a place that can be given a singular name (e.g., a name for a building, a meeting, or an occasion). Consequently, the "just thisness" of an object can include the accountable here-and-now presence of a "this" or "it" that does not already stand for a named and verifiable thing.[46]

Illustrative instances of such usage can be found throughout the tape recording of three astronomers on an occasion that later came to stand as the first "discovery" of an optical pulsar. Take the following fragment:

Disney: . . . I won't believe it 'till we get a second one.
Cocke: I won't believe it until we get the second one and until th – the thing has *shifted* somewhere else.[47]

An eavesdropping analyst who is well informed about the context of the expressions can, of course, attempt to clarify Disney's and Cocke's usage by ascribing meanings to some of the indexical expressions they use:

Disney: . . . I *[Michael Disney]* won't believe it *[that we have made an important discovery]* until we get a second one *[another spike on the oscillograph screen, during a repeat of the observation under comparable conditions].*
Cocke: I *[John Cocke]* won't believe it *[that we have made an important discovery]* until we get the second one *[a similar spike on the oscillograph screen, when we repeat the observation under comparable conditions]* and until the thing *[the spike on the screen]* has shifted somewhere else *[to appear in a different sector of the screen].*[48]

If the translations proved defensible, they could serve the purpose of reconstructing the logic of the exchange (i.e., Disney's and Cocke's methodological rationales for "suspending belief in the evidence at hand until further tests are made"). What such a classic exercise would miss, however, is that the indexical expressions, with all their "vagueness," are used, and used intelligibly, by the speakers, without an apparent need to translate them.

[46] Harvey Sacks, "Omnirelevant devices; settingod activities; 'indicator terms'", transcribed lecture (February 16, 1967) in Sacks, *Lectures on Conversation*, Vol. 1, pp. 515–22. See Chapter 5 for a more elaborate discussion.

[47] H. Garfinkel, M. Lynch, and E. Livingston, "The work of a discovering science construed with materials from the optically discovered pulsar," *Philosophy of the Social Sciences* 11 (1981): 131–58; quotation from p. 154 (transcript slightly simplified).

[48] Ibid., p. 154 (transcript slightly simplified). Garfinkel and his colleagues (pp. 135ff.) use the term "evidently vague it" when discussing a temporally available appearance of the object.

Indeed, the indexical expressions supply the same materials the translator uses when (re)constructing their sense in more elaborate terms.

Ethnomethodology's strange methodological policy to explicate "the rational properties of indexical expressions" by investigating the orderliness of singular occasions of conduct begins to make sense in light of instances like this. Such "rational properties" would not be revealed by a rational reconstruction of Cocke's and Disney's "methodological proposals"; rather, they would be evident in the production of the surface features of the dialogue. These "rational properties" include, for instance, the way that Cocke repeats Disney's pronomial expressions "it" and "a second one" to demonstrate an understanding of those terms without need for formulating their sense, and the way that both parties demonstrate that they are speaking of the same thing without having specified what that thing is.

To understand such rationality, one needs to do more than specify a stable referential background for the speakers' expressions. Cocke's and Disney's expressions may "fail to refer" to a verifiable entity in the world;[49] indeed, the way they speak of "it," "a second one," and "the thing" anticipates that very possibility. And yet their talk is sensible, and it is sensibly related to an immediate complex of activities, things, equipment, and horizonal possibilities. But to see how that is so, we must stay in tune with the "thises and thats" mentioned by the astronomers and made evident by the way they speak to each other. This is what haecceity means.

Ethnomethodology's approach to the rationality of indexical expressions is strange in relation to the prevalent tendency in the social sciences to assume that singular occasions of conduct are chaotic, random, messy, and disorderly until order emerges at the level of average tendencies, typical patterns, modeled structures, and methodologically "filtered" data. It is often said that scientific study is not suitable for understanding "unique" events because the scientific method can discern only underlying causes and general tendencies that are obscured by the contingent details of single cases.[50] Although ethnomethodology does not regard itself as an inductive mode of inquiry, its analytic program builds on the "social fact" that singular instances of conduct are intuitively recognizable and vernacularly describ-

[49] See Philip Kitcher, "Theories, theorists and theoretical change," *Philosophical Review* 87 (1978): 519–47. Kitcher's theory of reference requires the analyst to have independent knowledge of what a historical expression refers to. Consequently, an expression like *phlogiston* fails to refer (p. 531) because we now know that the word only seemed to have an actual worldly referent for those who used it in the eighteenth century.

[50] Ernst Nagel, for instance, asserts that scientific explanations "can be constructed only if the familiar qualities and relations of things, in terms of which individual objects and events are usually identified and differentiated, can be shown to depend for their occurrence on the presence of certain other relational or structural properties that characterize in various ways an extensive class of objects and processes." See E. Nagel, *The Structure of Science* (New York: Harcourt Brace & World, 1961), p. 11.

able; otherwise, how participants manage spontaneously to produce mutually coordinated activities would be a complete mystery. This policy differs from a program of analysis that presumes that substantive "data" are "given" to the analyst as though with names attached, and it relies instead on the existential "fact" that any inquiry will find itself "thrown" into an already intelligible world, even when that world's intelligibility includes indistinct, arguable, and doubtful features. Consequently, there is no attempt in ethnomethodology even to imagine the possibility of a presuppositionless inquiry.[51]

The prevailing version of constructivism tells us that scientists confront a primordial chaos from out of which they construct facts. This conception of activity owes an odd debt to the logicist tradition in philosophy that gave rise to the much cited Duhem–Quine "underdetermination thesis." Like the inheritors of the Vienna circle's project, proponents of social constructivism agree that the "fact" that the existential intelligibility and "equipmentality" of the world precedes and surpasses any attempt at intellectual justification points to an unsolved epistemological problem.[52] For logical positivists and social constructivists alike, the "fact" that justifications "come to an end" by running up against an intuitively recognizable and massively ordered context of practical actions and scenic particulars is treated as a gap in our understanding. It is as though the absence of a "complete" justification implied the absence of intelligibility.

Rationalist philosophers of science seek to extend a filamentous tissue of intellectual justification to reach the uncanny presumption of an already intelligible world, and social constructivists argue that the essential impossibility of such a justificatory project means that our sense of "reality" is an ultimately unjustified and artificial construction. In both cases, the *préjuge du monde* is reduced to the terms of a reflective knowledge: in one case, the terms of an axiomatic derivation and, in the other, the terms of a rationally unjustified account or belief.[53]

In his brilliant remarks about "primitive natural science" (see Chapter 6), Sacks identified a structure of accountability (observation–description–replication) that, he argued, provided a naive grounding for early natural

[51] This recalls Heidegger's existential critique of Husserlian phenomenology (*Being and Time*, trans. J. Macquarrie and E. Robinson [New York: Harper & Row, 1962]).

[52] Even, and especially, some of the more radical ("reflexive") constructivist exercises give a skeptical cast to the "problem" of indexicality and reflexivity. See, for instance, Woolgar, *Science: The Very Idea*, pp. 32–33. For a rebuttal to the skeptical position, see Wes Sharrock and Bob Anderson, "Epistemology: professional skepticism," pp. 51–76 in G. Button, ed., *Ethnomethodology and the Human Sciences* (Cambridge University Press, 1991).

[53] For further discussion of this point, see Jeff Coulter, *Mind in Action* (Oxford: Polity Press, 1989), chap. 2; and G. Button and W. W. Sharrock, "A disagreement over agreement and consensus in constructionist sociology," *Journal for the Theory of Social behaviour* (forthcoming).

historical and natural philosophical investigations. For Sacks, the life-world can be described in such a way that others can "go out and see" what the description says. This mundane miracle points the way not to the reality of a natural world so much as to the analyzable achievement of concerted descriptions. As CA's research program developed, Sacks's initial investigations of the work of speaking *intersubjectively* (of "saying what others can already see") became a technical program for investigating the work of talking together. Garfinkel's persistent call for radical studies of the primordial structures of accountability points in a less well-established direction. Rather than accepting the miraculous "achievement" of a world known in common and treating it as a basis for initiating a new natural observational science of human behavior, Garfinkel sought to make that achievement perspicuous.

Instructed actions and *Lebenswelt* pairs

We are not yet out of the foundationalist woods. Although Garfinkel's and his students' recent studies of work in various natural scientific and mathematics disciplines can be read as extensions of a more general program for investigating instructed actions, Garfinkel has repeatedly insisted on the special character of the "discovering sciences" and mathematics.[54] His argument may recall Mannheim's much criticized "exclusion" of mathematics and (some) natural sciences from the sociology of knowledge, but there are significant differences that are worth pursuing at length.

To understand Garfinkel's argument, it might be helpful to begin by recalling his thematic interest in a "gap" between the "natural accountability" of the life-world and the formal "renderings" produced by bureaucratic functionaries and professional scholars. This gap is produced through a transformation of locally accomplished, embodied, and "lived" activities into disengaged textual documents. Very early on, Garfinkel gave an example of such a transformation in an unpublished manuscript entitled "The Parsons Primer" that he based on his dissertation and circulated during the early 1960s.

> During the war my uncle had occasion to go to a government office because he wanted an increase in his allotment of fuel oil. There he complained to a

[54] See Garfinkel et al., "The work of a discovering science," (pp. 121 ff., n. 12); Garfinkel et al., "Respecifying the work of the natural sciences"; Eric Livingston, *The Ethnomethodological Foundations of Mathematics* (London: Routledge & Kegan Paul, 1986). When attributing these studies to Garfinkel, I do not mean to suppress recognition of the fact that others (including myself) are listed as coauthors in these studies, but there is no ambiguity about Garfinkel's having authorized the proposals about unique adequacy and the distinctive character of the discovering sciences and mathematics. Livingston's studies substantiate and develop those themes.

> clerk that his allotment was insufficient. He had a long story with which to justify his request for an increase. He described his circumstances at home. It was cold in the house; his wife was unpleasant because it was so cold; there was that large dining room which was always hard to heat even when you could buy as much fuel oil as you could afford; he was living in a particularly cold part of town; the children were down with one illness after another, one giving what ailed him to the next, so there was no rest for anyone; and so on.
>
> After several minutes of this the clerk stopped him. "How large is your house?" The story started again describing how large the house was; how it had always been a burden; that his wife and not he had wanted the house. The clerk interrupted again. "Excuse me, how many rooms do you have in the house? How many square feet?" My uncle told him. "What kind of heater do you have?" and "What was your allotment last year?" And so it went. Out of the flow of material with which my uncle described his situation the clerk established about four or five points.
>
> The clerk understood of course that the situation as my uncle described it was a fix in which a person could be. But the clerk consulted the rules of office operation, and in terms of these rules, exemplified in the information that was asked for on the form that the clerk filled out, the clerk undertook the process of selection, of classification, and the rest such that the clerk came up finally with what from the standpoint of the administered form was "the case."
>
> There was one description of the social structures that my uncle furnished the clerk. The transformed description of my uncle's circumstances found in the form described a world which did not include complaining wives, or a house whose size and expense were regretted. Such features, though known to the clerk, were not relevant. Instead the clerk described a social situation which included instances of houses with certain square footages, with certain types of heaters, that would on the average produce certain units of heat over a unit period of time, with the expected result that some expected amount of a scarce commodity would have been used up by one instance of a "home owner."[55]

This story presents a vivid account of a clash between two incommensurable descriptions of social structures: the uncle's complaint, as enunciated to the clerk, and the clerk's documentation of the "case." Throughout the described encounter between uncle and clerk, the uncle's narrative is translated into an intelligible and defensible bureaucratic document. The clerk formulates a set of elements of "the case" that are relevant to and congruent

[55] From H. Garfinkel, "Parsons's primer – 'ad hoc uses'," unpublished manuscript, Department of Anthropology and Sociology, UCLA, 1960, chap. 2, pp. 2–3.

with an unstated organization of "cases like this" in the government bureaucracy. As a disengaged document, the case report anticipates a body of rules, criteria, justifications, and identities that have an arguable relation to the "homeowner's actual situation" while at the same time it systematically omits reference to the details of the daily household situation that the uncle vociferously tried to impress on the clerk. By applying the later-developed policy of ethnomethodological indifference to this story, we can substitute an entire series of identities for the relational pair [uncle's description/clerk's case record]: [jury deliberation/Bales's interactional analysis of the deliberation], [interview encounter/interview data], [course of a game's play/rational reconstruction of the players' strategies], [performance of an experiment/report of the experiment], and [lived-conversation/conversation analytic transcript].[56]

The analysts and officials who use the "renderings" constituted by such relational pairs tend to privilege the disengaged document's objective, analyzable, and formal properties, whereas Garfinkel points to a different order of asymmetry: the surplus relevancies, significations, and temporal parameters of the "descriptions of social structures" that are irretrievably lost as soon as the disengaged case report becomes the official record of the case. Moreover – and this is the key point – the *transformation* that is achieved from the rendering of the case is itself hidden whenever the case report becomes the relevant analytic datum. This erasure of surplus detail does not necessarily produce a general epistemological problem, although it can be a source of occasional disputes and methodological uncertainties. Instead, it is an unacknowledged part of virtually every analytic program for producing accounts of social structure.

In later work, Garfinkel focused on the act of "writing" as itself the source of a "gap" between the work of composing a text and the retrospectively analyzable properties of the resultant document. This was elucidated in a videotaped demonstration by one of his students, Stacy Burns.[57] Burns produced a videotape that framed a typist's hands at an electronic typewriter keyboard. The tape documents the typist's hands working at the keyboard while her voice gives a running commentary of "what she is doing" as she composes the text. The typed document is shown unfolding on a sheet of paper positioned in the carriage while the typist strikes a sequence of keys, crosses out and restarts a passage, and pauses between letters while consid-

[56] The brackets [ ] are a notational convention for identifying an ethnomethodologically achieved identity: an internally produced, used, and *glossed* relation between "account" and "lived work." See Harold Garfinkel and Harvey Sacks, "On formal structures of practical actions," pp. 337–66, in J. C. McKinney and E. A. Tiryakian, eds., *Theoretical Sociology: Perspectives and Development* (New York: Appleton-Century-Crofts, 1970).

[57] Demonstrated in Garfinkel's seminar, Department of Sociology, UCLA, 1980.

ering aloud what to do next. The videotape thus frames a distinctive "pair" of intelligible documents: (1) a "real-time" video sequence of typing, complete with hesitations and commentary, and (2) a typed page that can be read, copied, and analyzed independently of the real-time sequence. On the videotape, the typed page can be seen as the product of a course of work, but when the page is read as a disengaged text, its coherent semiotic features implicate a different order of "authorship." The completed sentences stand as documents of a coherent set of "ideas," "intentions," "grammatical competencies," and so forth, which no longer display the local history of production documented by the videotape. The written text's analytic features do not document the singular "hesitations," "interruptions," and "second thoughts" made evident by the tape.[58]

In Garfinkel's terms, the two documents stand in a relation of "asymmetric alternation": One document (the videotaped sequence) enables the recovery of the other (the text on the page), but not vice versa; the written text's analyzable field no longer retains a trace of the surplus details of typing. This is not to deny that the videotape itself is a disengaged document of the "lived work of writing," since a similar asymmetric alternation is internal to that pair. This simple exercise clearly demonstrates the kind of reduction that is accomplished whenever social scientists use reports, archives, transcripts, codified data, and other such documents as representations of a practice. Athough such "renderings" enable social scientists to amass data bases from which they can make coherent and defensible analyses, their use produces a "gap in the literature" or a "missing what" that is recreated by the very existence and fact of "a literature."

The fact that social and natural scientists alike produce, rely on, and take for granted transformations of their local practices into disengaged "accounts" of those practices has become a familiar theme in social studies of science. Latour, for instance, is noted for arguing that the coherence of a scientific field – of a network of reproducible techniques, cumulative bodies of writing, and coherent domains of "natural" entities and relationships – is established and maintained through chains of "literary inscription."[59] For Latour, iterable traces and records of lab work encompass an entire process

[58] As Derrida points out ("Signature, event, context," *Glyph* 1 [1977]: 172–97), the "orphaned text" is far from unintelligible. Although it would be absurd to figure that one would need to observe the writer "in the act" of writing in order to understand the written text, Garfinkel's point is that a distinctive order of intelligibility – and one that is definitely part of "writing" – is opened up by inspecting the lived work of writing.

[59] Bruno Latour and Steve Woolgar, *Laboratory Life: The Social Construction of Scientific Facts* (London: Sage, 1979; 2nd ed., Princeton, NJ: Princeton University Press, 1986); Bruno Latour, "Visualization and cognition: thinking with eyes and hands," *Knowledge and Society: Studies in the Sociology of Culture Past and Present* 6 (1986): 1–40; Latour, *Science in Action;* and Latour, *The Pasteurization of France,* trans. Alan Sheridan and John Law (Cambridge, MA: Harvard University Press, 1988).

of making phenomena visible, calibrating instruments and scales, describing and measuring, triangulating different measurements and readings, developing models and arguments, and disseminating reports. He argues further that the stable formats provided by texts, scales, graphs, diagrams, photographs, and the methods for reproducing and disseminating them enable scientists to transcend the local circumstances of any single observation while incorporating selected inscriptions into an expansive field of power/knowledge. Latour invokes what he calls the "immutable" and "mobile" properties of mechanically reproduced texts to explain the growth of science and the stabilization of its knowledge.

From an ethnomethodological standpoint, however, such an explanation still retains a "gap in the literature." This gap pertains to the way that any trace, sign, inscription, representation, graphism, and citation is originally, and always, an expression, indication, icon, or moment of articulation in some practice. This is not to say that the meaning of such "indexicals" depends on one or another reader's subjective interpretation; rather, it is that such meaning is bound to the felicity conditions supplied by the *instructive reproducibility* of a practice. Latour recognizes that the development and distribution of a practice (of a discipline in Foucault's sense) works hand in glove with the dissemination of "immutable mobiles," but he negotiates his "semiotic turn" by assuming that a structural analysis of textual "statements" can recover the transformation of "science in action" into stabilized sociotechnical facts.[60]

Despite Latour's effort to distance his semiotic explanation from previous sociological solutions to the problem of order, in the end he follows a familiar analytic path. First, he defines the lived work of laboratory science as a practice of manipulating signs, of writing and reading inscriptions, representations, traces, statements, and texts. Consequently, science in action (or "laboratory life") becomes a matter of constructing and deconstructing formal properties of sign systems, and from this it follows that a formal program of semiotic analysis should be able to recover the relevant moves in a situated construction, translation, and deconstruction of inscriptions, traces,

[60] Latour ("Will the last person to leave the social studies of science please turn on the tape-recorder," *Social Studies of Science* 16 [1986]: 541–48) recommends a "semiotic turn" as an antidote to the severe limitations he finds in ethnomethodological studies of laboratory work. There is, of course, more than one way to make such a turn, and one could argue that Latour is recommending a semiotic "right turn" that borrows heavily from the formalist and structuralist traditions in semiotics rather than from the deconstructionist wing. For Latour, iterability and syntagma provide a solution to the production of a stable transituational meaning, whereas Derrida rephrases the Heideggerian questioning of the "identity" of the thing with itself by questioning the identity of the text with each iteration of "its" autonomous legibility. Although the iterability of the trace resists any effort to reduce the text to a particular speech act, it guarantees nothing about the formal stability of the text or its intelligibility.

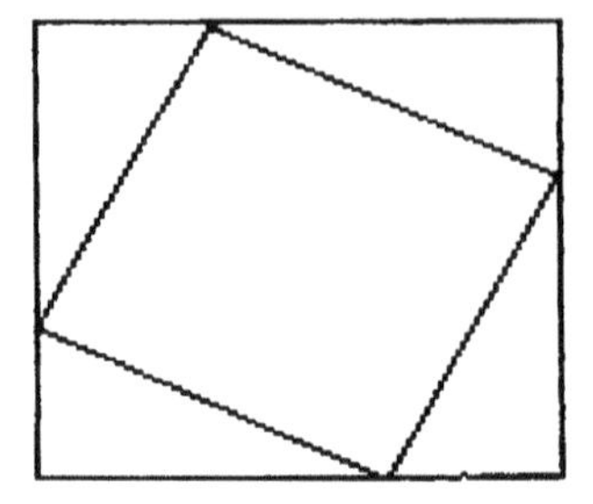

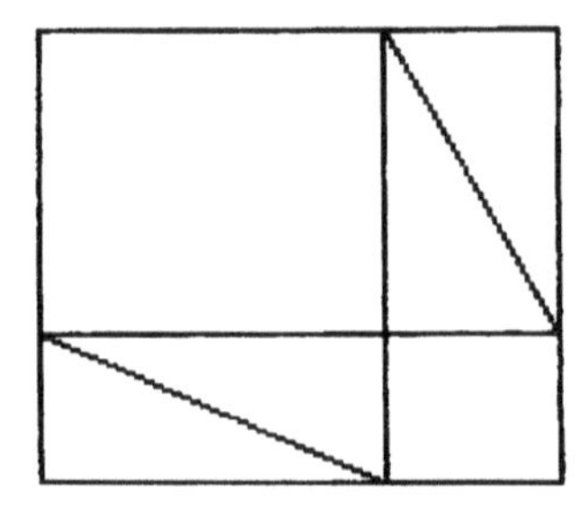

Figure 7.1

and texts. The transition from life-world to system thus comes down to the construction of analytically transparent chains of signifiers that subsume local practices in delocalized networks.

Ethnomethodology challenges this semiotic approach on at least two fronts, first by denying the *formal* equivalence between literary representations and life-world activities and second by treating the *practical* equivalence between literary representations and lived activities as itself a local and reflexive achievement that betrays any attempt to recover it with a general program of formal analysis. Instead of moving "ahead" into a model-building enterprise that presupposes the practical achievement of an identity between semiotic elements and the performative implications of those elements, ethnomethodologists make a sustained effort to investigate how particular actions are instructively reproduced.

Garfinkel does not suggest that ethnomethodology can fill the "gap" in the social studies of science literature by describing the preliterary contents of disciplinary life-worlds; instead, he argues that the gap arises as a necessary, intelligible, defensible, and unavoidable product of lay and professional sociologies. Although the gap points to the absurdity of a totalistic account of practical actions, it does not necessarily present an insoluble problem for the particular practices that ethnomethodologists study. This is especially clear in Garfinkel's and Eric Livingston's treatment of the *Lebenswelt* pair in mathematics. A simple demonstration of this can be seen in Figure 7.1, a visual proof of the Pythagorean theorem (the square of the hypotenuse of a right triangle is equal to the sum of the squares of the other two sides).[61]

In his discussion of the example, Livingston invites readers to try to "discover" how the geometric figure constitutes a "proof" of the theorem, and he observes that those who are unfamiliar with the particular proof are

[61] This "Chinese proof" is demonstrated in Eric Livingston, *Making Sense of Ethnomethodology* (London: Routledge & Kegan Paul, 1987), p. 119. Livingston and Garfinkel give a more extensive treatment of it in a collaborative paper. See Eric Livingston and Harold Garfinkel, "Notation and the work of mathematical discovery," unpublished paper, Department of Sociology, UCLA, 1983.

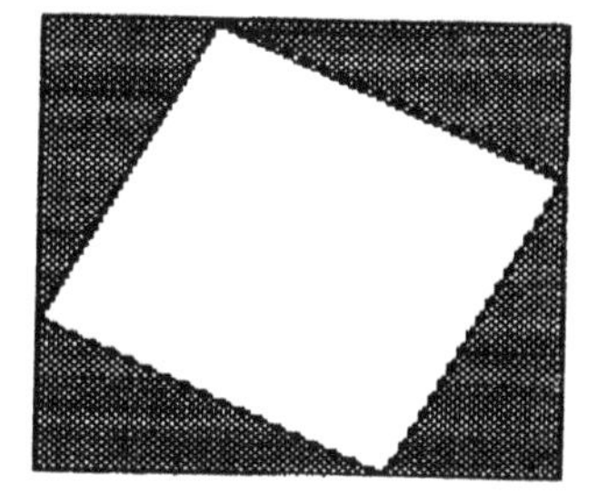
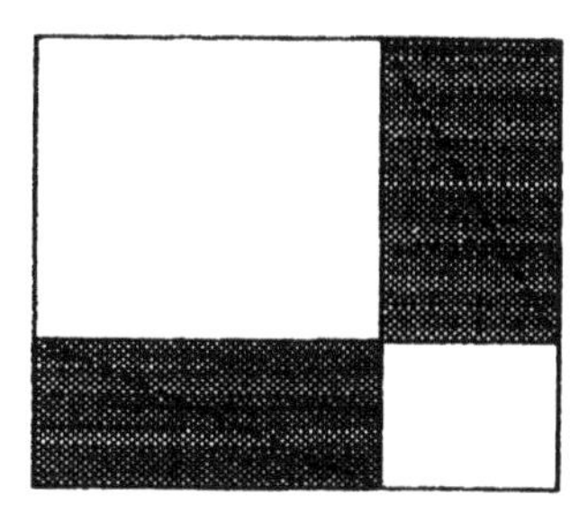

Figure 7.2

likely at first to feel bewildered. But once they are given a set of instructions, aided by various notational devices that indicate how to find selected equivalencies and ordered relationships in the figure, readers can begin to figure out how the constituents of the proof-account act as elements of the proof. Livingston first points out that the square on the left is "dissected" into four right triangles and an embedded square. By inspecting the figure and assuming various unquestioned geometric relations (such as the axiom that the interior angles of a triangle add up to 180 degrees, so that the two acute angles of a right triangle add up to 90 degrees), we can begin to see that the four triangles in the figure on the left have equivalent angles and sides. We should then be able to see that the figure embedded in the larger square is itself a square composed on each side by the hypotenuse of one of the four triangles. Livingston then leads us to see that the square framing the right-hand side figure "can be considered as the same square, but one in which the triangles have been repositioned to occupy the places that are shown" (see Figure 7.2).[62]

The trick now is to see that the two unshaded squares in the square on the right-hand side of Figure 7.2 occupy the same surface area as the embedded (unshaded) square in the square on the left. Their equivalence is established by the difference between the area of the larger square and the four equivalent triangles in both sides of the figure. The proof is discovered by realizing that the two unshaded squares on the right are the squares of the two sides of the triangle and that the unshaded square on the left is the square of the hypotenuse. The important point, however, is not whether the particular instructions that I have just given adequately show how the figure acts as a proof.[63] Rather, the point is that as soon as the figure is seen as the proof, an ensemble of equivalent and analytical relations among sides, angles, and areas internally support one another in accordance with the theorem. When

[62] Livingston, *Making Sense of Ethnomethodology,* p. 120.

[63] One could, for instance, use scissors to cut out the inscribed figures and then stack and fit them together in a kind of palpable proof.

this happens, the proof account emerges as a "precise account" of any of the various ways that it can be construed as instructions for performing the proof, and at that point, the proof:

> takes on a transcendental, objective, accountable presence. The proof is seen as already being in the proof-figure itself. It has a substantial, "massive" presence. It appears to have an endless depth of proof-relevant, discoverable details; these appear to be available from different perspectival viewings of *the proof's* various aspects. It withstands its repeated interrogation and is seen to be the cause and source of all inquiries concerning it. It is accountably and analyzably a proof of the Pythagorean theorem.[64]

As I read this – and as I encourage readers to understand it by figuring out the proof through the instructively guided course of work[65] – Livingston is saying neither that the transcendental "truth" of the proof account is the "cause" of the lived work of proving nor that it is merely a retrospective illusion that the proof account appears to be "the cause and source of all inquiries concerning it." Instead, he is insisting that both elements of the *Lebenswelt* pair – the proof account and the lived work of proving – are necessary for an adequate understanding of the proof as such. Otherwise, the proof account becomes an empty textual figure about which many things can be said, but without recovering how it stands in and as the proof. The reader's "discovery" of how the proof statement (Figure 7.1) stands as a "precise description" of the work of proving the theorem is achieved through the lived work of proving (an actual course of reading, sketching, puzzling about, handling, and reconfiguring the textual elements supplied in the proof account), and it is not enough to say that a reader must "understand" or "interpret" the proof in order to see the figure as an adequate set of instructions, because the point of the demonstration is to specify just what such "understanding" or "interpretation" might include as a course of embodied mathematical work.

The *Lebenswelt* pair in mathematics may seem akin to other such "pairings" of documentary renderings with the lived work of some activity, but Garfinkel conjectures that "there exist, but only discoverably, and only for the natural sciences, domains of lebenswelt chemistry, lebenswelt physics, lebenswelt molecular biology, etc. just as there exists the discovered domain of lebenswelt mathematics" and, further, that "lebenswelt domains cannot be demonstrated for the social sciences" nor can they be demonstrated for

[64] Livingston, *Making Sense of Ethnomethodology,* p. 119.

[65] Readers who find my instructions confusing or incomplete are encouraged to consult Livingston's text (ibid.). This and other examples are also discussed in Dusan Bjelic, "The praxiological validity of natural scientific practices as the criterion for identifying their unique social-object character: the case of the 'authentication' of Goethe's morphological theorem," to appear in *Qualitative Sociology* 15 (1992): 221–45.

various actions performed in accord with rules in games, manuals of instructions, freeway signs, occupational codes of ethics, contracts, and the like.[66]

Accordingly, the clerk's "case record" of the encounter with Garfinkel's uncle presumably would count only ironically as a "precise description" of the course of action from which it was extracted. Only ironically could the record be said to withstand repeated interrogation and to be seen as "the cause and source of all inquiries concerning it." The fact that the case report might later be treated as a documentary basis for denying, and justifying the denial of, the uncle's request for an increased allotment of fuel oil does not foreclose the likelihood of what Livingston calls "different perspectival viewings" of the document's formulation of the uncle's situation. Simply put, the documented denial of the uncle's request could itself become a source of complaint about the arbitrary, rigidly bureaucratic, and unsympathetic imposition of authority by the officials involved. If we were to say that a particular proof statement in mathematics is supported by an arbitrary imposition of power by the "officials" in charge of enforcing the dictates of mathematics, this would be far more startling and perhaps unintelligible as a conventional way to speak of mathematical proofs.

One can easily imagine, however, that ethnomethodologists would not want to accord special status to the way a chemist follows the instructions in a lab manual or a mathematician works out the proof of a theorem in Euclidean geometry. Why should such actions be essentially different from more "ordinary" instructed actions like following an overland route marked on a topographic map or preparing "aubergines à la Boston" from a list of ingredients and sequence of steps in a cookbook?[67] Ethnomethodological indifference – as well as Bloor's "symmetry" postulate – would seem to legislate against prejudicing the case with categorical assumptions about the special nature of mathematical proofs and experimental practices.

Viewed in that light, Garfinkel's contrast between the discovering sciences and the social sciences seems to juxtapose some familiar demarcationist themes with an unusual distinction between *bricolage* expertise and textual practices.

> The social sciences are talking sciences, and achieve in texts, not elsewhere, the observability and practical objectivity of their phenomena. This is done in literary enterprises through the arts of reading and writing texts, by administering compliance documents, and by "shoving words around." . . . Social sciences are *not* discovering sciences. Unlike "hard sciences" they cannot "lose" their phenomena; they cannot undertake the search for a phenomenon as a problem to be solved, finally be unable to do so, and thus

[66] Garfinkel et al., "Respecifying the work of the natural sciences," p. 128.

[67] According to Craig Claiborne (*The New York Times Cook Book* [New York: Harper & Row, 1961], p. 377), this recipe is "somewhat tedious to prepare but the game is worth the candle."

> have "wasted time"; they do not know the indispensability of bricolage expertise; and these are never local conditions of their inquiries and theorizing.[68]

Elsewhere, Garfinkel mentions other distinctive themes, such as the "unforgivingly strict sequences" of laboratory experiments and the fact that "an issue can get settled" among scientists and mathematicians.[69] This, of course, sets Garfinkel's program at odds with influential arguments in the sociology of scientific knowledge about the predominance of literary inscription, the social closure of controversies, and the flexibility of scientists' practices and arguments. His emphasis on the (merely?) textual character of social science inquiries seems to assume a contrast between the humanities and the natural sciences that has been contested by contemporary treatments of science as a literary genre that incorporates "fictional" rhetorical tropes and narrative frameworks. Moreover, his denigration of "literary enterprises" implies a rather narrow conception of textuality compared with the conception used by many sociologists and historians who employ literary theoretic concepts to describe heterogeneous modes of textual representation, such as the construction of specimen materials and instrumental traces, the organization of museum collections and diorama, and the composition of various kinds of visual documents.[70]

Although Garfinkel's proposals concerning the discovering sciences may seem to express a retrograde adherence to scientific realism and naturalism, I would argue against such a reading. He does not invoke the "reality" of an independent natural world in order to explain the possibility of making discoveries, losing the phenomenon, and so forth; instead, he contends that science writing and mathematics texts are part of materially embodied forms of life. By treating "the work of a discovering science" as an irreducibly embodied achievement, Garfinkel opposes reductions of scientific practice to bodies of ideas, formulas, rules of method, bibliographic networks, and theoretical and metatheoretical commitments. Without denying the role of rules, equations, and other formalisms, he argues that it is necessary to situate such formulations within embodied practices.

Although one might be inclined to object to Garfinkel's strict refusal to

[68] Garfinkel et al., "The work of a discovering science," p. 133. Unlike Garfinkel, Derrida treats writing as a *bricoleur's* practice that is shared by scientist, engineer, and literary scholar alike. See Jacques Derrida, "Structure, sign, and play in the discourse of the human sciences," pp. 247–72, in R. Macksey and E. Donato, eds., *The Structuralist Controversy: The Languages of Criticism and the Sciences of Man* (Baltimore: Johns Hopkins University Press, 1970).

[69] Garfinkel et al., "Respecifying the natural sciences," pp. 4, 33.

[70] See, for instance, Joseph Gusfield, "The literary rhetoric of science: comedy and pathos in drinking driver research," *American Sociological Review* 41 (1976): 16–34; Michael Mulkay, *The Word and the World: Explorations in the Form of Sociological Analysis* (London: Allen & Unwin, 1985); and Haraway, *Primate Visions*.

entertain the idea that the social sciences might have their own embodied practices,[71] as I understand it, the point of this refusal is to avoid presuming the transparency, context independence, and compulsive force of rules, models, texts, signs, and formulas. This sets Garfinkel at odds with the many social scientists who contend that a "merely local" understanding of social actions is not "enough" and that we must devise representations of social systems of action that transcend the limitations inherent in particular actors' life-worlds.[72]

Ethnomethodology's indifference to such injunctions has the effect of relativizing the foundational claims in the social sciences by questioning the adequacy of representational programs that place the social analyst "outside" the life-world and at the center of an integrated field of signs. A literary assemblage of signs – a schematic account of normative categories, a Habermasian typology, a semiotic map, a narrative structure, a collection of coded interviews, a set of cognate or homologous concepts, a body of rules defining a "speech-exchange system" – becomes the medium through which the analyst ranges freely across a landscape of diverse competencies in order to gather together the elements of a *modeled* social order that stands for an actual system.

By calling a halt to the analytic movement from singular expressions to delocalized semiotic schemas, Garfinkel suspends a preliminary requirement of virtually every established program in the social sciences, including much of what rides under the banner of ethnomethodology. And by saying that the social sciences "do not know the indispensability of bricolage expertise; and these are never local conditions of their inquiries and theorizing," he suggests that the social sciences *fetishize the sign* by disregarding the embodied production and interactional use of textual renderings. Even when we grant this, however, there seems to be little justification for distinguishing *Lebenswelt* pairs in the "natural sciences" from those that have a "discoverable" role in such local, embodied, and materially situated practices as auto mechanics, cooking, tea ceremony, and criminal forensics.

From what I said earlier about "language games," it may seem reasonable to assume that the language games of, for example, molecular biology are not bounded by a closed set of disciplinary skills and/or a corpus of knowledge but that they are permeated by an immense variety of discursive

[71] See Harold Garfinkel, "Can the contingencies of the day's work in the natural sciences be used to distinguish them as discovering sciences from the social sciences and humanities?" unpublished proposal, Department of Sociology, UCLA, 1989.

[72] See my discussion in Chapter 1 of criticisms by Jürgen Habermas, *The Theory of Communicative Action,* vol. 1: *Reason and the Rationalization of Society,* trans. Thomas McCarthy (Boston: Beacon Press, 1984); Anthony Giddens, *The New Rules of Sociological Method: A Positive Critique of Interpretive Sociologies* (London: Hutchinson, 1978); and Pierre Bourdieu, *Outline of a Theory of Practice,* trans. Richard Nice (Cambridge University Press, 1977).

and embodied practices, some of which seem no more or less rigorous, distinctive, and precise than tuning an engine or preparing a soufflé. Moreover, standards of exactness, strict sequentiality, and tolerance for variation are assessed "for all practical purposes" in laboratory work just as they are in various other crafts, literary arts, and household activities. Sequences of laboratory work are not always "unforgivably strict," nor, as many studies in sociology of science have documented, are issues necessarily "settled" in the natural sciences without lingering dispute (perhaps mathematicians do, after all, "come to blows").

Whether or not we accept Garfinkel's account of *Lebenswelt* pairs in mathematics and (some) natural sciences, it should be clear that he does not aim to exclude them from "sociological" interest; quite to the contrary, he identifies these "rigorous" practices as particularly interesting ethnomethodological phenomena. Recalling the discussion of Wittgenstein's number-series example in Chapter 5, we can appreciate that there is no need to suppose that in order to understand that a rigorous practice is a "social achievement," we need to take a skeptical position toward the practice. Nor does such an understanding necessitate an explanation of how members of a community of practitioners come to "share a belief" in the rigor and reliability of their practices, as though such "belief" was somehow groundless or arbitrarily imposed. Instead, Wittgenstein's demonstration respecifies what "rigor" and "certainty" can mean in a practical universe divested of immaterial and disembodied mentalities. The rigor of (some) concerted practices is not an artificial rigor, in contrast with a "real" rigor; for Wittgenstein it is the only kind of rigor *we* can have.

Also following Wittgenstein, one need not associate "unforgivably strict sequences" and "precise descriptions" only with the practices in the natural sciences and mathematics. The various language games associated with measuring, counting, depicting, observing, describing, and so forth have a "home" in daily life no less than they do in science. Of course, highly complicated and "rare" skills and intuitions are cultivated in the specialized disciplines (e.g., in molecular biology, in which ingredients routinely are "handled" in extremely minute quantities), but this does not justify a categorical distinction between the rigorous *Lebenswelt* pairs in the natural science disciplines and more flexible and ad hoc "documentary methods" performed in the social sciences and daily life. We would want to avoid the common tendency to treat "science" as a synonym for the most effective and virtuous qualities of practical reason.

Although a categorical distinction between the discovering sciences and other practical actions may be unwarranted, Garfinkel's and Livingston's studies nevertheless raise a challenge to constructivist sociology of science that is not based on a commitment to metaphysical realism or rationalism.

Livingston, for instance, does not try to explain the rigor of mathematical proofs. To the extent that his textual demonstrations succeed, they place readers in a phenomenal field in which a series of textual maneuvers provides further testimony to the proof's univocal and yet reflexive account of the action. The "test" of Livingston's version of mathematics is provided by whether or not readers discover what he says to hold for their reading. That is, he succeeds to the extent that the reader's struggles – both the struggles to "get" the proof and to "resist" its unequivocal lesson – become grammatical "contents" of the mathematical language game that the proof account describes. Although various pathways can articulate how the proof statement stands in an enactment of the proof, these pathways of "choice" show themselves not to be freely chosen individual strategies or reluctant acts of compliance to an orthodox "opinion." For ethnomethodologists, such a game is no less social than enacting a ritual ceremony or following military orders, but its grammar is distinctive and worthy of study in its own right.

Toward an investigation of primitive epistopics

At long last, I am ready to outline a "program" of investigation that combines ethnomethodology's treatment of ordinary practical actions with the sociology of scientific knowledge's interest in the "contents" of scientific practices. The investigations I have in mind will concern the primitive structures of accountability that make up the instructable reproducibility of social actions. I assume that this is not a single phenomenon. Judging from existing studies in ethnomethodology and the sociology of science, different configurations of skills, purposes, instruments, texts, materials, routines, and modes of agency are likely to be discovered in a highly dispersed and discontinuous field of practices. The starting point for such investigations is the "epistopics" – the discursive themes that so often come up in discussions of scientific and practical reasoning: observation, description, replication, measurement, rationality, representation, and explanation. The epistopics provide foci for classic epistemological and methodological discussions, but they are no less relevant to vernacular inquiries. To paraphrase Garfinkel, they are vernacular themes that "went to college and came back educated."

There is, of course, nothing new about proposing a (hostile?) takeover of epistemology's topics. The strong program in the sociology of science attempted a thoroughly "sociological" rewriting of epistemology, and Garfinkel and Sacks took up classic issues in the philosophy of the social sciences and sought to respecify them as ordinary social accomplishments. But as I have argued, such takeovers have been subverted by efforts to secure analytic grounding for one or another program of research by invoking mythological conceptions of science. The attractions and pitfalls of scientism have been no less prominent in ethnomethodology and the sociology of scientific knowledge than in the more conventional areas of sociology. The problem is that many social scientists can see no legitimate alternative.

In light of the pressing need for a comprehensive account of "systems" of human activity and of normatively grounded programs of evaluation and remediation, nothing short of a rigorous and determinate science seems to be in order. However, it is one thing to work under the conditions of an established science and another to wish that such conditions were already in place. In the absence of such conditions, it might seem worthwhile to take stock of the vernacular topics that a science attempts to professionalize. Although a full-blooded social science is not necessarily in the offing for such an investigation, it does allow an examination of how and why the social sciences become "stuck" in their efforts to transcend ordinary practical reasoning. The research program I have in mind can be outlined as follows:

1. Begin by taking up one or more of the epistopics. The epistopics have a prominent place in the large literatures in the history, philosophy, and sociology of science, but in this case our aim will be to break out of the academic literature by searching for what Garfinkel has called "perspicuous settings": familiar language games in which one or another epistopic has a prominent vernacular role. So, for instance, although there are many interesting and erudite discussions of "observation" in the philosophy of science, "observation" has no less prominent a place in the practices, written and oral instructions, and reports in numerous other organized activities, some of which are quite humble and ordinary. The academic literature provides a relevant background for beginning such investigations, insofar as a long history of scholarly treatments and argumentative positions establishes the initial significance of the epistopic. Although for the program I am outlining, the literature cannot be disregarded – it does, after all, supply a current situation of inquiry – the academic conversation will be continued by other means than an explication of the classic literature.
2. Search for primitive examples. In this book I have focused on science, and there is no question that the epistopics have a role in the sciences. We should not forget, however, that scientists do not have exclusive rights to observation, description, measurement, truth telling, or the performance of "unforgivably strict sequences." Science and math (however we might define those activities) do provide clear cases for the investigation of particular epistopics, but this does not mean that the way that particle physicists make observations should provide the paradigm for all observation.[73] Indeed, the practices of particle physicists

[73] Shapere, for instance, seems to assume without question that the techniques physicists use to establish that they have "observed" something provide a basis for making universalistic statements about observation as such. See Dudley Shapere, "The concept of observation in science and philosophy," *Philosophy of Science* 49 (1982): 231–67.

might not be the best case with which to begin a study of observation. Unless one were already trained in physics, the vernacular language, methods of calculation, and technical skills that go into "making an observation" in particle physics would be practically inaccessible. Moreover, descriptions of such observations would have to be presented in simplified and misleading ways in order to make sense to most readers in the sociology, history, and philosophy of science.

Although Garfinkel's unique adequacy requirement of methods may seem to require a thorough mastery of a practice before it can be analyzed ethnomethodologically, Wittgenstein's account of language games suggests a different picture. For Wittgenstein, we already know in some sense how to count, calculate, infer, measure, observe, describe, report, follow instructions, and so on. This does not mean that an effort to "reflect" on our understanding of these themes will bring our competencies under examination. Instead, Wittgenstein devises ways to make our activities *perspicuous* by, among other things, devising "primitive language games" in which a particular word or linguistic activity becomes prominent. In his investigations of mathematics, for instance, Wittgenstein generally uses simple examples, such as counting a series of cardinal numbers. Although counting may be trivially simple, it is nonetheless a bona fide mathematical operation; furthermore, it is one that "is an important part of our life's activities . . . a technique that is employed daily in the most various operations in our lives." As such, we learn to count "with endless practice, with merciless exactitude; that is why it is inexorably insisted that we shall all say 'two' after 'one,' 'three' after 'two' and so on."[74] Wittgenstein justifies his use of simple examples from arithmetic with the following remarks:

> I can as a philosopher talk about mathematics because I will only deal with puzzles which arise from the words of our ordinary everyday language, such as "proof," "number," "series," "order," etc.
>
> Knowing our everyday language – this is one reason why I can talk about them. Another reason is that all the puzzles I will discuss can be exemplified by the most elementary mathematics – in calculations which we learn from ages six to fifteen, or in what we easily might have learned, for example, Cantor's proof.[75]

3. Follow the epistopics around and investigate actual cases in detail. An ethnomethodological transformation of Wittgenstein's approach would be to search for "naturally occurring" primitive language games and to investigate their performance in detail. For example, in order to illuminate the practical

[74] Wittgenstein, *Remarks on the Foundation of Mathematics,* ed. and trans. G. E. M. Anscombe (Oxford: Blackwell Publisher, 1956), pt. 1, sec. 4.

[75] Wittgenstein, *Lectures on the Foundations of Mathematics,* ed. Cora Diamond (Ithaca, NY: Cornell University Press, 1976), p. 14.

organization of "counting," it might be worthwhile to search out such recurrent, routinely enacted, familiar, observable, and comparable "games" as training a child to count; counting cards in a game of blackjack; figuring out one's "income" for tax purposes; counting clients served by a government bureaucracy, or conducting a "count" of inmates at a prison. Each instance brings into relief an organization of relevancies that does not readily come to mind when we consider counting as an arithmetical operation; at the same time, however, a study of each case unquestionably extends and differentiates what it means to count. No single case stands for all others, but the "conceptual binding" among the cases supplied by the epistopic (in this case "counting") lends general significance to the investigation. The investigation of each language game "says something" about counting.

4. Investigate each case in accordance with a unique adequacy requirement. Given what I have just said, Garfinkel's "unique adequacy requirement of methods" should not be understood as an admonition to learn the discipline investigated as a prerequisite for an ethnographic analysis. Although it does warn one away from the privileges of grand theory and speculative criticism, as I understand it, the requirement has to do with a method for demonstrating what a description says about a practice by enabling readers to see what is said by entering into the phenomenal field of that practice.

What Livingston says, for example, about the role of notation in a mathematical proof is demonstrated by giving the reader a proof statement and a set of instructions. Livingston's argument is bound to the reader's work, since the authority of his claims about the inexorability of following the proof is not furnished by his argument unless and until those claims are "discovered" in the lived work of proving that the reader performs, for example, on a scratch pad. Understood in this way, "unique adequacy" is rarely achieved in textual descriptions and demonstrations (and needless to say, it is not achieved in this volume).[76]

For epistopics like observation, measurement, and explanation, the task is thus to construct exercises in which readers are led to conduct observations, measurements, and explanations – or, at the very least, to follow along vicariously as the accomplishments of others are elaborated in detail – so

[76] Garfinkel (personal communication) identifies four studies in the entire ethnomethodological corpus that are "uniquely adequate" in this sense. These are Eric Livingston's demonstration of Gödel's proof (in *The Ethnomethodological Foundations of Mathematics*); Garfinkel and Livingston's paper, "Notation and the work of mathematical discovery"; Dusan Bjelic and Michael Lynch's paper, "The work of a (scientific) demonstration: respecifying Newton's and Goethe's theories of prismatic color," pp. 52–78, in G. Watson and R. Seiler, eds., *Text in Context: Contributions to Ethnomethodology* (London: Sage, 1992); and an unpublished study of Galileo's inclined plane experiment by Garfinkel, Britt Robillard, Louis Narens, and John Weiler.

that they are able to examine the relevant performances. For instance, Garfinkel and some colleagues "recreated" Galileo's inclined-plane experiments, not in order to reconstruct what Galileo must have done, but to exploit the relative simplicity and familiarity of that classic case in order to investigate the detailed performative features of conducting an experiment. For all practical purposes, "Galileo's experiment" became a primitive language game for investigating "making experimental observations."

5. Apply ethnomethodological indifference to the fact of the existence of science. When Sacks proposed that "science exists" (see Chapter 6), apparently he assumed that a generalized method was bound to that existence. And as I stated earlier in this chapter, Garfinkel and Livingston sometimes seem to give special *epistemological* status to the *Lebenswelt* pairs in mathematics and the natural sciences. What I have recommended here is to apply the policy of ethnomethodological indifference to the fact of the existence of science. By this, I do not mean to question that the existence of "science" is a recognizable fact in the modern life-world but, rather, to suspend judgment on whether the activities of scientists and mathematicians are epistemologically "special." Without denying that scientific and mathematical practices, no less than fixing a car or preparing a dinner, require specialized training along with a disciplined use of some commonplace skills and routines, I am recommending that we not assume that these rare and specialized competencies discriminate a uniquely coherent set of methods for making true observations, constructing unquestionable proofs, and achieving discoveries. In a way, this recommendation merges aspects of Bloor's impartiality postulate with Garfinkel's policy of indifference, but the result is different from what Bloor imagines. A "reflexive" ethnomethodology of science would not secure its method by analogy with the field studied; it would suspend the assumption that "science" provides a "ground" both for the fields investigated and the method of investigation.

An indifference toward science takes away much of the impetus for treating science and mathematics as special topics for the sociology of knowledge. One could figure that there is no reason to suppose a priori that the way professional mathematicians conduct "proofs" should illuminate a study of "proving," any more than a study of how children are instructed in elementary arithmetic, or even how employees at a licensed establishment assess documentary evidences of "proof of age" when serving drinks to customers, would help explain what "proving" can entail. But given the background murmuring of the vast bodies of classic literature in which the case is made for the special character of the "documentary methods" in science and mathematics, there is a point in taking up the study of "high-level" mathematics.

Because the policy of indifference does not lead us to *disbelieve* the arguments about the truth of mathematics, it is always possible that a study of mathematical proving could respecify what might be meant by the "truth" of mathematics.[77] Science and mathematics are occasionally relevant as perspicuous cases for investigations of observation, measurement, discovery, and the like, since these "epistopics" are featured explicitly in the practice. Although there is no reason to privilege the way that physicists conduct their observations, their practices can inform, in an interesting way, a study of the heterogeneous practices that are identified under the concept of observation.[78] This would not foreclose studies of various other modes of observation in and out of the sciences; in fact, one would want those cases also.

6. Use a "normal science" methodology. This is not Kuhn's "normal science," as it derives from an offhand remark made by Noam Chomsky in a debate with a sociologist.[79] Chomsky presented a critical argument about the way the "mainstream" U.S. press covers international events and conflicts. In his talk he made a number of cross-national and historical comparisons, and afterward a sociologist commentator questioned whether his account followed appropriate "methodological" canons for the selection of comparable cases. Chomsky claimed in his rejoinder that no special knowledge of sociology or of its methodology was necessary for his purposes; rather, he asserted that he practiced "normal science" when he presented and documented his argument. By this, I take it that he meant "nothing fancy," that his method was one of juxtaposing (arguably) comparable cases, citing testimonies and reports, drawing out common themes, noting relevant discrepancies and trends, and appealing to common intuitions and judgments.[80]

"Normal science" in this sense uses ordinary modes of observing, describing, comparing, reading, and questioning, and its constituent activities are

[77] This, I take it, is what Garfinkel and Livingston claim about their studies of mathematics and the sciences. Livingston's book *The Ethnomethodological Foundations of Mathematics* is the primary document on which that claim is based.

[78] See, for instance, Trevor Pinch, "Towards an analysis of scientific observation: the externality and evidential significance of observational reports in physics," *Social Studies of Science* 15 (1985): 3–36.

[79] The Eastern Sociological Association's annual meeting, Boston, April 1990.

[80] A similar eschewal of sociology is given by Freeman Dyson in the following quotation: "My colleagues in the social sciences talk a great deal about methodology. I prefer to call it style. The methodology of this book is literary rather than analytical. For insight into human affairs, I turn to stories and poems rather than to sociology." See Freeman Dyson, *Disturbing the Universe* (New York: Harper & Row, 1979), quoted in Bernard Barber, *Social Studies of Science* (New Brunswick, NJ: Transaction Publishers, 1990) pp. 254–55. Dyson supposes that without a professional "methodology," he must rely on stories and poems. The attraction of Chomsky's "normal science" is that it does not equate science with such methodological restriction. I would not want to equate Chomsky's remark at the conference with a proposal for a research program, and I doubt whether he would have it cover his linguistics research.

expressed in vernacular terms. From the standpoint of an idealized scientific observer, this may be a disappointing methodology, because it offers little that could end dispute on a controversial subject or provide an authoritative basis for normative judgments that would override "the prejudices of common sense." This normal science offers an analysis that is thoroughly "contaminated" by native intuition, vernacular categories, and commonsense judgments. And in Chomsky's case, it furnishes a set of instruments for fashioning a politically contentious argument. Although normal science offers no foundation for a scientific sociology, it serves very well for the kind of investigation I have in mind, because a more "technical" approach might dazzle us, thus distracting our attention from the primitive epistemic phenomena that first must be understood in their "natural" settings.

I recommend "normal science" not in order to appeal to common sense but to maintain an indifference toward the special epistemological status associated with social science methodologies. Since the epistopics are both thematic objects and analytic "instruments," to be uniquely adequate (in the sense outlined earlier) any analysis must be subject to a kind of "double transparency." The language games examined, for example, of "describing the appearance of an object or giving its measurements," must be transparently recognizable to readers, and that transparency must be made thematic. This radical reflexivity of accounts is a question not of "observing oneself observing" with all of its regressive implications but of bringing the transparency of an action under examination by composing descriptions that enable an "adequate" reproduction of that action.

In order to partake of this double transparency, a description must both enable the practical reproduction of an intuitively recognizable action and provide a notational index of the transparent details of that action's performance. Both of these transparencies are collapsed into a single textual object (such as in Figure 7.1), but when enacted over the course of a reading they become distinctively "instructive" temporal moments. The reference point for this primitive natural science is not a universal consciousness or a specialized community of experts but an immense and varied set of competencies that "we" already have available but that are amenable to further instruction and explication. By recommending normal science, I do not mean to suggest a "light" or "easy" approach to ethnomethodology; instead, I mean to shift the burden from an a priori general methodology to the singular demands of a "uniquely adequate" way of coming to terms with perspicuous settings. This is by no means a light demand.

7. Relate the "findings" back to the classic literatures.[81] The epistopics

[81] This is the final point in the program that I have outlined here, but it is not necessarily the end of the game. What I have proposed is a way to get started from within the confines of an academic field. A more "advanced course" in ethnomethodology would pay more attention to the possibilities of developing "hybrid" disciplines from out of the praxiological exchange

are collecting rubrics, but particular "findings" about their situated enactment are likely to hold differentiating and therapeutic implications for classic epistemological and methodological versions of observation, measurement, and the like. By "differentiating implications" I mean the kind of news that Wittgenstein delivers when he asserts that "what goes to make the reproduction of a proof is not anything like an exact reproduction of a shade of colour or a hand-writing."[82] Although as suggested in Chapter 6, we may speak of the instructable reproducibility of actions as an elementary social "molecule," Wittgenstein warns us that the reproduction of a proof from a proof statement is not "the same" as, for example, reproducing a play from a script or a symphony from a score. Each of these *Lebenswelt* pairs is generated on a "home terrain." By itself, this is not a very profound lesson, but it becomes profound in the way that Wittgenstein critically relates the lesson to the long-standing puzzles in philosophy.

Similar lessons are in store for those in various fields of study who take an interest in epistemological matters. A mere inventory of the many ways that observation, representation, and the reproduction of social structures are accomplished under different circumstances can create immense trouble for any general theory, methodology, or epistemology that presumes, for example, that what we call "representation" is a single kind of process or that "the reproduction of social structure" can be encompassed by a particular scheme of learning or internalization. This may not seem like much. But it may seem more promising if we imagine that our social theorists, rather than being latter-day Newtons, are cosmologists who use categorical machineries analogous to the medieval fourfold table of earth, air, fire, and water. For any topic they investigate, their task is to settle "which one it is." By proposing to differentiate the theoretical categories, I am not attempting to divide the social equivalent of "fire" into a finer set of distinctions, but to search for a more adequate framework.

I have recommended a kind of empirical investigation of the epistopics, but rather than leading to more precise definitions of the central terms in epistemology, the effect seems to be a matter of *displacing* the framework of epistemology. Rather than trying to define, for example, "representation" or "measurement" by comparing empirical cases and showing what they have in common, the program I have outlined seems more suited to subverting all efforts to build general models and to develop normative standards that hold across situations. But even though such negativism may have a certain anarchistic appeal, it misses something that has been implicit in the way I

between ethnomethodological research and the practices studied. In order to give more serious consideration to this possibility, I would have to run well ahead of where I have taken this book (cf. Garfinkel et al., "Respecifying the work of the natural sciences").

[82] Wittgenstein, *Remarks on the Foundations of Mathematics*, pt. 3, sec. 1.

have characterized the epistopics. I have not said that the epistopics are "mere empty words," nor have I denied the generality of their use (even though I have questioned the possibility of giving adequate general definitions of them).

It might be said that terms like representation and observation are "usefully vague."[83] As constituents of a lingua franca, they permit a way of talking that glosses over deep and mutual misunderstandings; they provide interdisciplinary conversation starters, nominal bindings, verbal passage points, and literary escape hatches. When used as elements of a general theory, the epistopics invite disaster when their useful vagueness degenerates into "mere" vagueness. But when considered as topical places to begin and end a discussion of scientific practices, their very vagueness becomes an indispensable resource. Recalling that for ethnomethodology, "indexicality" is far from an epistemological problem, the indexical roles of the epistopics in the language games of the sciences become more than a source of vagueness and indeterminacy. They become topics for investigations of the "rational properties of indexical expressions."

When social scientists announce new programs, they often foster the illusion that every reader will be persuaded to drop what he or she has been doing in order to take up the kind of empirical study that is now called for. In ethnomethodology, a more jaundiced view is more sensible: It is "not for everybody." Harvey Sacks once began an undergraduate class by telling the students who showed up on the first day that they should spend a couple of hours viewing a videotape of an ordinary conversation (he supplied the tape and facilities) before deciding whether the class was "for them." Perhaps this was a device to cut down on the number of students enrolled in the class, but it acknowledged explicitly that Sacks's program of study was likely to resonate only with those students whose prior "preparation" in life attuned them to the phenomena of interest. In my understanding, Sacks's was a very honest policy for recruiting members, but for social scientists who are committed to a "universalistic" science (or politics) it must seem disappointing or worse, since ethnomethodology holds out no immediate prospect for solving social problems, fostering revolutionary change, rectifying the errors of common sense, or gaining a more panoramic view of how biography is connected to history.

I would be crazy to believe that the program I have outlined will be taken up by many fellow travelers. It simply does not fulfill the purposes of promoting a normative social science, organizing a politicized attack on a technoscientific hegemony, or enlightening the masses with expert knowl-

[83] This is akin to what Garfinkel termed the "specifically vague" character of general sociological concepts.

edge. So what good is it, and why would anyone be interested in doing it? I cannot give a straightforward answer to this question, and I am tempted to dismiss it abruptly or flippantly by saying, for instance, "If you've read this far and still haven't figured it out, what more can I tell you?" or "Ethnomethodology is for those who are already into it."[84] But this would fail to acknowledge the seriousness of the question. An indirect way to approach it would be to reflect on the way that the "classic" topics of epistemology are kept alive as both grand themes and methodological worries in the human sciences, and as such they are subject to interminable discussions and countless technical remedies.

The attractions of ethnomethodology's approach can perhaps be most clearly realized when a student (or a veteran of decades of academic study) has already reached an impasse in his or her investigations: when, for instance, a student of science studies becomes tired of the way that debates in the field keep coming back to the same endless arguments between realism and constructivism; when a quantitative sociologist has reached the point of concluding that no amount of technical refinement will satisfactorily resolve the problems of valid correspondence between measures and social phenomena; when a proponent of "discourse analysis" becomes frustrated with the way that classic definitions of "signs" and "meanings" provide such insubstantial guidance when one tries to conduct textual analysis; or when a proponent of conversation analysis concludes that the latest findings in the discipline seem sterile in comparison with the promise the field once held out.

A postanalytic study of science might best be thought of as a postgraduate course in the human sciences, not in the sense of conveying a body of information that is more difficult or specialized than would previously have been learned, but in the sense of requiring a different kind of preparation – a combined familiarity and frustration with classic academic approaches to the most fundamental and interesting topics in the social sciences and humanities curricula. Although the kind of study I have recommended does not promise to solve the problems or break the impasses that have arisen in previous discussions of these topics, it does promise a way to "inspect" them from a different angle.

[84] The latter answer paraphrases something that Garfinkel once said in a seminar at UCLA (ca. 1976). It actually is not a bad answer.

Conclusion

Ethnomethodology and the sociology of science have begun to develop radical alternatives to the classic versions of science promoted in the history, philosophy, and sociology of science, but their potential has been subverted by familiar epistemological tendencies. Both programs are riven with internal contradictions, unfinished programs, half-baked ideas, and interminable squabbles, and I cannot hope to set things straight merely by writing a book, but perhaps I have succeeded in indicating where the problems lie and what can be done to clarify them. Again and again, these research programs have inhibited their radical potential by trying to secure a vantage point that enables a sociological analyst to remain seemingly outside the vernacular language and epistemic commitments of the communities studied.[1] The various analytic positions I have discussed and criticized include the following:

· Mannheim's general nonevaluative total conception of ideology.

· Bloor's program for a reflexive "scientific" program of explanation.

· Latour and Woolgar's search for an analytic language that is uncontaminated by the "terms of the tribe."

· The protoethnomethodological distinction between research "topics" and methodological "resources."

· The conversation analytic distinction between vernacular intuition and professional analysis.

· Garfinkel's unique adequacy requirement, construed as a method for recovering the "core activities" in each scientific discipline.

In each case, an effort is made to set up a program of analysis that gains independent access to the way that members observe, describe, explain,

[1] *Radical* is a much-abused term in social studies of science, as it can mean either or both a politically critical stance toward Western science and/or an opposition to positivist, realist, or rationalist metaphysics. The "radicalism" professed by ethnomethodologists is none of these. Rather than seeking to problematize or change the "root" causes or grounds of the alleged condition of modern science, ethnomethodologists attempt to displace the unified theoretical and methodological edifice of sociology while dissolving the "problem" of social order into myriad local practices. The incendiary effects of ethnomethodology for the most part have been limited to debates within the social and communication sciences. Garfinkel's work, as suggested by the theme of the "routine grounds" of everyday activities (*Studies in Ethnomethodology* [Englewood Cliffs, NJ: Prentice-Hall, 1967], pp. 35 ff.), also proposes a radical genealogy of social order.

represent, or otherwise engage in practical actions. In each case, social science models and methods offer ways to analyze and explain the "essential contents" of scientific and ordinary activities. This is done by converting practical activities into detachable configurations of signs that can be integrated with collections of similar cases, descriptions of contexts, simulacra, maps, cases, archival records, and other texts. Such programs of analysis become problematic as soon as we take into account what I called the "primary lesson" from ethnomethodology (see Chapter 4): There can be no intelligible theoretical position "outside" the fields of practical action studied in sociology. Even "inductive" or "empiricist" programs of social analysis that have no apparent debt to an overarching theory use collections, archives, and codes that reconfigure local expressions into variously decontextualized representations.

So what is the alternative? A groundless science of practical action that entirely collapses science into common sense? A hyperinductive sociology that relies on a "direct" apprehension of members' life-worlds unaided by mediating documents? An abandonment of the field of social theory? No. What I recommend is a programmatic *amnesis:* a "forgetting" of the dream to build a general science of society. In place of it, I propose a kind of "primitive natural science" (a "normal science" with no disciplinary pretensions) that takes up familiar epistemic themes with a deliberately "underbuilt" methodology. While recommending such a program, I would advise that we forget all anticipations of a scientific future. The primitive natural science that I have in mind is by no means a natural philosophy destined to become a natural science. Instead, it remains open to the possibility that an examination of "epistopics" (nominal themes in an epistemic lingua franca) like observation, description, and explanation will convince us that a natural observational science of human behavior is unlikely ever to be invented.

Among the things to be forgotten is the idea that sociology is an "underdeveloped science" that only needs more time (or an infusion of capital and technology) before it will become effectively "industrialized." In response to this tendency, we should consider the counterpossibility that the more industrialized (i.e., rigorous, well funded, standardized, cumulative, public-policy relevant, and hierarchically administered) sociology becomes, the less interesting and more oppressive it will be. Likewise, we should forget epistemology; that is, we should forget "metatheory" and "theory of knowledge" as prerequisites for building a social science. Not only that, we should forget "knowledge" as an adequate way of formulating the entire "content" of a science. Much of what goes under the heading of "knowledge" in science studies can be decomposed into embodied practices of handling instruments, making experiments work, and presenting arguments in texts or demonstrations. Much in the way that "observation" acquires a distinct, and perhaps more limited, role when it is formulated as a matter of getting "some bit of equipment to exhibit phenomena in a reliable way,"[2] so "knowledge" becomes more tangible – and less monolithic – when translated into various practical activities and textual productions.

To many readers, these proposals may seem to burn down the sociological house while leaving no legitimate dwelling for those of us who are sincerely committed to devising a strong and progressive alternative to what currently passes for social and cultural knowledge. Such a conclusion would blow things entirely out of proportion. Many of the arguments I have used in this book have been around for a long time, and they have done little to deter analytic philosophers and social scientists from pursuing traditional modes of argument and scholarship. It would be unrealistic, indeed grandiose, for me to expect to persuade very many sociologists to abandon the dream of a science of society, and furthermore, I would not want to convince them that their research practices (e.g., interviewing, using statistics, consulting archives, analyzing transcripts, etc.) are worthless.

Given what I have said here, I would not want to argue that an absence of secure epistemological "grounding" implies the absence of intelligibility or practical utility. To say that the social sciences are "merely practical" or "merely literary" enterprises begs the question, What else can they be? At the same time, I do maintain that we should be loath to bolster sociology's prospects by drawing analogies with more successful sciences. My suggestion to "forget science" therefore means: Forget trying to act – or trying to convince others that you are acting – in accordance with some general epistemological scheme. This advice applies most specifically to the research programs I have discussed here, because the sociology of scientific knowledge and ethnomethodology explicitly propose to examine epistemic topics. The lesson that observation, representation, replication, measurement, and the like are "locally organized" applies no less to the aims and methods of social scientific investigations than it does to the lay and professional activities described and explained through such investigations. But far from suggesting that epistemic activities are ubiquitously "problematic," this lesson encourages us to examine how these activities are accomplished *in situ.*

Postanalytic science studies

This book has been almost entirely polemical and programmatic, and readers may be inclined to ask, Why not simply get on with the empirical work? Both ethnomethodology and the sociology of scientific knowledge are polemical

[2] Ian Hacking, *Representing and Intervening* (Cambridge University Press, 1983), p. 167.

fields with a high ratio of program to accomplishment,[3] and it is easy to grow impatient with interminable debates about "epistemological" and "metatheoretical" issues. Although I could always try to dismiss such hankering for empirical research as "naive" while assuming the academic high ground of theory and textual criticism, I am inclined to agree that this book will be insufficient, merely programmatic, and little more than a promissory note, unless it is followed up by some kind of empirical research. In a later work I plan to undertake such a task, but much of what I have proposed here has already been done. For the most part, I have suggested a way to extend initiatives that are already present in a fairly large body of studies.

As mentioned in Chapter 7, the program I have outlined here is largely a negative one. While admitting this, I also would like to argue that such negativity can be "therapeutic" or even "emancipatory," in light of the current state of the social sciences. If we assume that the social sciences are on the wrong track, then negativity has a place even when no "right track" has yet been identified.

In this book I have criticized various analytic moves through which social scientists authorize their versions of ordinary and scientific practices, and I have suggested how a "postanalytic" program of study might take up investigations of familiar epistemic themes. The *post-* prefix suggests an affiliation with the poststructuralist and postmodernist approaches that are now so popular in the humanities and social sciences. Although I would like to avoid the sweeping claims about massive historical "eras" that so often are made by analysts of the "postmodern condition," there is one sense in which I find *post-* appealing. *Post-* differs from *anti-* by suggesting a temporal (dis)placement "after," rather than an opposition to, the term that follows the prefix. Direct opposition and inversion are replaced by "free play." A postmodern architecture plays itself off in various ways against modernist styles while retaining an ironic affiliation to an earlier genre.

By advocating postanalytic science studies, I am not repudiating analysis but suggesting a retrospective relation to already accomplished analyses. Such a position is endemic to the sociology and history of science as well as ethnomethodological studies of science. Studies of science presuppose their subject,[4] along with an established set of topics associated with it: theory,

[3] Stephen Turner, *Sociological Explanation as Translation* (Cambridge University Press, 1980), p. 4.

[4] Even when sociologists argue that science is problematic – in the sense of being essentially indistinguishable from "nonscience" – they do not question the social, practical, or rhetorical facticity of science. Indeed, by taking up the topic of science they acknowledge the prior existence of "it." The difficulty of making a single adequate definition of "science" does not mean that "it" cannot be recognized, performed, and discussed in detail.

observation, description, replication, measurement, experiment, rationality, representation, and explanation. Sociological and historical analyses themselves require a relation to a previously established subject, and under the policy of *docta ignorantia* the "knowledge" or "belief" that is analyzed is presumed to be familiar and yet somehow unknown or unexplicated. Otherwise, analysis would have a purely deictic relation to its subject: "Look! See! Here it is!"

For a Baconian program, scientific analysis is contrasted with the unclear, prejudiced, interested, and partial knowledges that are endemic to the tribal community and its routine relations. Sociologists face a difficult problem when they try to investigate the endogenous "beliefs" held by one or another "tribe" of scientists, because any incommensurability between the member's vernacular interests and beliefs and the sociologist's more comprehensive explanations of the sources of those beliefs sets up a competition over who is entitled to speak on behalf of "science." Although sociologists of scientific knowledge profess impartiality toward the correctness of particular natural scientific theories and facts, the analytic claim that inherently controversial or indeterminate features of those theories and facts are shut away in a "black box" (a kind of collective unconscious reopened by sociohistorical investigation) directly conflicts with the "tribal belief" that those theories or facts are accepted because of their superiority to any "reasonable" alternative.

Because sociology is widely assumed (even by many sociologists) to be a weak and underdeveloped discipline (or, less charitably, as "mere common sense" cloaked in jargon), any competition between practitioners' and sociologists' accounts of the "contents" of particle physics or biochemistry is likely to be resolved "on scientific grounds" in favor of the relevant natural scientists. Short of assuming that sociology (or social history) is a "superscience" capable of comprehending specialist scientists' claims and subsuming them within a larger explanatory picture, it might be reasonable to consider whether it is possible to investigate scientific activities without claiming scientific authority.

I have argued that such an "ascientific" approach has been developed by ethnomethodology's exemplary "extension" of Wittgenstein's philosophical investigations. Wittgenstein conducted investigations that he claimed were neither explanatory nor grounded in a scientific method but that relied on the intuitive familiarity of ordinary language to members of a community of users. He did not propose a linguistics but a way of demonstrating how "reasoning" is embodied in public uses of a common language. Ethnomethodologists extend Wittgenstein's grammatical investigations by devising procedures for supplying our "members' intuitions" with perspicuous instances of actual usage in commonplace as well as more specialized circumstances.

This investigatory procedure was sidetracked in conversation analysis when proponents of that field imagined that they had secured an empirical science. Rather than treating the collection, transcription, and systematic inspection of tape-recorded conversation as aids to a reflexive explication of public understandings embedded in commonplace activities, some prominent conversation analysts confused the issue by contrasting their professional analyses with vernacular intuitions. Thereafter, the conversation analysts hitched their fates to the success or failure of a new social scientific program, and that program began to diverge from ethnomethodology although it still provides a ready case for study.[5]

A postanalytic ethnomethodology begins with an ironic appreciation of the claims of analysis and of the Schutzian contrast between the attitude of scientific theorizing and the natural attitude of the "unreflective" member.[6] The irony refers to an appreciation of how a "radical" attempt to investigate the endogenous production of order in a prescientific life-world became yet another claimant for the privileges of a social science. To speak of this irony is to express an "amused" acknowledgment of ethnomethodology's failure to produce a natural observational science of human behavior: "Of all people, why would the ethnomethodologists think they could have done *that!*" Far from predisposing a rejection of ethnomethodology or a cynical detachment from the radical initiatives that were once lively for ethnomethodologists, the irony can motivate a wary attempt to revive those older initiatives. This does not necessarily mean going back to a "classic" or "fundamental" text or program, since there can be no coherent foundation for ethnomethodology. Instead, what may be needed is an infusion of new life into the agenda.

In this book I suggested such an infusion by pairing ethnomethodology with a critically worked-over sociology of scientific knowledge. Although sociologists of science have tended to bolster their analyses with large doses of scientism (or, in some variants, semiosis – a metaphysical inflation of the sign), their historical case studies and ethnographies have effectively chal-

[5] Given the historical linkage between ethnomethodology and CA, combined with the fact that CA is by no means a facile approach to discourse, CA is a relevant and challenging case for ethnomethodological investigation. Since it is closer to hand than, say, physics, CA might seem to be a much more likely candidate for "hybridization" with ethnomethodology. However, the potential for (re)building a "hybrid" ethnomethodology/CA is complicated by the fact that this hybrid (represented by Garfinkel's and Sacks's collaboration in the late 1960s) has already been bypassed, and it would need to be revived through a critical engagement within an existing and fractious community of ethno/CA practitioners (many of whom have their own designs on such a hybrid).

[6] The mode of "irony" here is not of the sort that makes moral capital out of the difference between what someone says and what they actually do (e.g., the ironic theme found in many social studies of science to the effect that scientists do differently than their research reports say). Instead, it is more of a retrospective gloss, acknowledging without condemnation how a well-intentioned struggle successfully recreated a variant of what it once battled against.

lenged the honorific versions of the "men" of science and their practices that once gave such strong impetus to the parasitic methodologies promoted in the social sciences. Along with various (often misguided) postmodern attacks on the edifice of epistemology, the success of the sociology of science provides a historical circumstance for reviving a radical ethnomethodology. Whereas Schutz and the early generation of ethnomethodologists once had little alternative but to try to justify their research in "scientific" terms, this may no longer be necessary. It is now possible to suggest investigations without first plotting out a methodological foundation and without suffering the withering accusation of being "unscientific." It is now possible (though perhaps tendentious) to say that such accusations are grounded in a mistaken conception of science.

Science as both orderly and ordinary

A postanalytic ethnomethodology, like the various other post-Enlightenment projects that have rapidly become an interminable topic of conversation in the social sciences and humanities, faces the question of how to get "ordinary life" out from under the shadow of science and scientific rationality. Although as mentioned earlier, the *post-* prefix suggests something other than vehement opposition, all too often this is ignored in contemporary critiques of "Western science." An opposition to science on behalf of subjugated knowledge, narrative knowledge, marginalized discourse, commonsense reasoning, or despised "irrationality" tends to frame the problem in the outlines of the now "outmoded" program for a unified science.

Such political – epistemological opposition treats "science" as a coherent rationality, and scientists as metaphysicians who hold conceptions of nature and of their own activity that are dominated by a positivistic epistemology. This picture dissolves when it is proposed that in practice and *in situ* the engineer is a species of *bricoleur* and the scientist is an ethnomethodologist. The injunction from ethnomethodology and the new sociology of science is "Stop talking about science! Go to a laboratory – any laboratory will do – hang around for a while, listen to conversations, watch the technicians work, ask them to explain what they do, read their notes, observe what they say when they examine data, and watch how they move equipment around!" Although such an experience can raise innumerable doubts about how a social scientist can hope to identify, let alone to explain, what goes on in the thickness of the technical routines of another discipline, it should be sufficient to answer the question, "Do you see anything other than *bricolage,* ordinary discourse and situated actions?" The specters of science and technical reason are likely to dissolve into myriad embodied routines and diverse language games, none of which may be uniquely "scientific."[7]

It is often argued that the ethnomethodological dissolution of science and of all of the classic inquiries that have been raised about it is the product of a methodological mistake. Almost immediately in the wake of the first laboratory studies a counterinjunction was raised: "Step back from the confusing and trivial details of particular laboratory projects; turn off the tape recorder; look at interorganizational linkages, corporate and military sponsorship, the rhetoric of scientific texts, interorganizational networks, transepistemic communities, and long spans of history."[8]

The established topics and problems of an older sociology of science – bibliographic networks, interorganizational linkages, norms, culture, institutionalization – returned for revisionist modes of analytic reconstruction. Presumably, it is mistaken to look for science by examining the seemingly chaotic details of particular laboratory projects, because science resolves itself only at a more global level of analysis.[9] In science studies it has become commonplace to claim that close observations of "some people at work" cannot come to terms with how that work is raised to the status of "innovation," lowered to that of "failure," or simply ignored. We are often told that

[7] When described specifically, techniques such as "fixation by vascular perfusion" are unquestionably unique in a number of ways: unique to a set of disciplines that employ one or another variant of the technique, and uniquely articulated in reference to a project at hand. This does not make the technique "unique to science" *as such,* and variations on the technique may also be used as a constituent of "industry".

[8] This injunction was stated most forcefully by many of the proponents of the "new" sociology and anthropology of science. See, for instance, Karin Knorr-Cetina, "The ethnographic study of scientific work: towards a constructivist interpretation of science," pp. 115–40, in K. Knorr-Cetina and Michael Mulkay, eds., *Science Observed: Perspectives on the Social Study of Science* (London: Sage, 1984); Latour, "Give me a laboratory and I will raise the world," pp. 141–70, in *Science Observed;* and Latour, "Will the last person to leave the social studies of science please turn on the tape-recorder?" *Social Studies of Science* 16 (1986): 541–48.

[9] In social studies of science it has become conventional to say that enough visits to laboratories have been made by Latour, Knorr-Cetina, and others and that now we must take up the challenge of explaining the "micro" actions of scientists by referring to events and social structures "beyond the walls" of the laboratory. See, for instance, William Lynch and Ellsworth Fuhrman, "Recovering and expanding the normative: Marx and the new sociology of scientific knowledge," *Science, Technology, and Human Values* 16 (1991): 233–48. Such proposals tend to reinstate the sort of multivariate analysis that was once rejected by ethnomethodologists and sociologists of scientific knowledge, and they ignore one of the principal motives for engaging in a "close encounter" with scientific practices. The point is not to tie "observations" of laboratory scientists to a more comprehensive "framework" of sociological investigation; among other things, it is to shake up the rather tired assumptions about science, methods, observation, and explanation that are entrenched in sociology. This is not to deny that laboratories depend heavily on funding, public support, and so forth. See, for instance, Chandra Mukerji, *A Fragile Power: Scientists and the State* (Princeton, NJ: Princeton University Press, 1990); and Michael Dennis, "Accounting for research: new histories of corporate laboratories and the social history of American science," *Social Studies of Science* 17 (1987): 479–518. Rather, it is to insist on a different orientation to practical action than can be gathered within the terms of a comprehensive explanation.

it is necessary to look "outward," to navigate through networks of events beyond the laboratory walls. The aim of such outward-looking projects is to explain how the chaos of the laboratory is translated into the order of science at the scale of a historically and institutionally contingent process of "construction."[10]

It is easy to understand the theoretical, pragmatic, and professional rationales for retreating from a direct ethnographic engagement with the "messy" and "disordered" practices of laboratory work in favor of established modes of historical or institutional analysis. For many sociologists' purposes it is enough to define the laboratory and its practices as "messy," in contrast with the "purified" accounts of method and logic in what Latour calls "ready-made science."

The contrast between this mess and the purified products of scientific research nicely supports the idea that the social construction of science operates at a level of organization "beyond" the laboratory, but the characterization of messy laboratory work itself begs numerous questions. Is there not order in the mess?[11] Does "social construction" imply a free, deliberate, and unconstrained "invention" and "fabrication"? Does the demonstrable "fact" that science, like any other organized practice, involves the use of indexical expressions, *bricolage* expertise, ad hoc practices, improvisation, persuasion, plausibility judgments, tinkering with equipment, and so forth, deter us from saying that scientific enterprises produce stable and reproducible facts and highly reliable procedures? Once it is agreed that absolute certainty is not a meaningful standard for assessing scientific procedures or scientific facts, to say that actual scientific practices fail to live up to that standard no longer reveals anything. It makes more sense to say that the alleged standards fail to "live down" to what scientists do. The critical target, if there is one, is a waning horizon of *theological* doctrines regarding science, and not the rationality, efficacy, orderliness, and stability of actual scientific practices.

Far from ending the debate about the rational and naturalistic foundations of scientific inquiry, laboratory studies and the new historiography of science helped shift the terms of the rationality debate from the domains of reason, cognition, and logic to those of practice, writing, and instrumentation. It is doubtful that even the most detailed laboratory studies have described, let

[10] The most influential summary statement along these lines is Latour's *Science in Action* (Cambridge, MA: Harvard University Press, 1987).

[11] Bruno Latour and Steve Woolgar (*Laboratory Life: The Social Construction of Scientific Facts* [London: Sage, 1979; 2nd ed., Princeton, NJ: Princeton University Press, 1986], p. 33) propose that the order of scientific facts is constructed from an initial condition of "chaos." Their account is akin to a pragmatist version of perception (opposing an initial "blooming buzzing confusion" to a constructed perceptual order), except that they stress collectively accomplished and textually inscribed modes of construction. Their account of the initial condition of chaos, however, is entirely metaphysical, as it requires us to imagine that lab scientists begin their projects in a world that is not already infused with meaning, intelligibility, and familiarity.

alone explained, the "content" of natural science researches, and it is also doubtful that such studies can argue with any confidence that "actual" scientific practices cannot justify themselves "internally." Critics of the "new" sociology of scientific knowledge often argue in favor of some role for "nature" and "cognition" in the determination of scientific discoveries, as well as for the importance of practical solutions to the problems of theory-ladenness and indeterminacy raised in skeptical treatments.[12] These proposals sometimes rehash older lines of argument against philosophical relativism, and typically they give short shrift to the diversity and specificity of studies in the sociology of science, but they do raise some serious doubts about the claims made about the "messiness" of actual laboratory practices.

What might be meant by "actual" practices is not always obvious, but it is clear from the way the term is used in laboratory studies that "actual" scientific practices are material, embodied, real-time accomplishments, as opposed to after-the-fact accounts. There is a certain naiveté in this characterization, since it ignores that sociologists compose after-the-fact accounts whenever they describe "actual" scientific practices and that other after-the-fact accounts are concrete features of the "material presence" of those practices. Nevertheless, participants in science studies research have had little trouble finding places where "actual" scientific practices are performed. Such places are alive with embodied practices in which instruments and specimens are palpably present; they are sites where speaking, writing, and reading are performed as real-time constituents of scientific projects. The question remains: Is the life-world of scientific practice inherently chaotic and disorderly?

Characterizing the science workshop as a practical and epistemic morass nicely suits a program that seeks to explain the "apparent" order of scientific facts and scientific descriptions by searching for explanatory factors "beyond" the laboratory. But if the empirical findings of laboratory studies are themselves challenged as confused, incoherent, and largely programmatic, then it might seem premature to subsume those findings within more ambitious models of "the larger technoscientific field." Until such a challenge is successfully turned aside, one can always conclude that the sociology of science, though challenging previous versions of scientific observation, description, explanation, and the like, has been unable to secure scientific authority for its own observations, descriptions, and explanations.

Rather than recommending that the sociology of science fall back on a picture of science and of nature that both limits its domain of inquiry and lends authority to its method, I have argued that Wittgenstein and ethno-

[12] See Stephen Cole, *Making Science: Between Nature and Society* (Cambridge, MA: Harvard University Press, 1992); Allan Franklin, *Experiment Right or Wrong* (Cambridge University Press, 1990); and Peter Galison, *How Experiments End* (Princeton, NJ: Princeton University Press, 1987).

methodology offer a novel way to avoid the antinomies of the realist–constructivist debate. As I stated in Chapter 5, ethnomethodologists investigate the "rational properties of indexical expressions and indexical actions." This implies that far from being a chaos from out of which order is constructed, the locally organized and reflexive details of actual conduct in a laboratory are orderly and describably so.

A reluctance to settle for an account of local actions as messy and disorderly was expressed in an early article by Harvey Sacks, in which he disputed the widespread idea in the social sciences that it is pointless to study "naturally occurring" human conduct without first isolating relevant variables and controlling for extraneous sources of variation.[13] Sacks argued instead that singular conversations are "finely ordered," that such order was to be found "at all points" in the temporal production of interaction, and, most important, that the intelligibility of singular actions was endogenous to their production as socially organized actions. Accordingly, it would make no sense to begin an account of technical actions in a laboratory (or in any other recognizable setting) by proposing that such work is responsible for transforming a primordial chaos into order. For Sacks and ethnomethodology, any observation or description of human conduct is bound to a densely ordered configuration of actions in which that conduct takes place; there is no exit from the accountability of the singular "moves in the game" through which "actions" and "contexts" reflexively attain specificity.

Although I have advocated ethnomethodology as a way out of some of the culs-de-sac faced by the sociology of knowledge, I have been careful not to recommend ethnomethodology as a nascent scientific program capable of generating privileged analyses of epistemologically relevant phenomena. The pressure to build a social science can be immense, and it is understandable why enthusiasts for novel research programs would want to build on an abstract version of science in order to authorize and legitimate their work. Given the often-demonstrated fact that the guardians of scientific sociology are always ready to charge heretical movements with being "unscientific," it is difficult to avoid the tendency to invoke the authority of scientific method in order to withstand that charge. In this book, however, I have recommended a suspension of the use of general definitions of science and scientific methodology as presumptive "grounds" of a social science research program. Although it may at first appear to be a hopelessly "weak" position to take – an inquiry without methodological foundations and with no claims to scientific authority – I would argue that it is just what is called for in light of the questioning of the idea of a unified science that has occurred in all of the social sciences and humanities disciplines.

[13] Harvey Sacks, "Sociological description," *Berkeley Journal of Sociology* 8 (1963): 1–16; Sacks, "On sampling and subjectivity" (Lecture 33, spring 1966), pp. 483–8 in Sacks, *Lectures in Conversation,* vol. 1, G. Jefferson, ed. (Oxford: Blackwell, 1992).

Name index

Subject index

For EU product safety concerns, contact us at Calle de José Abascal, 56–1°, 28003 Madrid, Spain or eugpsr@cambridge.org.

www.ingramcontent.com/pod-product-compliance
Ingram Content Group UK Ltd.
Pitfield, Milton Keynes, MK11 3LW, UK
UKHW042145130625
459647UK00011B/1198